Anthropogenic Climatic Change

Anthropogenic Climatic Change

M. I. Budyko and Yu. A. Izrael, Editors

Translated from the Russian by the Authors

The University of Arizona Press
Tucson

The University of Arizona Press

(∞) This book is printed on acid-free, archival-quality paper.
Manufactured in the United States of America.

96 95 94 93 92 91 6 5 4 3 2 1

Library of Congress Cataloging-in-Publication Data

Antropogennye izmeneni͡ia klimata. English.
Anthropogenic climatic change / M.I. Budyko and Yu. A. Izrael, editors : translated from the Russian by the authors.
p. cm.
Translation of: Antropogennye izmeneni͡ia klimata.
Includes bibliographical references.
ISBN 0-8165-1122-5
1. Climatic changes—Environmental aspects—Soviet Union. 2. Ecology—Soviet Union. I. Budyko, M. I. (Mikhail Ivanovich) II. Izraėl′, I͡U. A. (I͡Uriĭ Antonievich) III. Title.
QC981.8.C5A59 1991
551.6—dc20 91-3722
CIP

British Library Cataloguing in Publication data are available.

CONTENTS

FOREWORD

The U.S.S.R. report of the State Committee for Hydrometeorology and the Academy of Sciences offer a provocative and controversial vision of future climates of the Earth. Evoking the traditional concept of geological thinking that the past is prologue to the future, they argue that certain geological intervals can serve as models of the climate of the future. The Holocene optimum (5000–6000 years ago), the last interglacial (some 125,000 years ago), and the Pliocene (some several million years ago) represent a sequence of time, each somewhat warmer than the previous one. By analogue, these past intervals represent a hypothetical sequence of future climate changes.

A major portion of this book develops the theoretical ideas for this hypothesis and summarizes the patterns of climate change for these intervals. Although Soviet authors recognize that many factors contribute to climate change, they argue that the change in CO_2 levels is the dominant one.

The implication of these analyses is that the future greenhouse world (climate change induced by man-made activities) may be largely favorable, at least by these reconstructions by Soviet scientist for a large part of the northern hemisphere. This has led the principal author, Michael Budyko, to argue that international measures to reduce the emissions of greenhouse gases are not justified.

At the root of this hypothesis are several fundamental scientific questions which Soviet authors address and which readers will want to ponder and analyze themselves. Central to this discussion is the relative roles and contributions to climate change of orbital changes and atmospheric CO_2 changes. Many geologists and paleoclimatologists will also want to review carefully the discussion on the paleoclimate reconstructions. All readers will benefit from the exposition of Soviet thinking on climate change and the extensive Soviet literature referenced in the text.

Alan D. Hecht
Environmental Protection Agency
Washington, D.C.

PREFACE

It has been widely assumed by many nonscientists that modern global climate is more or less stable and no basis exists for expecting any noticeable changes in the near future. A few scientists at the end of the nineteenth century hypothesized that the climate might be affected by the increase in atmospheric carbon dioxide content resulting from combustion of fossil fuels. Their hypotheses were neither acknowledged nor supported by the scientific community.

The question of anthropogenic climatic change attracted the attention of governmental agencies for the first time in the USSR, when the staff of the Hydrometeorological Service (now the USSR State Committee for Hydrometeorology and Environmental Monitoring-Goskomgidromet) in 1961 recognized the possible development of anthropogenic warming and decided to organize a systematic study of man's impact on global climate. That same year, Academician Ye. K. Fedorov and corresponding member of the USSR Academy of Sciences M. I. Budyko conducted the All-Union Conference on the Problem of Climatic Modification by Man in Leningrad (Gal'tsov, 1961). It should be noted that other countries initiated similar studies much later. The studies sponsored by the USSR Hydrometeorological Service developed the first realistic prediction of impending climatic change (Budyko, 1972).

Scholars from several scientific institutions in the Soviet Union representing many scientific disciplines are now engaged in research on anthropogenic climatic change. The interdisciplinary nature of this problem necessitates this broad approach to study of anthropogenic climatic changes. The advances made between 1961 and 1985 have been discussed at five all-union conferences on anthropogenic climatic change at which the participants were representatives of scientific institutes and various organizations studying this problem. In 1975, the Goskomgidromet founded in the State Hydrological Institute (Leningrad) a department for investigating climatic changes and the hydrologic cycle of the atmosphere. This department is coordinating the major research on this problem in the Soviet Union.

Since the anthropogenic climatic changes occurring today are global in nature, it is obvious that they should be studied on an international basis. Consequently, over the past ten years a series of scientific conferences on anthropogenic climatic changes for scientists from socialist countries have been held in the USSR. Since 1972, the United States and USSR have cooperated successfully in studying climatic changes with major emphasis on man's impact on climate. Within the framework of this cooperation, about ten scientific conferences have been held and several joint scientific reports and findings published on various aspects of climate change. Particular attention should be given to the conclusions reached at the 1982, 1984, and 1987 Soviet-American Conferences of experts regarding imminent climatic changes resulting from economic development (Effect of Increased Carbon Dioxide..., 1982; Anthropogenic Climatic Change, 1984; Causes of Modern Climatic Change, 1987; see also Budyko and MacCracken, 1987).

Since the mid-1970s the problem of anthropogenic climatic change has received much attention from various international organizations, particularly the World Meteorological Organization (WMO) which, has prepared a series of findings on this topic and sponsored the World Climate Program.

In 1979 the First World Conference on Climate was convened in Geneva under the auspices of the WMO. Four of the 25 sponsored reports were presented by Soviet scientists, Ye. K. Fedorov, G. I. Marchuk, I. P. Gerasimov, and Yu. A. Izrael'. In his introductory report Ye. K. Fedorov said, "Future climatic changes are unavoidable. They will become noticeable and perhaps irreversible in the next several decades... It is thus obvious that some strategy should be worked out, i.e., a system of preplanned actions that would enable mankind to avoid the negative consequences of possible climatic changes...) (Fedorov, 1979).

The Conference adopted the Declaration addressed to all countries of the world, indicating a future significant global climatic change due to economic activity. Since anthropogenic climatic changes will be beneficial in some regions and disastrous in others, complex social and technical problems will arise on an international scale. Unprecedented efforts will be necessary to solve them, to organize research on climatic change and create new forms of international cooperation.

The proceedings of the 1985 (Villach, Austria) conference on anthropogenic climatic change convened by the WMO together with the United Nations Environmental Program (UNEP) and the Council of Scientific Unions (ICSU) should be noted. This conference issued a statement addressed to the governments of all nations. It states that in several decades, the change in chemical composition of the air will bring about a greater warming than that which occurred during recent centuries. This statement notes the fallibility of the current practice of basing important decisions on the assumption that climatic conditions of the recent past will continue in the future. It is consequently a matter of the utmost urgency to assess the climatic conditions expected in the near future.

Among foreign nations, the United States has organized the most extensive studies of anthropogenic changes in global climate; an example is the report published by a commission of the National Academy of Science (Energy and Climate, 1977). A year later a national climate program bill was passed in the United States emphasizing the need to considerably improve methods of predicting future climatic changes. Several scientific reviews on the problem of anthropogenic climatic changes have been published in recent years in the United States by commissions of the National

Academy of Science and by federal agencies (Carbon Dioxide and Climate..., 1979; Carbon Dioxide and Climate..., 1982; Changing Climate, 1983; MacCracken and Luther, 1985, etc.).

Studies on the problem of anthropogenic climatic change carried out in England, West Germany, and Sweden should also be mentioned.

Data from earlier studies by Soviet scientists on anthropogenic climatic change have been presented in several monographs (Budyko, 1974, 1980; Karol', Rozanov, and Timofeev, 1983; Byutner, 1986; Vinnikov, 1986). However, these monographs gave more coverage to the results of studies by their authors and did not fully review other works on this problem. This book is the first of its kind, presenting the results of anthropogenic climatic change studies made by Soviet researchers.*

*These results were discussed at a meeting of the Coordination Council on the Atmosphere of State Committee for Science and Technology, which recommended that data on future climate be taken into account in solving various national economic problems.

Anthropogenic Climatic Change

CHAPTER 1

THE PROBLEM OF ANTHROPOGENIC CLIMATIC CHANGE

1.1. Causes of Anthropogenic Climatic Change

Carbon dioxide. Arrhenius (1896, 1908) and Callendar (1938) were the first to assume that the burning of greater amounts of coal, oil, and other kinds of fossil fuel would result in a noticeable increase in atmospheric carbon dioxide. They believed that this change in the chemical composition of the atmosphere would cause an increased greenhouse effect and a rise in lower air layer temperature.

Although the calculations of Arrhenius and Callendar about future climatic change later proved to be inaccurate, their investigations were far ahead of contemporary science. The results of their investigations received scant attention, however, possibly because of the universal viewpoint that the ocean was capable of rapidly absorbing any additional amount of CO_2 entering the atmosphere from any source.

It is typical that the first of many reviews (Energy and Climate 1977) published abroad on the CO_2 problem since the late 1970s did not mention the studies of Arrhenius and Callendar. This indicates that their works were forgotten, and that at that time the problem of the carbon dioxide effect on contemporary climatic change was completely novel for many scientists.

Organization of systematic observations of atmospheric CO_2 concentration by sufficiently precise instruments was very important in studying the laws governing variation in atmospheric CO_2 content. These observations began during the International Geophysical Year (1957–1958). At the end of the 1960s, analysis of the resulting data revealed a year-to-year increase in the mass of atmospheric CO_2 , and the rate of this increase was essentially identical at distant stations (Hawaii, Alaska, etc.). It was established that the amount of atmospheric CO_2 increases annually by about 0.3% of its total content. This amounts to about half the CO_2 released into the atmosphere from burning of fossil fuel (Bolin and Bischof, 1970; Inadvertent Climate Modification, 1971).

The first estimate of the expected change in atmospheric CO_2 concentration that was confirmed by subsequent observations was made in the early 1970s based on a very diagrammatic analysis of impending variations in the carbon cycle (Budyko, 1972). The major hypothesis of this study was the assumption that after 1970 the rate of increase in fossil fuel combustion would gradually decrease. On the basis of this hypothesis it appears that the concentration of atmospheric CO_2 will double in the middle or second half of the twenty-first century. This agrees well with the conclusions of early studies employing detailed carbon cycle models that began to be developed at the end of the 1950s (Revelle and Suess, 1957; Bolin and Erikson, 1959).

The CO_2 balance in the atmosphere and anthropogenic effects on the balance are discussed in detail in Chapters 3 and 4; therefore, here we simply mention that, according to observational data, the volumetric CO_2 concentration in the atmosphere has increased from 315 ppm in 1958 to more than 350 ppm in 1988. There is no doubt that the rise in CO_2 concentration began long before the institution of systematic observations. Consequently, it is very important to know the atmospheric CO_2 content in the preindustrial era. Based on the calculations described in Chapter 3, we can conclude that this concentration was about 280 ppm in the mid-nineteenth century.

By the mid-1980s the CO_2 content had thus increased by 20 to 25%. The majority of early studies assume that the increase in fossil fuel consumption in the coming decades will taper off, while the portion of anthropogenic CO_2 emission into the atmosphere will increase. As a result, the rate of increase in atmospheric CO_2 content will change comparatively slowly in the next few decades. There is no doubt that there will be a 50% increase in the first half of the next century as compared with the preindustrial era. It is quite probable that the CO_2 concentration will double in the second half of the twenty-first century. As subsequent chapters of this book will show, this change in the chemical composition of the atmosphere will have a very significant effect on the climate and life of photosynthesizing plants. We also have reason to believe that the present increase in CO_2 content already is influencing global climate and the biosphere as a whole. The work of E. K. Byutner (1986) is one of the recent surveys on anthropogenic change in atmospheric CO_2 content we should mention. We discuss this problem in Chapters 2 through 4.

Trace gases in the atmosphere. Man's impact on the chemical composition of the atmosphere is not confined to an increase in CO_2 concentration. This influence is causing the atmospheric concentration of other gases to rise as well. They, just as CO_2 , intensify the greenhouse effect and contribute to a temperature increase in the lower air layers. These gases include methane (CH_4), nitric oxides, ozone, and others.

The concentration of methane in the atmosphere coming from swamps, deep cracks in the earth's crust, and some other natural sources is not high (about 1 to 2 ppm). The current amount of atmospheric methane is increasing rapidly, both as a result of greater agricultural production (particularly the expansion of abundantly irrigated rice paddies) and of climate warming in permafrost zones.

Among nitric oxides, N_2O and NO_2, with concentrations of 0.3 ppm, are the most important for climatic changes. A considerable quantity of nitric oxide is released into the atmosphere during manufacture of mineral fertilizers and as a result of some other types of industrial activity.

There is reason to believe that man's activity is causing the ozone (O_3) concentration in the troposphere to increase, and this should also intensify the greenhouse effect in the atmosphere.

In addition to the aforementioned gases that were in the atmosphere in small quantities before industrialization began, modern air contains trace gases originating solely from anthropogenic sources. Of these, the most important are chlorofluorocarbons (CFC) that enter the air when some paints dry, when the liquid used in refrigeration units evaporates, etc. The amount of CFC in today's atmosphere is minute (of the order of 10^{-10} of the atmospheric volume), but it is increasing rapidly, and that increase could exert a pronounced effect on the climate.

Although the question of the climatic effect of an increased amount of trace gases was raised about ten years ago (Ramanathan, 1976; Yung et al., 1976; Wang et al., 1976), it was only clarified recently that this influence on global warming could be comparable to the CO_2 effect. This problem is discussed in detail in the monograph by I. L. Karol', V. V. Rozanov, and Yu. M. Timofeyev (1983), in a number of other reviews, and in Chapter 5 of this book.

Atmospheric warming. Increased energy production that causes additional warming of the atmosphere is among the other factors influencing the change in climate and due to economic activity. All energy consumed by man is ultimately converted into heat, and the majority of this heat is an additional energy source for the Earth, contributing to an increase in its temperature.

Of all the existing components of contemporary energy consumption, only hydraulic power and the energy contained in wood and agricultural products represent the conversion of solar energy absorbed annually by the Earth. The expenditure of these types of energy does not alter the thermal equilibrium of the Earth nor cause any additional warming.

However, these types of energy constitute only a small fraction of all the energy consumed by man. Other sources, such as coal, oil, natural gas, and nuclear energy, are a new heat source independent of present solar energy conversions.

There are estimates of the amount of heat generated as a result of mankind's economic activity. This amount per unit area of the Earth as a whole is small, equal to about 0.01 W/m^2. This value is two orders greater for the most developed industrial regions (tens and hundreds of square kilometers in area), reaching 2 to 3 W/m^2. In large metropolitan areas (tens of square kilometers), this value is one or two orders greater, i.e., up to tens and hundreds of W/m^2. One can just calculate how this additional heat generation influences the mean surface temperature of the Earth (Budyko, 1962).

In the case of a 1% change in the energy received from the Sun, the Earth's mean surface temperature would vary by 1.5°C. If it is assumed that the present heat production by man is about 0.006% of the total radiation absorbed by the Earth-atmosphere system, then the corresponding increase in mean temperature would be approximately 0.01°C. This is a comparatively small amount, but since anthropogenic heat sources are not uniformly distributed over the Earth's surface, the temperature increase in certain regions could be considerably greater.

The same paper (Budyko, 1962) noted that in the absence of atmospheric circulation, the temperature might increase on the order of 1°C in the most developed industrial regions and by tens of degrees in large cities, which obviously would make life there impossible. Atmospheric circulation considerably moderates the corresponding temperature elevation; the smaller the area where additional thermal energy generation is concentrated, the greater the moderating effect.

Flohn (Inadvertent Climate Modification, 1971) has made a detailed analysis of consumption of energy that serves as an additional heat source for the atmosphere. He established that in the center of New York and Moscow the amount of anthropogenic heat released is several times greater than the influx of solar energy. In a number of smaller cities and in the most industrially developed regions measuring tens of thousands of square kilometers, the additional heat released equals 10 to 100% of the solar energy influx. This, of course, leads to a higher temperature in the cities, where it is usually 2°to 3°C warmer than in the suburbs.

In many countries hundreds of thousands of square kilometers in area, the additional heat influx amounts to 1% of the solar energy.

Flohn's data confirm the above conclusion that not only in large cities but in considerable territories of industrial regions, energy generation by man is an important climate-modification factor.

The aforementioned study (Budyko, 1962) noted that a several percent annual increase in energy consumption over a period of 100 to 200 years will result in a drastic rise in air temperature over the entire planet. Since the growth of global energy consumption has slowed down lately, the possibility of our reaching the "thermal barrier" on the global scale is attracting little attention now. It may be that this question will not really take on any practical value until the middle or end of the twenty-first century, but we cannot rule out the possibility of a rapid increase in global energy consumption, especially if the sources of that energy become even more accessible. A case in point is the problem of thermonuclear fusion. If practical application of the fusion reaction is achieved, the question of reaching the "thermal barrier" may become critical. It therefore cannot be ignored as a vital aspect of the problem of anthropogenic climate change. In addition, in the immediate future direct heating of the atmosphere as it relates to climatic change will be a source of concern solely for cities and limited areas with a high concentration of power-consuming industries.

Other factors. Among other anthropogenic factors that could exert some influence on global climate is a possible increase in the mass of anthropogenic aerosol in the atmosphere. As noted in section 1.3, natural variations in the amount of aerosol particles in the lower stratosphere have a definite, albeit rather limited influence on climatic change. The possibility of a similar effect by an increase in anthropogenic aerosol on global climate has been discussed in many studies; the recent conclusion has usually been that such an effect cannot be very significant. One reason for this conclusion is that most anthropogenic aerosol is concentrated in the lower troposphere, where the lifetime of the aerosol particles is only several days because they are quickly removed from the atmosphere by precipitation and vertical air movements. These conditions cause anthropogenic aerosol to be distributed mainly over limited regions with the most intensive economic activity which occupy a small part of the Earth's surface.

Previous studies (Budyko, 1974, and subsequent reports) have noted that irrigation of arid regions can have some effect on the mean temperature of the lower air layer. According to monitoring data, the Earth's surface albedo can be decreased by approximately 0.10 when arid regions are irrigated. The albedo (reflectivity) of the Earth-atmosphere system, which under cloudless conditions is a function of scattering and absorption of solar radiation in the atmosphere, will vary somewhat less, in this case by approximately 0.07. Assuming that irrigation takes place mainly in regions with few clouds, the latter value may be used to estimate the influence of irrigation on the rise in mean global temperature. In this case the relative area of irrigated land (0.4% of the Earth's total surface area) and the dependence of mean temperature on the albedo of the Earth-atmosphere system should be taken into account. Climate theory studies have established that a 0.01 decrease in albedo results in a 2°C rise in mean temperature. Keeping this in mind, we find that irrigation elevates the mean temperature by about 0.05°C. Although such a temperature increment is significant, additional heating of the atmosphere due to expansion of irrigated areas will be considerably less than the warming produced by other anthropogenic factors.

Besides irrigation, the construction of reservoirs can affect the mean surface air temperature. The creation of reservoirs in regions with vegetation cover decreases the mean albedo to the same extent as the irrigation of arid regions. But since the largest artificial reservoirs have been constructed in regions with comparatively wet climate, with more or less considerable cloud cover, the albedo of the Earth-atmosphere system is changed less in this case than over irrigated regions, where the cloud cover is sparse. Also, since the total area of artificial reservoirs is considerably smaller than that of irrigated lands, their influence on the Earth's mean surface temperature turns out to be comparatively slight.

The possible change in mean air temperature due to anthropogenic effects on the vegetation cover has been discussed in the article by Sagan, Toon, and Pollack (1979). They list the various forms of man's impact on the vegetation cover and evaluate the resulting changes in global albedo and mean global surface air temperature. They believe that man's destruction of the savannahs and their conversion into desert (which took several thousand years and encompassed an area of 9.10^6 km^2, thus increasing the Earth's surface albedo by 0.19), as well as the felling of tropical forests (7.10^6 km^2, increasing the albedo by 0.09), were of paramount importance. These are the main processes that have caused the global albedo of the Earth-atmosphere system to increase by 0.006, which should have led to an approximately 1°C drop in mean surface air temperature.

They also assume that during the last 25 years the albedo has increased for the same reason, resulting in a 0.2°C drop in temperature. Those authors believe that the temperature changes they discovered explain the Little Ice Age as well as the temperature drop of the 1950s and 1960s.

Although the estimates obtained by Sagan et al. are somewhat exaggerated (the actual albedo variations evidently were less than the values used in the calculations), the effect that they suggest the albedo variations had on temperature probably did occur. It must be kept in mind, however, that utilization of wood from the cleared forests was accompanied by release of a considerable amount of CO_2 , which caused an increase in the mean air temperature. As a result, the effect the destruction of the vegetation cover has had on air temperature change has been compensated for to a considerable extent.

Climatic conditions are affected to some degree by the urbanization process, whereby former forest and crop areas become covered by buildings, asphalt pavement, and so on. Sagan, Toon, and Pollack (1979) also estimated this effect, stating that this process has covered an area of 1.10^6 km^2, thus decreasing its albedo from 0.17 to 0.15. The associated decrease in global albedo was $2.5 \cdot 10^{-5}$, which raised the mean surface air temperature by approximately 0.005°C.

Considering all the ways in which human activity can affect the Earth's albedo, we must conclude that in the immediate future it will not have any effect on global climate that can compare with the change induced by the increase in concentration of CO_2 and of a number of trace gases in the atmosphere.

1.2. Climatic Sensitivity to Anthropogenic Factors

Methods for estimating climatic sensitivity. The central issue in studying anthropogenic climatic change is determination of climatic sensitivity to different factors.

To clarify this question, it was very important to introduce the concept of climatic system, which incorporates those components of the Earth's outer shell in which physical and chemical processes directly affecting climate occur. This shell includes the atmosphere, the hydrosphere, the cryosphere, and the upper, comparatively thin, layer of lithosphere. Major external factors influencing climate are the influx of solar radiation into the outer boundary of the atmosphere, the lithosphere-atmosphere-hydrosphere gas exchange, and Earth's surface relief.

This definition of the climatic system is very general. The climatic system may be considered more limited as applied to particular problems of studying climatic changes, which simplifies the solution to these problems.

Thus, the most important factors for the relatively short-term anthropogenic climatic change taking place now are the physical and chemical processes in the lower layers of the atmosphere (troposphere and lower stratosphere) and in the upper layer of the ocean. Most of the cryosphere involving vast land glaciations is not considered a part of the climatic system under consideration, whereas sea ice and snow cover on the continents are a substantial part. It is obvious that in this case man's economic activity should be considered an important external factor of climatic changes.

Only models of climatic theory were used initially to study climatic sensitivity to anthropogenic effects. It was believed that they would permit a fairly accurate and detailed investigation into the sensitivity of the climatic system to the effects of external factors. Although the use of climatic models has enabled us to make considerable advances in solving this problem, they are still far from yielding precise answers to the myriad questions about expected anthropogenic climatic changes. For this reason, rather soon after it was found how difficult it is to determine a number of the climatic system sensitivity parameters using model calculations, empirical methods began to be employed. These proved to be very fruitful.

These methods are based on relationships obtained by studying data from meteorological observations made while the world network of stations existed (i.e., in the last 100 years) and on patterns revealed by analysis of climatic conditions existing in the geological past.

The feasibility of using two or three independent methods to estimate climatic system sensitivity to external effects has greatly enhanced the validity of these estimates when there is fairly good agreement between the results. This approach was proposed in the late 1970s (Budyko et al., 1978) and is one of the most important features of the anthropogenic climatic change studies made within the framework of the Soviet Climatic Program. We will discuss the results of applying this method to solving the major problems in studying climatic system sensitivity to anthropogenic factors.

Data on variations in mean global temperature of the lower air layer, which is a tangible measure of the energy balance of the climatic system, generally are used to evaluate climatic system sensitivity. It was found in the 1960s that outgoing longwave radiation is linearly related to the temperature of the lower atmosphere layer (Budyko, 1968). Consequently, the anomaly of mean global temperature is proportional to the anomaly of outgoing radiation, and the latter determines the sign of mean temperature change with time and the rate of this change, when there is constant influx of solar radiation absorbed by the Earth.

A number of studies (they are discussed in more detail in subsequent chapters) have established that, during climatic fluctuations, regional changes in air temperature and other meteorological elements are rather closely related (although not always unequivocally) to changes in mean global temperature. Therefore, in studying the effects of different factors on the climatic system, estimates of the influence of these factors on the mean temperature of the lower atmospheric layer are very important. This is determined either for the Earth as a whole or just for the Northern Hemisphere, where the network of meteorological stations is much denser than in the Southern Hemisphere.

Climatic reaction to variation in atmospheric gas composition. The first fairly detailed theoretical study of the climatic system sensitivity to the increase in atmospheric CO_2 (Manabe and Wetherald, 1967) showed that the mean temperature of the lower air layer is almost a logarithmic function of CO_2 concentration and that the temperature will rise by 2.4°C if the CO_2 content is doubled. The authors of that work made calculations for mean global conditions and ignored the influence of latitudinal and longitudinal variations in meteorological elements on climatic sensitivity. That study, as well as other theoretical research cited below, determined climatic sensitivity by comparing two steady states of the climatic system.

Analogous calculations were later made on the basis of general atmospheric circulation models, in which the spatial distribution of meteorological elements was taken into consideration, thus making it possible to factor in the effects of snow and ice cover on climatic system sensitivity (Manabe and Wetherald, 1975, 1980; Hansen et al., 1979, 1984; Manabe and Stouffer, 1980). G. I. Marchuk and his colleagues (Marchuk, 1979; Dymnikov et al., 1980; and others) have made a significant contribution to the theory of climate. Estimates of climatic sensitivity to a change in CO_2 concentration have also been obtained using various simplified models (Budyko, 1974; Ramanathan et al., 1979; Mokhov, 1981; and others).

The first two columns of Table 1.1 show the results of determining ΔT_C, which is equal to the difference between the mean global temperature of the lower air layer with doubled CO_2 concentration and with its preindustrial value obtained in climatic theory model calculations.

Table 1.1. Change in Air Temperature (°C) with Double CO_2 Concentration, According to Data of Different Researchers.

Simplified Climatic Theory Models	General Atmospheric Circulation Models	Modern Climate Changes	Climate Changes in Geological Past
1. 2.4	5. 2.9	12. 3.3	15. 3.5
2. 2.5–3.5	6. 3.9	13. 2.0–3.0	16. 3.4
3. 3.3	7. 3.5	14. 2.1–4.2	17. 2.8
4. 3.5	8. 2.0		18. 3.0
	9. 2.0		
	10. 3.0		
	11. 4.2		

Note. 1, 5, 10: Manabe and Wetherald, 1967, 1975, 1980. 2, 12, 15, 16: Budyko, 1974, 1977, 1979, 1980. 3: Ramanathan et al., 1979. 4: Mokhov, 1981. 6, 7: Hansen et al., 1981, 1984. 8: Manabe and Stouffer, 1980. 9: Dymnikov et al., 1980. 11: Hansen et al., 1981. 13, 14: Vinnikov and Groysman, 1981, 1982. 17: Climatic Effects of Increased Atmospheric Carbon Dioxide..., 1982. 18: Budyko, Ronov, and Yanshin, 1985.

It is obvious that these estimates of the parameter ΔT_C differ only slightly and that their mean value is close to 3.0°C. The same value was accepted as the most valid in surveys on the CO_2 effect on climate (Carbon Dioxide and Climate, 1979, 1982; Budyko, Vinnikov, and Yefimova, 1983; Bolin et al., 1986).*

*Additional information on using climatic theory models to estimate the effect of CO_2 on air temperature is presented in Chapter 6.

Another approach to determination of climatic system sensitivity is based on data regarding modern climatic changes obtained from meteorological observations.

Monitoring data in lower air layer temperature from the world network of meteorological stations can be used to calculate the mean Northern Hemisphere temperature for each year of the last 100 years (earlier, the number of meteorological stations was insufficient for a more or less accurate determination of the mean hemispheric temperature). Although these monitoring data are more limited in the Southern hemisphere, making them difficult to use in calculating mean global temperature, it is clear that the variations in mean temperature in the Northern and Southern hemispheres are comparable.

The physical mechanism of modern climatic change can best be explained by comparing the secular change in air temperature anomalies for the Northern Hemisphere with the anomalies of direct solar radiation under clear sky shown in Figure 1.1 (Budyko, 1972). The temperature and direct radiation anomalies in this case are smoothed over 10-year running periods.

It is easy to see the definite similarity between curves 1 and 2 in Figure 1.1. Both curves have two peaks, one referring to the late nineteenth century and the other to the 1930s. But there are some differences: the first peak is more pronounced in the secular change in radiation than that of temperature. The similarity between these curves allows us to assume that variations in radiation caused by instability of atmospheric transparency are an important factor in climate fluctuations. Calculations that confirmed this hypothesis were made to clarify this question.

Returning to Figure 1.1, we note that the far smaller temperature increase during the first radiation peak as compared with the second can be attributed to the influence of thermal inertia in the climatic system (mainly, thermal inertia of the upper ocean layer).

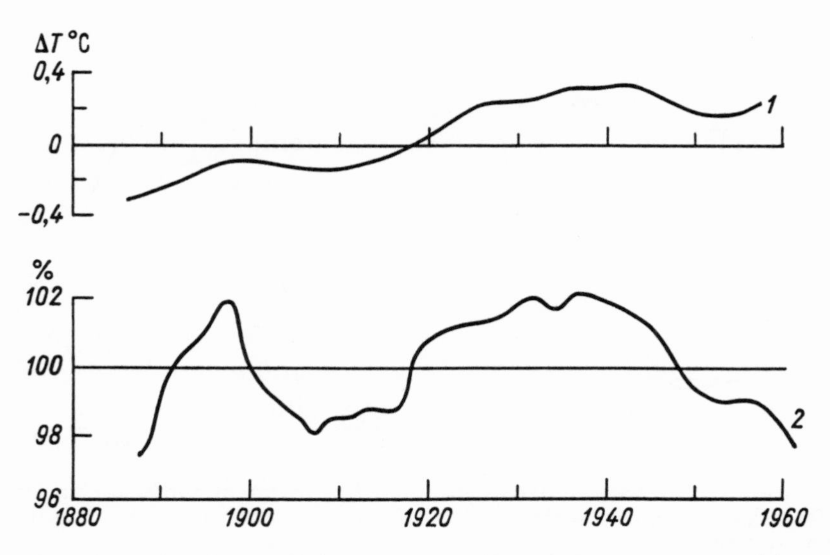

Figure 1.1. Secular variations in anomalies of air temperature (1) and direct solar radiatuon under clear sky (2).

Figure 1.2 compares the secular change in mean air temperature in the Northern Hemisphere smoothed over 5-year running periods with temperature variation due to increase in CO_2 concentration (Budyko, 1977). As this figure shows, the monitoring data agree fairly well with the trend of rising temperature caused by higher CO_2 concentrations if the effect of anomalies in atmospheric transparency on temperature variations is taken into account. To use the data on atmospheric transparency for the last 10–15 years is difficult owing to considerable air pollution at many radiation stations.

This conclusion can be confirmed by different means. One is to calculate ΔT_C according to temperature anomaly data at the beginning and end of the monitoring period, with time intervals having approximately the same atmospheric transparency values. This calculation yielded $\Delta T_C = 3.3°C$ (Budyko, 1977). Analogous calculations were made later in greater detail (Vinnikov and Groisman, 1981, 1982),

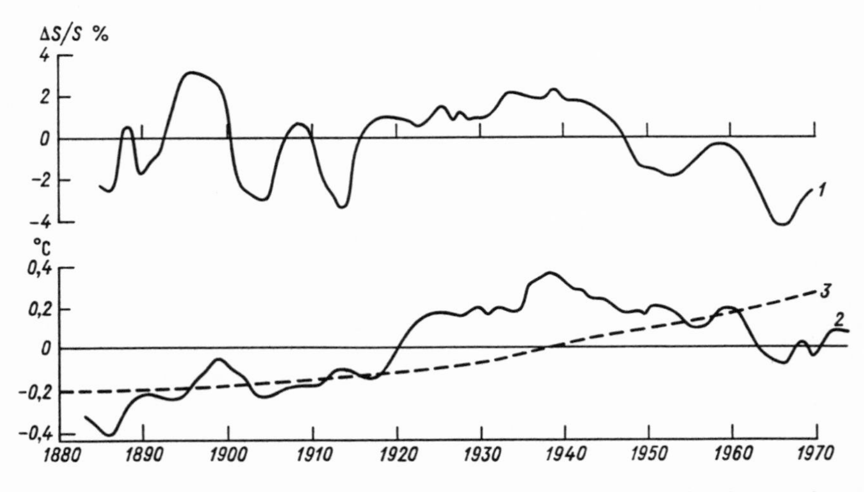

Figure 1.2. Secular variations in anomalies of direct radiation under clear sky (1), mean surface air temperature smoothed over running 5-year periods (2) by observational data and temperature variations due to higher CO_2 concentration (3).

yielding a value of ΔT_C 2.1 – 4.2°C in different versions of the calculation.

A third method for estimating climatic sensitivity to variations in CO_2 concentration is based on paleoclimatic evidence. The idea of this method is to use data on climatic conditions during warmer epochs in the geological past as analogs of the climate in the near future that is changed as a result of increased CO_2 concentration in the atmosphere. This approach can be especially fruitful if the warm climates in the past as well as expected warming in the future can be attributed to an increase in the CO_2 content in the atmosphere.

Detailed investigations of the atmospheric chemical composition in the geological past that confirm this assumption have been made in the last decade (Budyko and Ronov, 1979; Budyko, Ronov, and Yanshin, 1985; etc.). Mean CO_2 concentrations at the end of the Phanerozoic taken from the latter study are presented in Table 1.2. It is evident that at the end of the Phanerozoic a trend of diminishing CO_2

content in the atmosphere dominated; however, this decline was irregular. During all the epochs of this time interval, CO_2 concentration was higher than the 0.028% of the nineteenth century. Data on climates of the late Neogene, when the Earth's topography differed less from the present than earlier epochs, may be especially useful as the analogs of future climatic conditions.

Table 1.2. CO_2 Concentration (%) at the End of the Phanerozoic

Time Interval	Duration of Time Interval (10^6 yr)	CO_2 Concentration
Late Cretaceous	101 - 67 = 34	0.178
Paleocene	67 - 58 = 9	0.076
Eocene	58 - 37 = 21	0.120
Oligocene	37 - 25 = 12	0.032
Miocene	25 - 9 = 16	0.076
Pliocene	9 - 2 = 7	0.045

One should remember, however, that past climatic conditions differed from the modern climate not just in terms of higher CO_2 concentration. Estimates of the influence of these factors on mean global temperature of the lower air layer are shown in Table 1.3.

Using empirical data on mean temperatures for the late Phanerozoic, one can solve the inverse problem, i.e., find the value ΔT_C that describes the climatic system sensitivity to variations in CO_2 concentration. The calculation results in Table 1.1 yield values of ΔT_C of 2.8 to 3.5°C.

Here are data from a similar calculation in *History of the Earth's Atmosphere* (Budyko, Ronov, and Yanshin, 1985).

Table 1.3. Difference Between Mean Air Temperature (°C) in the Geological Past and the Present

Time interval	ΔT_s	ΔT_α	$\Delta T'$	ΔT
Late Cretaceous	−0.6	3.2	7.7	10.3
Paleocene	−0.4	3.0	4.0	6.6
Eocene	−0.3	2.8	6.0	8.5
Oligocene	−0.3	2.8	0.3	2.8
Miocene	−0.1	2.5	4.0	6.4
Pliocene	0.0	1.6	1.8	3.4

Note. ΔT_s is the temperature difference due to solar radiation variation, ΔT_α to albedo variation, $\Delta T'$ to CO_2 content variation. The value ΔT is calculated from the data on CO_2 content given in Table 1.2, taking into account the logarithmic dependence of temperature differences on CO_2 content and assuming that $\Delta T_c = 3.0$°C. The total temperature change $\Delta T = \Delta T_s + \Delta T_\alpha + \Delta T'$.

Figure 1.3 depicts the values of $\Delta T'$ as a logarithmic function of CO_2 concentration for several epochs of the Cenozoic era. In this case $\Delta T' = \Delta T - \Delta T_s - \Delta T_\alpha$, where ΔT is determined from empirical data, while ΔT_s and ΔT_α are taken from Table 1.3. After computing the slope angle of the straight line drawn through the points in Figure 1.3, we obtain $\Delta T_C = 3.0$°C. The good agreement between the estimates of climatic sensitivity to variations in CO_2 concentration, obtained by these three independent methods, becomes obvious if we compare the values of ΔT_C calculated according to the climate theory models (the first two columns in Table 1.1) with those found by empirical methods (the third and fourth columns). This conclusion is of

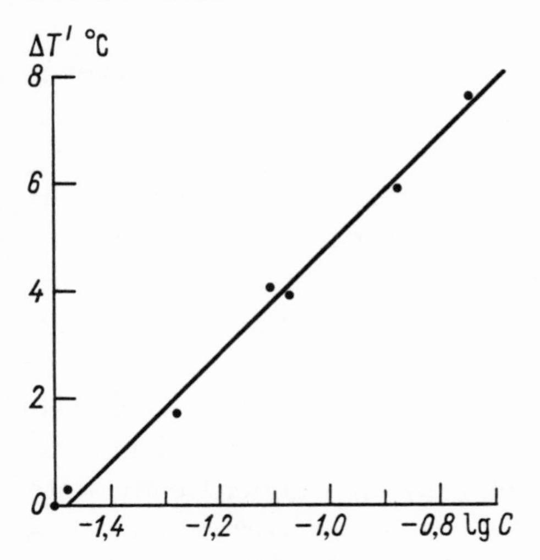

Figure 1.3. Determination of ΔT_C -parameter.

paramount importance for substantiating the validity of estimates of future climatic conditions based on the value of ΔT_C obtained in the aforementioned studies.

Here we note the possible use of these data on past climatic conditions to verify the theoretical conclusion that the temperature increment is a logarithmic function of atmospheric CO_2 concentration. The data in Figure 1.3 confirm the existence of this dependence. This figure is plotted on a semilogarithmic scale. The proximity of the empirical data, represented here as points to a straight line, corroborates the logarithmic relationship between the variables considered.

Let us now consider climatic system sensitivity to a gradual increase in CO_2 concentration. It is obvious that the variation in mean air temperature accompanying a gradual increase in atmospheric CO_2 content should be somewhat less than this variation calculated as the temperature difference for the two steady states, for the elevated and preindustrial values of CO_2 concentrations. This decrease will be caused by thermal inertia of the upper layers of the ocean.

There are empirical data showing that the transience of the climatic system has relatively little influence on anthropogenic temperature increase. As noted below, study of the influence of atmospheric transparency fluctuations on the mean temperature of the lower layer concluded that a 1% decrease in the "meteorological" solar constant (i.e., the influx of radiation into the upper boundary of the troposphere) by stratospheric aerosol lowers air temperature at the Earth's surface by 1.2°C (Budyko, 1974). This calculation was based on average data for 30 years. The most reliable climate theory models show that under steady-state conditions, this difference in heat influx corresponds to a drop in air temperature by 1.5 – 2.0°C (Budyko, 1980). The thermal inertia of the climatic system for this time interval therefore reduces the mean temperature change by approximately 20–40%.

Empirical data thus show that, for the typical 30-year period of variation in climate-forming factors, thermal inertia of the oceans reduces by approximately 30% the temperature change that would have occurred in an inertialess climatic system.

These calculations show that most of the current anthropogenic temperature increase is due to the change in chemical composition of the atmosphere over the last several decades. It can be demonstrated that a similar situation will also occur in the near future. Therefore, there is reason to believe that the rise in mean global temperature should be approximately 30% less than that estimated for steady state.

A number of theoretical studies have discussed the influence of thermal inertia of the climatic system on anthropogenic temperature increase. Some have obtained higher estimates of the moderation of temperature variation than the indicated value and some have obtained lower estimates.

This difference can be attributed to their essential dependence on the parameterization of heat-exchange processes in the ocean, which is not realistic enough in all studies. Those investigations based on the most valid oceanic heat-exchange models have yielded estimates of the influence of climatic system thermal inertia on temperature increase that do not differ from the aforementioned empirical calculation. The work done by Byutner (1983) and a number of other authors is representative of these studies.

The influence of thermal inertia on current climatic change is discussed in greater detail in Chapter 6.

The effect on climate of increases in the amounts of trace gases and CO_2 in the atmosphere is calculated in a similar manner since in both cases the enhanced greenhouse effect causes temperature of the lower air layer to rise. One-dimensional climatic models (the so-called radiative-convective models) are generally used to calculate the effect of trace gases, because empirical methods cannot be applied owing to the comparatively weak effect of trace gases on past climatic conditions.

1.3. Future Climatic Change

Detection of anthropogenic climatic change. In order to ascertain the validity of estimates of future changes in climatic conditions caused by human activity, it is very important to determine the anthropogenic climatic change that has occurred during the existence of the world meteorological stations network, i.e., in the last 100 years.

As already mentioned, anthropogenic climatic warming during that time resulted primarily from an increase in the CO_2 content in the atmosphere. Combustion of different kinds of fossil fuel has caused the atmospheric CO_2 concentration to rise 20 to 25% to date. Assuming that $\Delta T_C =$ 3.0°C and taking into account the logarithmic dependence of the mean lower air layer temperature on atmospheric CO_2

concentration,* we find that this rise in CO_2 mass should correspond to a 0.9°C elevation in mean global temperature. Since the rate of increase in atmospheric CO_2 has gradually increased during the last 100 years, the majority of anthropogenic CO_2 was released into the atmosphere in the last several decades. It can consequently be assumed that the influence of the climatic system thermal inertia should have caused an approximately 30% drop in temperature elevation. The warming in this case should be about 0.6°C (see Chapters 2, 6 and 9).

Detection of such change in mean global temperature would not be difficult if this temperature varied only under the influence of anthropogenic factors. However, previous investigations have established that the temperature change is influenced noticeably by natural factors, and as a result short-term natural temperature variations somewhat mask the longer-term anthropogenic temperature change.

We have already mentioned that mean global air temperature is affected by atmospheric transparency fluctuations, with temperature rising and falling along with transparency. A noticeable year-to-year variability in temperature also exists, caused mainly by instability in general atmospheric circulation. The influence of this year-to-year variability on temperature can be significantly eliminated by smoothing out the temperature data for 5 or 10 years. After this operation, the effect of atmospheric transparency variations on air temperature can be estimated and the influence of natural and anthropogenic factors on this temperature can be separated (Budyko, 1977).

Figure 1.2 gives data characterizing variations in the Northern Hemispheric mean surface air temperature averaged over running 5-year periods (curve 2) and the results of calculating temperature changes due to the increase in CO_2

---

*That is, assuming that the temperature change is proportional to logarithmic ratio of concentration to its initial value.

concentration (curve 3), all based on the assumption that a doubling of CO_2 concentration leads to a 3°C rise in mean surface temperature. The nonlinear dependence of higher temperature on greater CO_2 content is also considered.

The data in this figure indicate that, although the actual temperature variations differ from the calculated values, monitoring data confirm a tendency toward a higher temperature in the second half of the period under consideration as compared with the first. A case in point is that before 1920 the temperature never reached the mean value for the entire period, whereas after that year, except for a brief interval, it was above the mean value. It is also evident that the increase in CO_2 concentration could have raised the mean surface air temperature by about 0.5°C since the end of the nineteenth century. Although this is not a small amount, it is masked to a considerable extent by brief climatic changes associated with fluctuations in atmospheric transparency.

Analysis of the data in Figure 1.2 leads to the conclusion that during the period of instrument observations the Northern Hemispheric mean 5-year air temperature was a function of the increase in CO_2 concentration and atmospheric transparency fluctuations. This figure depicts direct radiation anomalies under clear sky from data of Pivovarova (1977) (curve 1) to evaluate the effect of the second factor. The values of the temperature anomalies are averaged for running 5-year intervals.

It is very interesting to compare curves 1 and 2 in Figure 1.2. The main difference between them is the tendency toward a rise in temperature (see curve 2) that is observed throughout most of this period. Such a trend is missing from the secular course of direct radiation, which is characterized by departures to both sides from the stable mean, while radiation decreases somewhat in recent decades.

Considerable similarity in the main features of the secular course of direct radiation and temperature is also quite obvious. It is easy to see that the major peaks and dips on curve 2 correlate with similar variations on curve 1, and the temperature changes usually lag a little behind the radiation fluctuations and are smoother.

The link between radiation fluctuations and long-term warming that began in the 1920s deserves particular emphasis. In 1914–15 radiation drastically increased and its positive anomaly persisted for a long time. A rise in temperature corresponding to the higher radiation level occurred in 1918–22. It should be noted that a further radiation increment in 1931–34 was followed by higher temperatures in 1935–38. The radiation minimum in 1953 resulted in a temperature minimum in 1956–57, the radiation maximum of 1959 was followed by a temperature maximum in 1960, and the radiation minimum in 1966 led to a temperature minimum in 1967. Noticeable increases and decreases in radiation, thus, have usually been accompanied by air temperature variations that lagged behind the radiation fluctuations by 1 to 5 years (most often by 3 years).

According to monitoring data, any change in radiation influx to the troposphere depends mainly on the mass of stratospheric aerosol, which, because of intense horizontal mixing, varies little within the confines of the stratosphere of each hemisphere. As a result, it is sufficient to have actinometric observational data (or aerosol concentration data) from a small number of stations to determine mean conditions of stratospheric transparency.

The influence of variations in atmospheric transparency and CO_2 concentration on air temperature can be studied using the given relationships.

The sign of difference between the temperature anomalies (see curves 2 and 3 in Figure 1.2) generally coincides with the sign of the direct radiation anomalies (curve 1). This demonstrates definitive influence of radiation fluctuations on the shorter variations in the lower-layer air temperature of the Northern Hemisphere.

It is interesting to compare the values of temperature anomalies caused by radiation fluctuations with the values of the radiation anomalies. According to the data in Figure 1.2, it can be established that the ratio of the temperature anomaly caused by radiation fluctuations to the radiation anomaly rises with an extension of the period during which the sign of the corresponding anomalies remains unchanged. Its mean value was 1.3% during the longest positive radiation anomaly (from 1920 through 1945). Calculations based on atmospheric optics equations show that this value corresponds to an approximately 0.2% increase in total radiation (Budyko, 1971). The temperature anomaly caused by atmospheric transparency fluctuations (equal to the mean difference between the values represented by curves 2 and 3) was 0.24°C during the period involved. The ratio of the second quantity to the first (ΔT_1) equals 1.2°C per 1% change in total radiation.

The series of studies published from 1977 to 1983 (Budyko, 1977, 1980; Vinnikov and Groisman, 1981, 1982; Budyko and Vinnikov, 1983) is valuable in detecting the anthropogenic climatic change. These studies are based on the following assumptions.

Mean air temperature anomalies can be considered to be the sum of three components. The first is the result of the effect of natural factors on climatic change: the second is usually called climatic noise. We emphasize that this term is arbitrary because this component is actually the sum of temperature changes induced by random atmospheric circulation fluctuations, errors in estimating all analysis data, and variations in the anomalies caused by those factors of natural, determined climatic changes, which are ignored in the calculations. The third component is caused by anthropogenic factors and should be found from analysis of data from observations of mean temperature variations.

The only factor of natural variations in global climate whose influence has been conclusively established in modern theoretical and empirical studies is atmospheric transparency fluctuations caused by variations in stratospheric aerosol concentration. Although it is not particularly difficult to calculate the effect of this factor on air temperature, it should be as accurate as possible. The most reliable results are based on monitoring data of atmospheric transparency fluctuations from existing actinometric stations. Use of these data is based on the assumption that the amount of tropospheric aerosol for the entire Earth and the Northern Hemisphere has changed little during the last 100 years. The transparency fluctuations consequently can be attributed mainly to unstable concentration of stratospheric aerosol. This assumption is confirmed by the known fact of the brief lifetime of tropospheric aerosol (about ten days), after which it is primarily removed by precipitation. By contrast the lifetime of stratospheric aerosol particles is measured in months and years.

Two criteria can be used to estimate the effect of CO_2 on mean air temperature variation: the mean air temperature increase in the last century (or a shorter period) caused by the increase in both CO_2 concentration and air temperature from doubling of CO_2 concentration (the second criterion can be obtained from the first). Both are well-known values from data other than from air temperature monitoring. It has already been stated that these temperatures equal approximately 0.6°C and 3°C, respectively.

Two methods can be used to determine these values from air temperature measurement data. The first computes the difference between temperature anomalies for various CO_2 concentrations, if these anomalies refer to similar atmospheric transparency conditions or if the effect of transparency fluctuations on the air temperature anomalies is taken into account by making the proper corrections. By using this method it is possible to detect the temperature anomaly caused by the CO_2 effect, if it is large enough as compared with the aforementioned "noise" level.

The second method is based on the energy-balance equation for the Earth-atmosphere system in a transient state. It incorporates the relationship between outgoing longwave radiation and CO_2 concentration. The parameters in this relationship are found from monitoring data on mean surface air temperature variations. This approach makes it possible to verify the hypotheses of the significance of individual climatic change factors.

The results of one of the aforementioned works (Budyko, 1977) indicate the possibility of determining the increase in the Northern Hemispheric mean temperature caused by the higher CO_2 concentration by comparing the mean air temperature anomalies for two time intervals: 1881 to 1890 and 1961 to 1970. Actinometric monitoring data established that, although the mean atmospheric transparency for both intervals was noticeably lower, its values were practically the same. The 0.5°C temperature increment observed at that time could therefore be attributed to anthropogenic factors. This assumption was confirmed by the almost exact coincidence between the measured temperature increase and the analogous increment computed by climatic theory models with allowance for the CO_2 concentration increase that had occurred.

K. Ya. Vinnikov and P. Ya. Groisman (1981, 1982) detected further mean air temperature variations in the Northern Hemisphere caused by a rise in atmospheric CO_2 concentration. These authors used a transient model of the Earth's energy balance that was based on climatic system thermal inertia, albedo variations due to atmospheric transparency fluctuations, the effect of atmospheric CO_2 concentration on outgoing long-wave radiation, and the albedo-temperature feedback.

This model can be used to find the ΔT_C parameter equal to 2.1 to 4.2°C, which is close to its theoretical estimates.

The analysis demonstrated that the anthropogenic rise in atmospheric CO_2 concentration led to a $0.4 - .06°C$ higher annual mean surface air temperature in the Northern Hemisphere at the end of the period covered compared with the mid-1880s. This warming sometimes increased or decreased depending on the atmospheric transparency fluctuations. The hypothesis that CO_2 has no effect on mean air temperature variations was refuted with over 99% probability (see also Chapters 7 and 11).

By analyzing all of the aforementioned studies, one can conclude that they reveal the current anthropogenic global climatic change very reliably.

We note that a similar conclusion can be drawn from an analysis of the results obtained by Miles and Gildersleeves (1977), Hansen et al. (1981), and some other researchers, although they themselves did not draw such a conclusion.

We can see that research on anthropogenic climatic change in other countries lagged somewhat behind the Soviet studies. Reports of meetings held abroad a few years ago stated that anthropogenic climatic variation had not been detected. Similar conclusions recently began to reflect the present situation better, although they still contain unsubstantiated reservations.

The problem of detecting anthropogenic climatic change is unclear to some researchers in other countries for the following reasons:

First of all, these researchers evaluate climatic sensitivity to the doubling of CO_2 concentration (parameter ΔT_C) only from climatic theory models, assuming that this parameter is within the range from the smallest to the greatest value produced by the model, i.e., ΔT_C = 1.5–4.5°C (Carbon Dioxide and Climate, 1979, 1982). This approach raises objections. First, one should use only the best available models to estimate the parameter ΔT_C , and ignore model calculation results with doubtful physical basis. Second, assuming all the models are equivalent, it is impossible to consider that the most probable range of the ΔT_C values equals the difference between its maximum and minimum estimates. Statistics make it clear that this range should be smaller than the indicated difference. The third and main objection concerns the statement that current empirical estimates of parameter ΔT_C are much more reliable than the model calculation results. The lower variability in empirical versus climatic model estimates, for example, reflect this.

Secondly, surveys make the clearly unsubstantiated statement that it is difficult to take into account the effect of atmospheric transparency fluctuations on mean temperature. It is clear that the errors in calculating the influence of atmospheric transparency variation on mean temperature fluctuations cannot be considered equal to the difference between all similar calculation results. There are several studies along this line that use absolutely invalid transparency characteristics, which lead to poor results.

Thirdly, estimates of the influence of climatic system thermal inertia on mean temperature variations can employ only those climatic models that are based on fairly reliable empirical data. It should be especially emphasized that the data on transport of some tracers from the atmosphere to the ocean cannot be used for this purpose. As Byutner (1983) showed, this is an erroneous approach, since there are different mechanisms for the transport of heat and tracers in ocean water. This is caused by interaction of surface and deep ocean layers as opposed to exchange of passive substances.

Lastly, some studies on detecting anthropogenic climatic variations "discovered" them by taking into account additional temperature-modifying factors that are actually meaningless. These factors usually include solar radiation, whose random variations allow the researcher to obtain any result desired. While the authors of some surveys correctly note the erroneous calculations, they incorrectly consider this to be proof of difficulties in solving this problem as a whole.

There is no doubt that all these questions will be cleared up in the near future, and that the problem of detecting anthropogenic climatic change will be considered completely resolved. This problem is additionally covered in Chapters 7 and 11.

Variations in mean global and latitudinal climatic characteristics. Future estimates of anthropogenic climatic change will be made in several stages. The first step is to calculate the factor variations that will exert the greatest effect on future climate. Previous reasoning considers the rise in CO_2 concentration to be the major factor. It is assumed that it will double the preindustrial level by the second half of the next century. An increase in the trace gas content intensifying the atmospheric greenhouse effect will also have a pronounced effect on global warming.

As mentioned in Chapter 3, the expected change in atmospheric CO_2 concentration can be calculated by predicting the rate of anthropogenic CO_2 influx into the atmosphere. These calculations are based on the basic scenario of future energy development covered in Chapter 2, in which the rate of industrial CO_2 emission will be 7.5 GT/yr in the year 2000 and 12.5 GT/yr in the year 2030 (it is currently 5 GT/yr). These calculations show that atmospheric CO_2 concentration will be 380 ppm in the year 2000 and 450 ppm in the year 2030.

Considering this and the fact that temperature of the lower air layer rises by 3°C when CO_2 concentration doubles, we find that the mean 1970 temperature will rise by 0.65°C in the year 2000, and by 1.4°C in the year 2030. These values are determined without regard for the effect of climatic system thermal inertia on temperature variations.

Thermal inertia (i.e., the influence on warming of specific heat and conductivity of the ocean surface layers) lessens temperature variations. The calculations show that given the current time dependence of CO_2 industrial emission rate, the ocean lessens the temperature growth by approximately 30%. The resulting change in surface air temperature caused by higher atmospheric CO_2 concentration will be 0.5°C in the year 2000 and 1°C in the year 2030 versus the 1970 level.

The higher concentrations not only of CO_2 but also of some trace gases, including methane, nitric oxides, chlorofluorocarbons (CFC), and tropospheric ozone, are exerting a greater influence on climate. Although the change in the amount of these atmospheric admixtures is small (10^{-6} to 10^{-9} of the atmosphere's volume), they have a pronounced effect on intensification of the greenhouse effect, thus causing an additional rise in temperature of the lower air layer.

The values of higher mean lower layer air temperature shown in Table 1.4 were found on the basis of climatic system thermal inertia (see Chapter 5).

Table 1.4. Expected Concentrations of Atmospheric Trace Gases and Their Influence on Rising Temperature

Substance	Admixture Ratio (10^{-9})			Temperature Increase (°C)	
	1970	2000	2030	2000	2030
Tropospheric ozone	30	50	80	0.11	0.38
Methane	1500	1900	2900	0.06	0.26
N_2O	295	340	390	0.11	0.22
CFC	0.17	1.80	6.0	0.12	0.45

This table shows that the higher quantity of minor admixtures will additionally raise the mean air temperature by 0.4°C in the year 2000 and by 1.3°C in the year 2030.

We can thus conclude that the mean temperature of the lower air layer should increase by 0.9°C in the year 2000 versus 1970, and by approximately 2.3°C in the year 2030. The first value is comparable to the temperature observations in the early 1980s. They indicate that the mean temperature then was several tenths of a degree above that in the 1960s and early 1970s.

Since the accuracy of estimating the influence of trace gases on mean temperature for the year 2050 is quite limited, one can presume that in the mid-twenty-first century, the contribution of CO_2 and trace gases to the intensified greenhouse effect will be approximately the same. It follows from this assumption that the probable rise in temperature by the mid-twenty-first century versus 1970 will be 2.5 to 3.5°C.

These conclusions are in fairly good agreement with the predicted temperature increment presented in the Soviet-American meeting report (Climatic Effects of Increased Atmospheric CO_2 ..., 1982) that contains Table 1.5.

Table 1.5. Variations in Mean Global Surface Air Temperature (ΔT)

Year	2000	2025	2050
ΔT°C	1–2	2–3	3–5

This table presents the higher temperature values versus the preindustrial period, which should be approximately 0.5°C above those cited earlier. The lowest values of temperature variation in this table are regarded as the most probable; the highest values correspond to the greatest possible warming.

The second stage in estimating future climatic conditions is to clarify spatial distribution of expected variations in meteorological elements. Since these data can be used to solve important practical problems, they should be gathered by various independent methods, in accordance with the idea stated in the beginning of this chapter.

This problem for mean latitudinal patterns of air temperature and continental precipitation has been discussed in two studies (Budyko and Yefimova, 1981; Budyko, Vinnikov, and Yefimova, 1983). The first obtained these distributions by two methods (from climatic theory models and from paleoclimatic data), and the second used three methods (the two aforementioned and from modern climatic change data).

Figure 1.4 shows the relative rise in mean annual temperature at different latitudes in the Northern Hemisphere. This figure demonstrates that all methods for determining relative temperature variations yield more or less similar

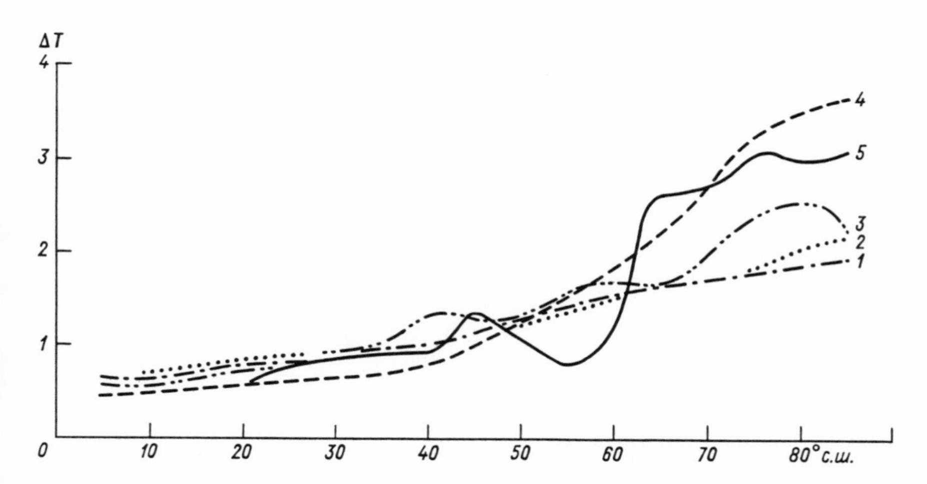

Figure 1.4. Relative change in mean annual surface air temperature at different latitudes of the Northern Hemisphere. 1–3, calculations by the general atmospheric circulation models (1, Manabe and Stouffer, 1980; 2, Wetherald and Manabe, 1981; 3, Manabe and Wetherald, 1980); 4, paleoclimatic evidence (Climatic Effects of Increased Atmospheric Carbon Dioxide..., 1982); 5, modern climatic changes (Vinnikov, Groisman, 1982).

results for latitudes below 60°N. For higher latitudes, empirical methods show a higher temperature sensitivity to the increase in CO_2 concentration compared to climate theory model results. This discrepancy hypothetically reflects an insufficiently accurate parameterization in the models of polar sea ice dynamics that significantly influence arctic air temperature.

Figure 1.5 presents the distribution of differences in total annual precipitation in various continental latitudinal zones where mean global temperature of the lower air layer rises by 1°C because of the higher atmospheric CO_2 concentration. It is evident from this figure that global warming

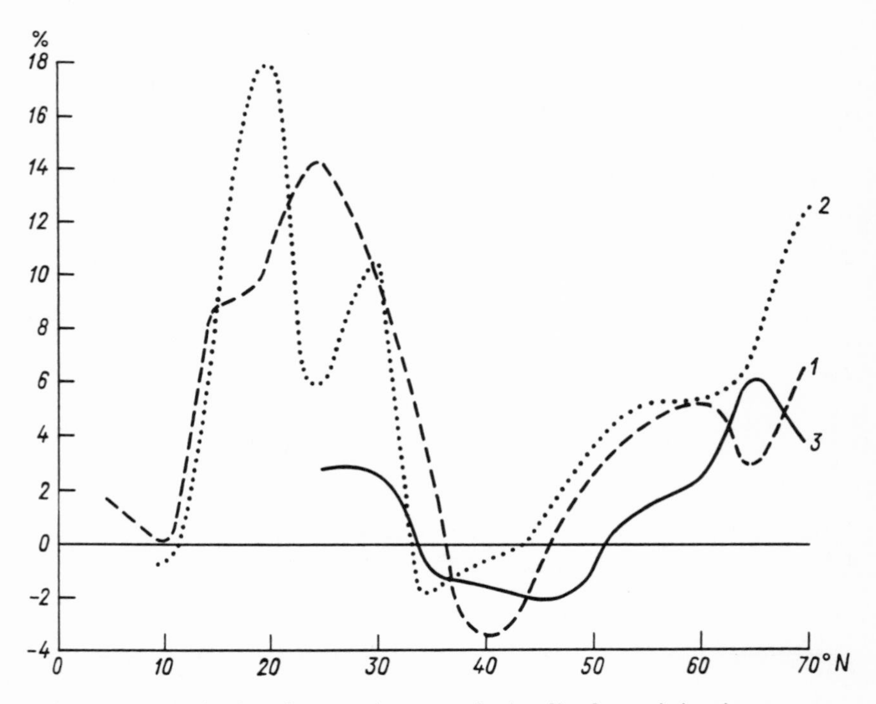

Figure 1.5. Relative changes in mean latitudinal precipitation on continents of the Northern Hemisphere with 1°C higher mean surface air temperature. 1, calculation by general atmospheric circulation model (Manabe and Stouffer, 1980); 2, Paleoclimatic evidence (Climatic Effects of Increased Atmospheric Carbon Dioxide..., 1982); 3, empirical data on modern climatic changes (Budyko et al., 1983).

causes the mean precipitation to increase noticeably in the 10–30°N zones and north of 50°N. Precipitation tends to diminish in the 30–50°N zone.

It became clear later that the spatial distribution of variations in precipitation during greater warming differs somewhat from that presented here (see Chapter 9). Nevertheless, satisfactory agreement between data on temperature variation and the amount of precipitation during moderate warming, found by various methods, shows that it is possible to make a sound scientific estimate of future climatic change. We have already mentioned that this estimate should refer not only to latitudinal zones but also to specific geographical regions. Recent studies have compiled maps of changes in a number of climatic elements resulting from the rise in atmospheric CO_2 concentration; however, each map was prepared by only one of the aforementioned methods, so it is difficult to determine their validity. Chapter 9 covers the possibility of constructing such maps by a complex method based on different approaches.

CHAPTER 2

DEVELOPMENT OF ENERGY PRODUCTION

2.1. Introduction

Energy consumption is a necessary condition of mankind's existence. Human energy consumption is 40–60 W for everyday life without heavy physical work. With a human energy efficiency of about 20%, our life support requires 2.4 to 3.6 kW $\times$ h or 8400–12,600 kJ per day. This energy enters the body when different kinds of food are eaten. Primitive man consumed precisely this amount of energy for his needs. All his comforts came from his use of muscle energy.

Civilization developed historically with steady economic growth and technological progress, higher satisfaction of physical, spiritual, and social demands of the people, and improvement in the means of production. Each stage required more energy consumption (Figure 2.1; Nuclear Power ..., 1982; Kuz'min and Stolyarevsky, 1984). It was always necessary to have energy to satisfy human needs, increase longevity, and improve living conditions.

The first jump in energy consumption occurred when man learned to obtain fire and use it for cooking and heating his dwellings. The specific (per capita) energy consumption in this historical period (approximately 100×10^3 years ago) rose several times compared with that of primitive man, reaching 25,000 kJ/day. Human muscle power and wood were the energy sources at that time. The next important stage occurred with the invention of the wheel, the discovery of methods for smelting metal and making alloys to produce various tools, and the development of forging.

By the fifteenth century, the man of the Middle Ages, using cattle, wind, and hydraulic energy to rotate water wheels and windmills, wood, as well as a small amount of coal, was already consuming approximately 120,000 kJ/day, i.e., ten times more energy than primitive man.

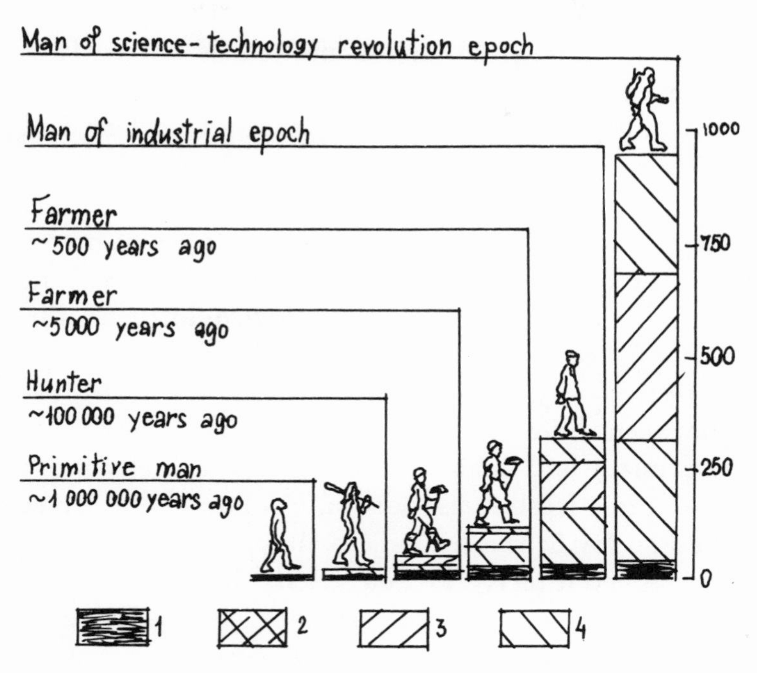

Figure 2.1. Growth of per capita (1000 kJ/day) energy consumption and changes in energy consumption structure in the course of human evolution. 1, food; 2, housework and services (including the work of institutions, trade, education, etc.); 3, industry and agriculture; 4, transportation.

The evolution of civilization occurs through invention of newer methods of energy conversion, and consequently through development of new energy sources. Energy consumption rose considerably when the steam engine was invented. As a result, per capita energy consumption in industrially developed countries reached 340,000 kJ/day by the mid-nineteenth century.

Man in the modern industrial society of highly developed countries uses about 10^6 kJ/day, i.e., 100 times more energy than primitive man. His material standard of living is consequently much higher and his life span is quadruple that of primitive man and double that of fifteenth century man.

Energetics of highly developed nations is currently a major sector of industry that consumes organic fuel (oil, natural gas, and coal), hydraulic power, and nuclear fuel as energy sources. Figure 2.2 presents a schematic example of the USSR unified energy balance structure in the early 1980s (Melent'yev, 1984). Three basic types of power plants—electrical, boiler, and so-called direct fuel use (engines and mechanisms, industrial technological furnaces, domestic heaters)—consume 40, 15, and 45% of total energy resources, respectively. It is important at this point to distinguish between different types of energy, particularly between the primary and final. The first refers to such energy carriers as organic fossil fuels, natural uranium, wind, river current, and so forth; the second, to those types of energy that are consumed directly, i.e., electricity, heat, and so on. Figure 2.2 presents approximate primary energy losses at the main conversion stages and the final energy transmitted in various forms to the national economy. These numbers demonstrate that today's energy consumption efficiency is about 36%. Its rise is the number-one national economic problem.

Energy is the basis for development of the fundamental branches of industry that determine progress in social production, since there is a direct relationship between power available per worker, automation level and electrification of the processes, on the one hand, and labor productivity, on the other hand. The rates of energy development in all industrially developed countries have exceeded those in other sectors. At the same time, energy is one of the sources of unfavorable influences on man and his environment. It influences different components of the environment: the atmosphere (oxygen consumption, emissions of gases, moisture, and solid particles), the hydrosphere (water consumption, construction of artificial reservoirs, discharge of polluted and heated waters, liquid waste), and the lithosphere

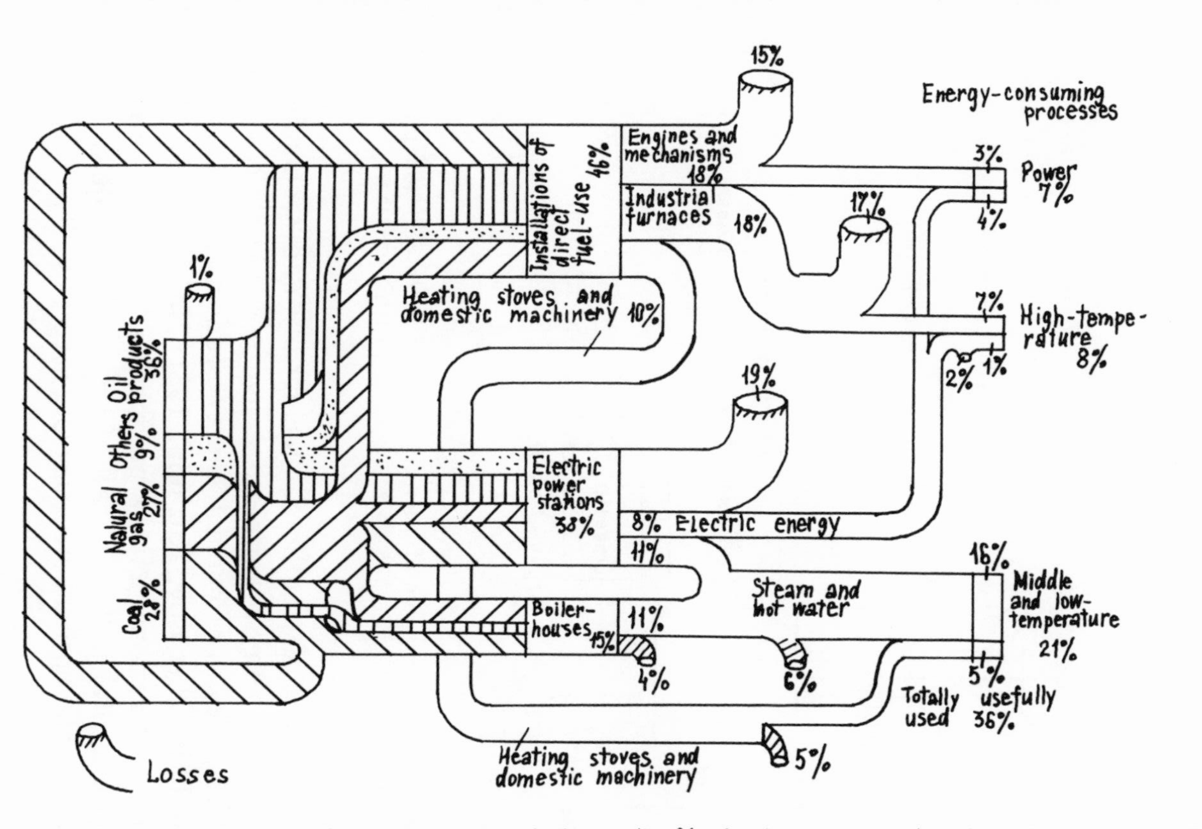

Figure 2.2. Structure of the USSR energy balance (in % of primary energy). The value is given of gainfully employed energy (36%) minus the expenditure for domestic needs.

(fossil fuel consumption, changing relief, surface and deeper emissions of solid, liquid, and gaseous toxins). The extent of this influence is determined mainly by the type of power plants. Figure 2.3 presents, as an example, the material balance of 2400 MW (elec.) coal thermal power plant (Makukhin, 1985).

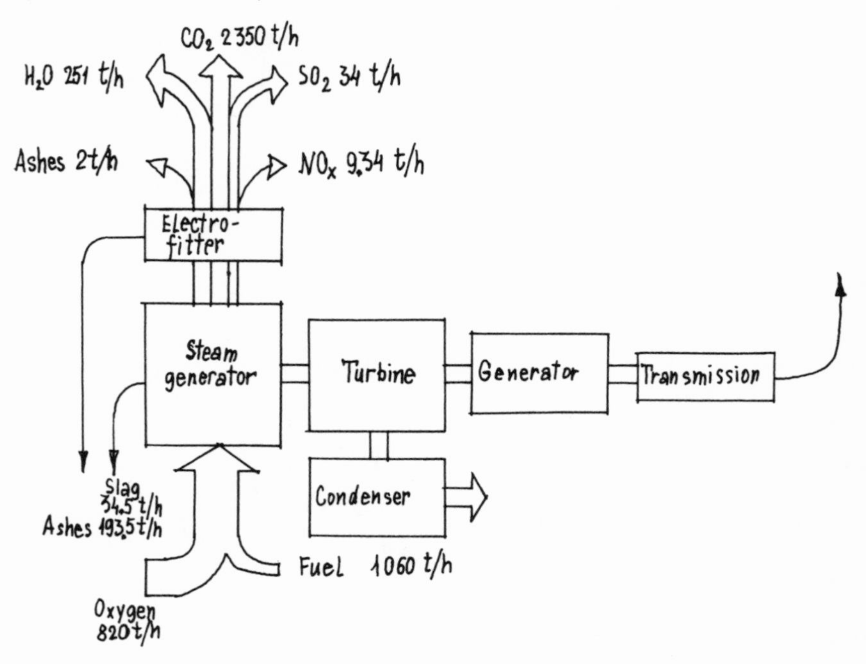

Figure 2.3. The scheme of operation of coal thermal power plant of 2400 MW.

Depletion of traditional, technically and economically accessible energy resources is a real possibility because of the large scales of current energy production and its predicted rise (Legasov, Kuz'min, 1981; Hafele, 1981). In a number of regions, energy production has caused greater environmental pollution. All of this raises the question of choosing the optimal structure of energetics.

2.2. World Energy Development

The average global primary energy consumption has increased exponentially in the last 100 years by about 2.2% per year (Hafele, 1981). As a result, the 1980 global primary energy consumption was 9.61 TW × year* (IIASA '83). Current global energy and fuel demand is met basically through fossil fuel (Hafele, 1981; IIASA '83 ..., 1983). A discussion of the evolution of the fuel and energy balance (Table 2.1) over the last 100 years will give us an insight into its modern structure. Data (Hafele, 1981) in Figure 2.4 show that in the mid-nineteenth century wood was the predominant form of primary energy resources; burning it met about 70% of the energy demand. The developing industry of the time increased coal consumption, gradually displacing wood fuel. The higher specific caloric value of coal versus wood that promotes large-scale use, transportation, and storage of energy led to basic technological changes in the energy producing system. Then, rapid development of plants and cars with internal combustion engines, which began in the early twentieth century, required a higher percentage of oil and natural gas in the fuel and energy balance. Oil and natural gas are particularly high-caloric fuels and they pollute the environment less than solid organic fuel. These advantages led to a gradual increase in the percentage of oil and natural gas in the global fuel and energy market (Figure 2.4), reaching 71.0% of total consumption in 1980 instead of 8.1% in 1920. The percentage of coal declined for the same period from 62.5 to 28.5%.

The current structure of the fuel and energy distribution is changing owing to the rapid depletion of fossil fuel reserves, primarily oil and natural gas. Oil and natural gas recovery experience shows that energy expenditures to extract them (energy outlays per unit of drilling operations)

---

*1 TW × yr = 10^9 kW × yr = 31.54 10^{18}J.

Table 2.1. Global Primary Energy Consumption

Energy Resources	1975		1980	
	TW × yr/yr	%	TW × yr/yr	%
Coal	2.26	27.5	2.56	26.6
Oil	3.62	44.1	4.26	44.4
Natural gas	1.51	18.3	1.78	18.5
Hydraulic power	0.50	6.1	0.59	6.1
Nuclear energy	0.12	1.5	0.23	2.4
Other*	0.21	2.5	0.19	2.0
Total	8.22	100.0	9.61	100.0

*Including biogas, geothermal resources, commercial use of wood fuel, etc.

is constantly growing, while the quantity extracted per unit of drilling operations is steadily declining. These trends are confirmed by U. S. oil production data (Hall and Cleveland, 1981) presented in Figure 2.5. As seen, the energy expended to produce oil in the beginning of the next century in the United States will exceed the energy that the oil recovered could produce. A similar situation is developing in other countries with oil refineries, including the Soviet Union.

Analysis of the changing structure of global energy production (see Figure 2.4) shows one very important trend in primary energy substitution: F is the change in percentage of a specific energy resource in the total consumption described by a logistic function of the time. Actually, in this case the behavior of function $F/(f-1)$ with time, describing the replacement of a specific fuel is determined by the relationship

$$\rho_n \frac{f}{F-1} = \alpha + \beta t, \tag{2.1}$$

where α and β are constants, and t is time. Equation 2.1 is a

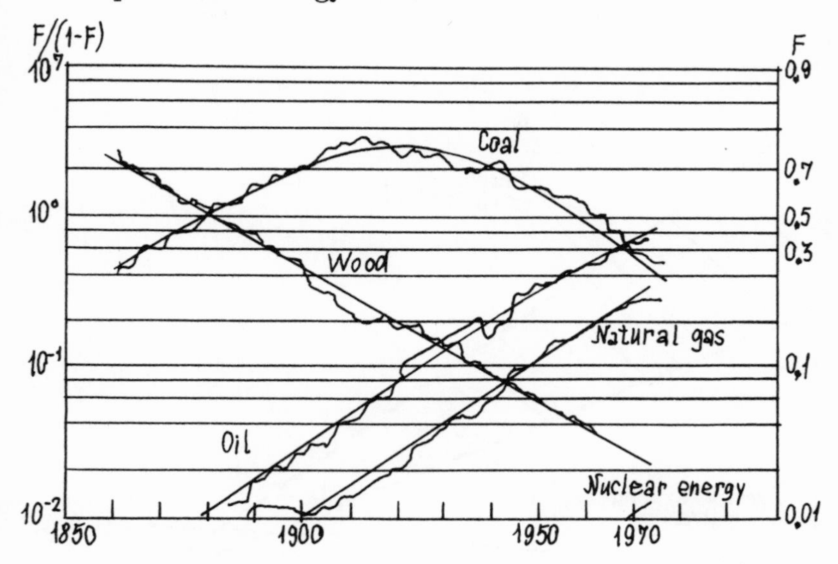

Figure 2.4. Change in structure of the global fuel and energy balance. Broken lines reflect actual dynamics of consumption, and smoothed lines, the dynamics of change in different energy resources; F is the share of the given energy source in meeting total energy demands.

straight line. Data on the dynamics of oil, natural gas, and wood consumption shown in Figure 2.4 obey this relationship. The results indicate a slow development of replacement. It took about 100 years for coal to comprise half of the total primary energy resource consumption. The same is true for oil and natural gas; it took the same amount of time for the percentage of oil consumption to reach the 50% level. Study of possible variations in the structure of fuel and energy production thus requires that long-term periods of energy development be considered.

The high rates of global energy consumption growth in the last 100 years have been caused by high rates of economic growth. It has already been mentioned that energetics to a considerable extent determines the level of economic development as a whole. Practically all aspects of life in any modern society are related to the amount of energy produced in various forms. This is statistically confirmed by Figure 2.6, which shows that global economic growth is

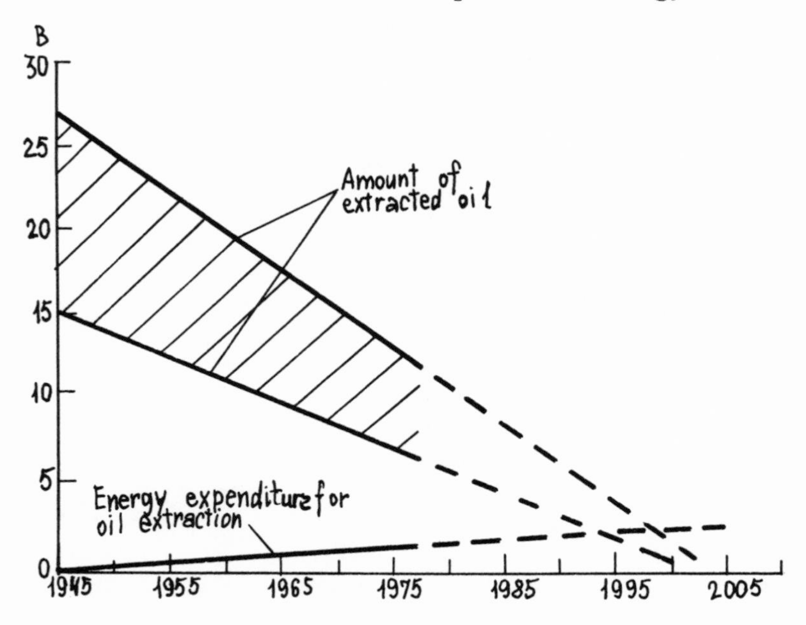

Figure 2.5. The amount of oil per unit of drilling operations obtained during its exploration and recovery in the United States (crosshatched area) and energy expenditure for its extraction. Dashed lines are the extrapolation of today's tendencies into the future. The intersection points show that the amount of energy expended per unit of extracted oil becomes equivalent to the amount of energy that can be produced from an extracted unit of oil; B is the number of barrels of light Arabian oil per 1 foot of drilling.

directly proportional to the rising energy consumption (Chant, 1981; Hafele, 1981).

At the same time, distribution of energy consumption is very uneven in various parts of the world: annual per capita primary energy consumption for 72% of the world's population is less than 2 kW × yr, for 22% it is from 2 to 7 kW × yr, and only for 6% it is about 10 kW × yr (Hafele, 1979). This figure is 0.2 kW × yr in over 80 countries, i.e., there is a 50-fold difference in per capita maximum and minimum consumption in the modern world. Figure 2.7 demonstrates the annual per capita primary energy consumption in different regions of the world (Hafele, 1981) before 1975. The global average estimate was 2.08 kW × yr in 1975.

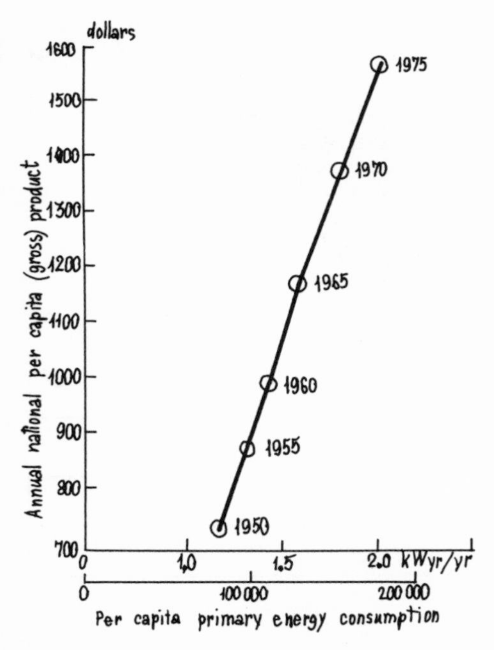

Figure 2.6. Ratio between GNP and per capita energy assumption for the world as a whole (GNP is estimated in 1975 U.S. dollars).

The need for industrial growth as the basis of prosperity of developing countries, where three-fourths of the Earth's population live and only one-fourth of the global energy is consumed, determines a great deal of the future rise in global average per capita energy consumption. In developing countries, the direct proportion between gross national product and energy consumption will be maintained while developed countries show new trends. Figure 2.8 presents the correlation between primary per capita energy production and rise in gross national production in the United States from 1910 to 1978 (Hafele, 1981). The quantity of primary energy expended per unit of gross national production (GNP) declined 11% in the United States in 1980 compared to its mean value for 1960–1980 (Hafele, 1981; Edvard and David, 1982; Kuz'min and Stolyarevskiy, 1984). U.S. national energy policy plans a further decrease in the

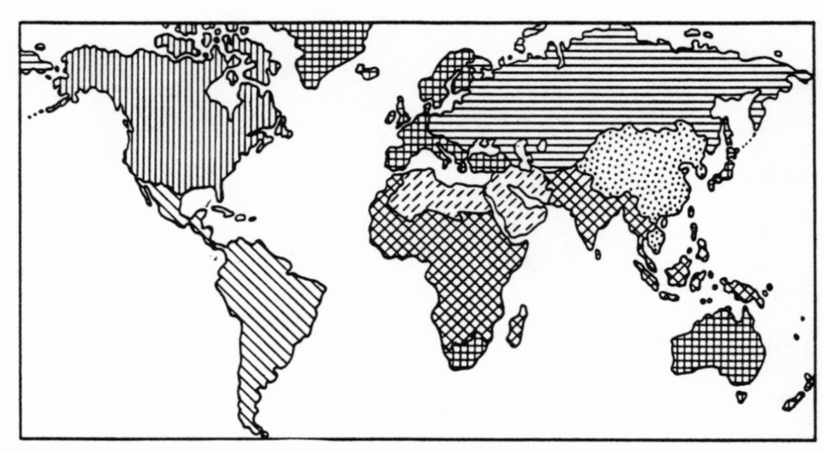

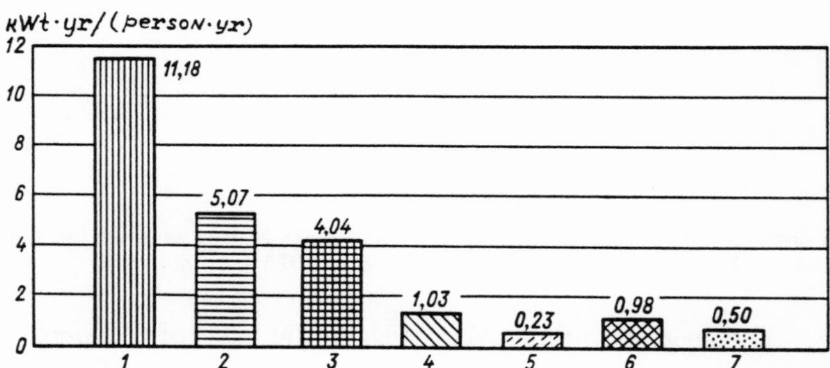

Figure 2.7. Annual per capita consumption of primary energy throughout the world in 1975. 1, North America (237 million people); 2, USSR and socialist countries of Eastern Europe (363 million people); 3, developed capitalistic countries of Western Europe, Japan, Australia, New Zealand, South African Republic, Israel, etc. (560 million people); 4, Latin America (319 million people); 5, Africa, excluding South and North Africa (1422 million people); 6, Middle East and North Africa (133 million people); 7, China and other socialist countries of Asia (over 1 billion people). The global average per capita energy consumption in 1975 was 2.08 kW $\times$ yr/yr and per capita GNP was 1565 dollars per year.

amount of energy expended per unit of GNP by another 19% by 1990. The USSR energy program also projects a decrease.

Figure 2.9 depicts the forecast of decreasing energy expenditures per unit of GNP for the United States, Canada, Western Europe, and Japan (Kuz'min and Stolyarevskiy,

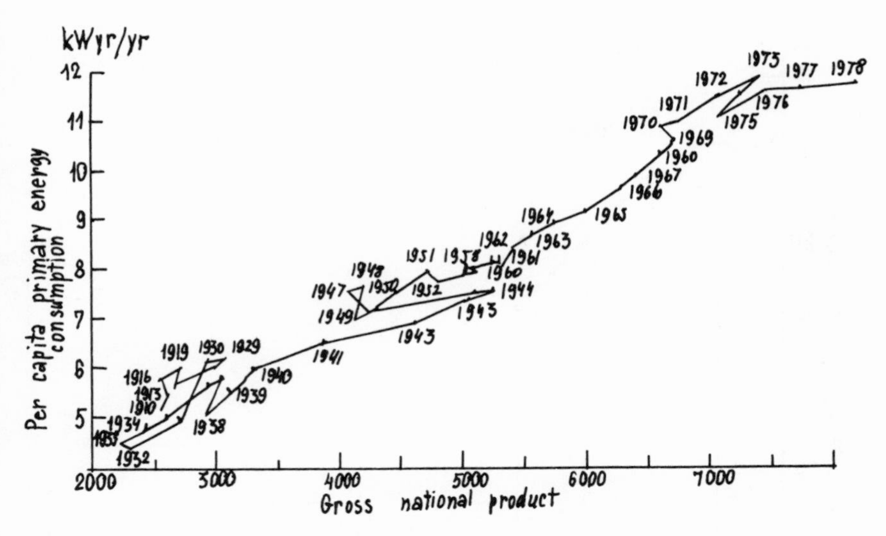

Figure 2.8. Ratio between GNP (dollars per year) and per capita energy consumption in the United States from 1910 through 1978.

1984; Edvard and David, 1982). Compared to the average 1960–1969 level it is expected that this estimate will drop by 34% by the year 2000. GNP in Japan more than doubled from 1974 through 1985, although energy demand increased by only 7 to 8%. Japan therefore produces twice as much goods and services with practically the same energy expenditures.

How can primary energy sources be conserved? First, efficiency of energy-producing and energy-consuming units can be improved, i.e., reduce direct heat losses into the environment. However, there is always a limit, which the efficiency of a thermal engine, for example, cannot exceed. Energy conservation generally means more efficient use of primary energy without decreasing the rates of economic progress. The second major means of energy conservation is therefore a switch to energy-conserving technologies. In transportation, for example, this means use of lighter materials, streamlining, developing more efficient engines, and microcomputer mounting of engine operation. In construction, this means use of insulation, computer control of heat

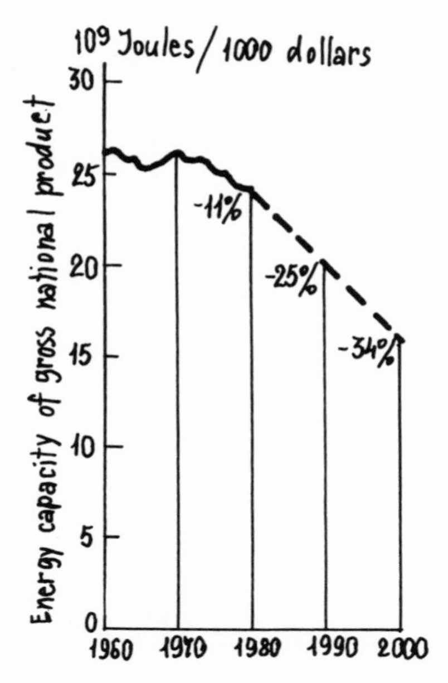

Figure 2.9. Statistical data (1960-1980) and international forecast (1980-2000) of changing energy capacity of GNP together for the United States, Canada, Western Europe, and Japan.

supply systems, broad application of dual-purpose cooling and heating pumps, and so on. In short, transition to new energy-conserving technologies in all sectors is the main avenue to energy conservation.

Table 2.2 presents data on reduction in energy expended per unit of GNP predicted here for the United States, Canada, and Western Europe (see Figure 2.9) for different sectors of the economy (Edvard and David, 1982).

Developed countries thus currently display a trend toward slower rise in per capita energy consumption while maintaining the growth rates of per capita GNP. It should also be mentioned here that population growth rates are slowing down in these countries, and as data in Figure 2.10 show, this is related to a certain extent to the high level of energy development. World experience therefore shows that factors are operating in highly developed countries to decrease energy consumption reaching asymptotically some essentially constant level (Babayev et al., 1984).

Table 2.2. Forecasted Decrease in Energy Expended for GNP in Different Economic Sectors Together for the United States, Canada, and Western Europe (in % of mean GNP in 1960–1969)

Economic Sectors	1980	2000
Heating	9	35
Transportation	5	31
Industry	16	43
Average for all sectors	11	34

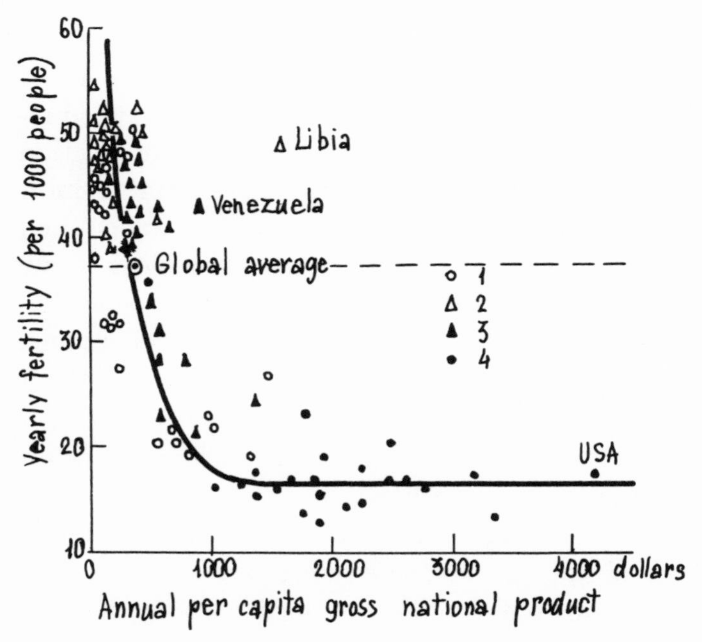

Figure 2.10. Ratio between annual per capita gross national income and birth rate in different countries of the world. 1, Asia; 2, Africa; 3, Latin America; 4, Europe, the USSR, and North America.

2.3. Long-Term Forecasts of Energy Development

Studies on defining a strategy for energy development are the current focus of attention. Problems of global energy development in the distant future (up to the middle or even the end of the next century) are being extensively worked out both in the Soviet Union and in other countries (Legasov and Kuz'min, 1981; Styrikovich, Sinyak, and Chernyavskiy, 1981; Lovins et al., 1981; Edmonds and Reilly, 1983a and b; IIASA'83 ..., 1983; WEC ..., 1983; Babayaev et al., 1984; Edmonds et al., 1984; Goldenberg et al., 1984; Keepin, 1984; Reister, 1984; Keepin et al., 1985). The following circumstances explain why extensive research is necessary.

First, correct solution to the problem of energy supply, based on an adequate estimation of the role of different energy sources in the future fuel and energy balance, is one of the decisive factors in subsequent economic growth both in the world as a whole and in separate regions and countries.

Secondly, the great financial, material, and labor intensity of energetics, and time-consuming development of new economically feasible technologies of energy production, leads to its great inertia. As the previous section showed, study of the history of energy development indicates that restructuring of the fuel and energy balance (i.e., transition from one energy source to another) takes many decades. There could thus be a considerable time span between the consequences of an energy policy (socioeconomic, ecological, etc.) and the actions to stimulate or prevent them. It is therefore very important to make a thorough and early analysis of the potential current or long-range effects of the modern energy policy.

Finally, the ecological capacity of the environment and reserves of certain energy resources were practically unlimited for the previous scales of development. Today we cannot rely on this, because the scale of energy consumption is so great that resource, hygienic, and ecological limitations have appeared simultaneously. Current global energy development is characterized by rapid depletion of the fossil fuel on which it is primarily based.

Numerous studies of long-range global energy development use various approaches, have a different degree of reliability, and are designed to solve various problems. Data in

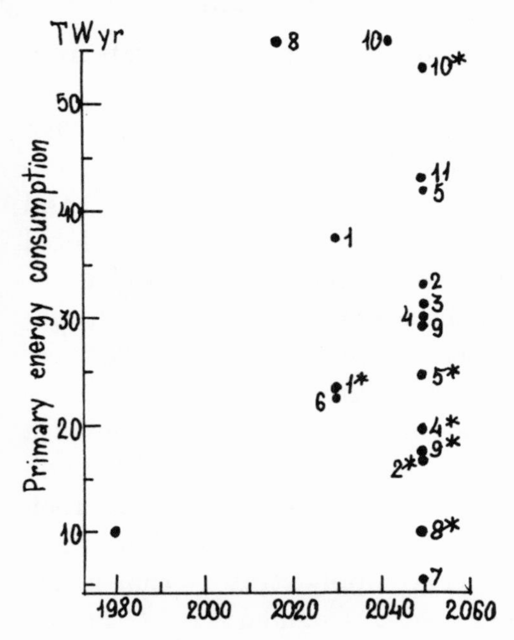

Figure 2.11. Projected energy consumption for the world as a whole in 2050 compiled by different authors and scientific groups. 1, Hafale, 1981; 2, Rose et al., 1983; 3, Seidel, Keyes, 1983; 4, WEC..., 1983; 5, Reister, 1984; 6, IIASA'83..., 1983; Lovins et al., 1981; 8, Nordhause, Yohe, 1983; 9, Edmonds, Reilly, 1983a; 10, Edmonds, Reilly, 1983b; Legasov et al., 1984. The upper projected estimate and lower (marked by asterisk) are indicated.

Figure 2.11 reveal the uncertainty in long-range energy consumption calculations. It should be mentioned that these data were obtained from models designed to solve different problems. One researcher in predicting development rates and global energy structure could be interested in estimating the future of nuclear power, whereas another could focus on ecological problems arising during development of a certain energy source. Both researchers will build models of energy development; however, these models and their results could be difficult to compare due to differences in the initial problems.

The report of the International Institute of Applied Systems Analysis (IIASA) "Energy in a Finite World" (Hafele,

1981) was a very elaborate study. It took 140 scientists from 20 countries 7 years to prepare it.

The main purpose of the IIASA study was to provide an objective analysis of different energy policies. Two detailed energy scenarios are presented, labeled "high" and "low," to cover all possible paths of evolution of the global energy system from 1980 through 2030. The world is divided into seven homogeneous regions, similar in main economic and energy factors (see Figure 2.7). The detailed scenarios describe an iterative procedure involving three models. Iteration begins with population and economic growth projections, and lifestyle parameters in an energy-demand model. These forecasts of energy consumption (from 1980 through 2030) are then introduced into a model that calculates the least expensive energy supply strategy that meets the expected consumption levels. Calculations of the optimal energy supply strategy are used to assess economic and environmental impacts. The findings help to correct the original economic projections, thus closing the iterative model loop. The entire procedure is repeated until internal consistency is reached.

The IIASA scenarios were recently criticized (Keepin, 1984) because the iterative modeling method was never used in practice, and the models were mainly a numerical demonstration of subjective projections by analysts. It was also found that the major results depend strongly on inaccurate estimates of resource cost. Analysis of sensitivity showed that the models have an unstable analystic structure: although the models include many detailed aspects of the energy system, they omit the exceptionally important variability of future energy costs (Figure 2.12).

Realistic conclusions were also drawn, however. The work done in the IIASA was one of the first attempts to analyze energy systems in global terms. Persistent efforts led to the collection of consistent and important results from all regions. The global potential of every major energy source

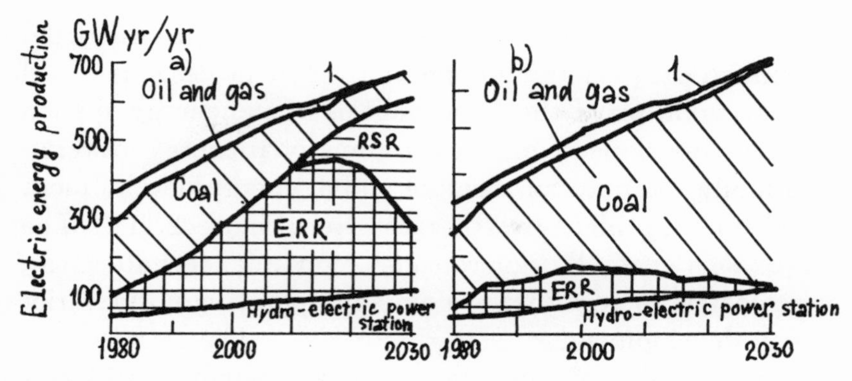

Figure 2.12. Predicted electricity production (1) in North America (Canada and the United States) obtained by the model developed in the International Institute of Applied Systems Analysis. Two possible scenarios are presented that differ in the contribution of various energy sources to electricity production. It is anticipated that the cost of electricity produced by nuclear power (scenario b) will be 16% higher than in scenario a.

was assessed. Finally, this was an effective program because it raised the level of professional knowledge and focused attention on the global nature of energy and related problems such as the greenhouse effect.

The IIASA recently developed a new scenario (1983) using an improved energy supply model, which foresees an expanded role for natural gas, but is otherwise similar to the "low" version of the previous scenario.

Of all the available long-range global energy development forecasts, we should discuss the universal one by Lovins et al. (1981). Analyzing the method used to make these projections we should first emphasize that the purpose of this study was not to predict the future but to describe the paths of social development that could significantly reduce total energy consumption without lowering the level of economic development. It was shown that this requires a drastic shift in emphasis in the global energy policy: a transition from a policy of producing more energy by involving higher amounts of energy resources in the fuel and energy balance to a policy oriented toward improving efficient use and maximum conservation of energy. The global energy development projection implying more efficient usage is based on extrapolation of the potentialities of the energy system in highly developed countries to the global energy system.

Finally, we note forecasts of global energy development based on very high growth rates (see Figure 2.11). The projections generally ignore some economic, technological, or ecological factors that are already operating and restrict energy development.

Predictability of economic, social, and scientific and technical development of society is currently not specific enough to make a detailed and accurate forecast of the development of any industrial sector or any aspect of human activity, including energetics. We are thus faced with a contradiction. On the one hand, a long-range scenario of energy development is extremely important, while on the other hand, those social and economic long-range plans using our scenarios that should already have been adopted could prove to be incorrect. To avoid this contradiction, so-called diagnostic projections are suggested (Legasov and Kuz'min, 1981; Babayev et al., 1984). They do not claim to be universal, i.e., make full consideration of all economic, social, demographic, ecological, and other factors. The purpose of diagnostic forecasts is to use general laws of development and the most substantiated forecasting concepts to evaluate the possible limits of energy growth and clarify their potential, for example, ecological impact. Thus, in a certain sense, the inverse problem is raised: how must energy develop to cause a certain ecological change? In other words, the diagnostic projection is confined to the framework of the method for analyzing energy strategy, in particular, estimating the efficacy of various measures to restrict or stimulate the development of a certain energy scenario.

2.4. Diagnostic Projections of Energy Development

Global experience shows that any country, after reaching

a certain level of development, cuts its per capita energy consumption and population growth rates. Construction of a diagnostic model for long-term global energy consumption growth (Babayev et al., 1984; Legasov and Kuz'min, 1981) is therefore based on the following assumptions:

Annual per capita energy consumption $W(t)$ rises constantly, but its increment rates decrease with time t, and in the final analysis, it begins to asymptotically approach some fixed level W_0 (the "saturation" level). To describe analytically this per capita energy consumption growth, a logistic function is used:

$$W(t) = W_0/(1 + e^{b-at}), \tag{2.2}$$

where a and b are free parameters; the global population changes with time $P(t)$ according to the demographic forecast made at the UN Conference on Global Population (Global and Regional Long-Range Estimates ..., 1974), that predicts the global population will stabilize at 10 to 12 billion by the end of the twenty-first century.

Thus, the annual primary energy demand $E(t)$ is thus determined from the following relationship:

$$E(t) = P(t)W(t). \tag{2.3}$$

Since the global population change (according to the previous assumption) is a specific function in our model, the predicted rise in energy consumption depends on the assumptions used in selecting values a, b, and W_0 of logistic function $W(t)$. It is difficult to determine the energy consumption saturation level W_0. Overall energy consumption should provide a fairly high standard of living. The most developed countries have reached the highest level of per capita energy consumption, 10 kW $\times$ yr/(man $\times$ year). Assuming that this value will be reached by the time the world as a whole is stabilized it is currently about 2.5 kW $\times$

yr/(man × year)), the total energy consumption by a population of ten billion people will be 100 TW × yr/yr. We recall that population growth and industrialization typical for the majority of developing countries (with the greatest population) generate trends toward higher energy expenditure per unit of GNP. Industrialization requires an additional per capita consumption of materials (metal, plastic, etc.), and, correspondingly, increases the per capita commercial energy expenditure. When the Earth's population is more than double present level, additional energy expenditures will be required to produce food, and soils with low fertility will have to be developed for this purpose. Therefore, energy expenditures will be needed to produce a large amount of fertilizers, and for land reclamation. Depletion of accessible mineral beds will lead to working of poor ore deposits (which amount to about 65% of total reserves) and deep fields. It will be necessary to increase energy outlays for mining and secondary recovery of rare elements. A higher population will require additional efforts and consequently, energy outlays to supply mankind with fresh water. Finally, there is the question of environmental protection. Environmental protection measures will also require energy expenditures; for example, for domestic and industrial effluent disposal and air purification. Use of hydrogen in future industry and transportation is a promising solution to ecological problems, but large-scale production of hydrogen as a heat carrier requires additional outlays of energy.

Calculation of the specific expenditures for all these needs allowed Hafele (1979) and Babayev et al. (1984) to draw the following conclusion: since the global population growth will stabilize to a great extent because of the developing countries faced with intensive industrialization, in order for all people on Earth to have the same living standard as the most highly developed modern countries, an increase might be needed in the mean global per capita energy consumption to 10–20 kW × yr/(man × year) (Hafele,

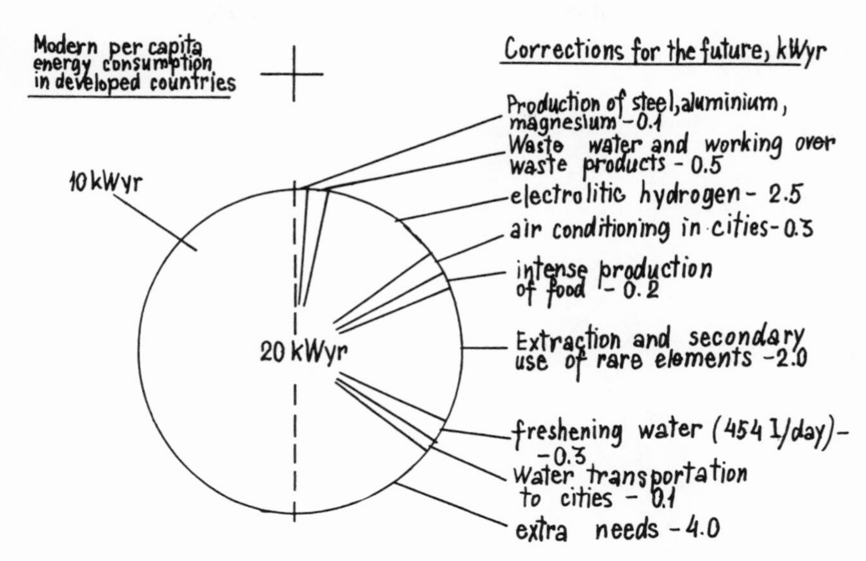

Figure 2.13. Level of stabilization of annual per capita consumption of primary energy.

1979; Legasov and Kuz'min, 1981; Babayev et al., 1984), or correspondingly, 100–200 TW × yr/yr for a population of 10 billion (Figure 2.13).

On the other hand, analysis of the saturation level W_0 must also take into account the aforementioned trends of decreasing energy outlays per unit of GNP. Based on a comprehensive consideration for these trends presented, for example, in Styrikovich et al. (1981), the conclusion is drawn that the primary energy consumed by the time the Earth's population stabilizes will be 60 TW × yr/yr. We adopt this minimum consumption projection as the base.

An analysis was made of the published scenarios for global energy consumption growth in order to plot long-term energy development (Hafele, 1981; Styrikovich et al., 1981; Kuz'min and Stolyarevskiy, 1984). The conclusion can then be drawn that data on energy consumption growth (Figure 2.14) are sufficiently substantiated approximately up to the year 2010 and can be the basis for defining the logistical curve parameters for further extrapolation of global

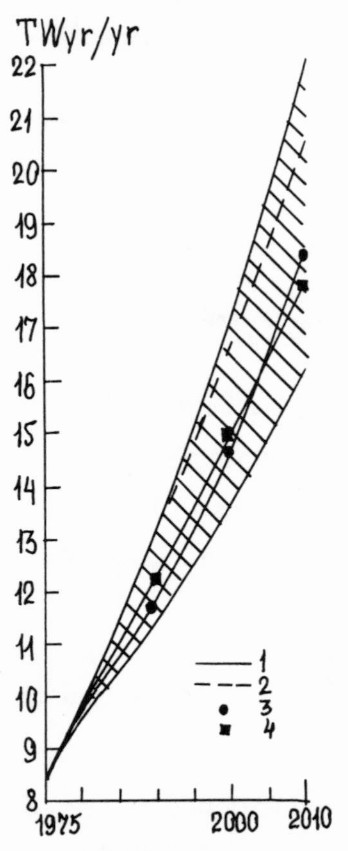

Figure 2.14. Predicted annual energy consumption up to 2010 compiled by different authors. 1, Hafele (1979); 2, Styrikovich, Sinyak, Chernavskiy (1981); 3, International Symposium of Energetics; 4, Kuzimin, Stolyarevskiy (1984).

energy consumption growth predictions. Figure 2.15 presents projected energy development to the year 2100, with the assumption that the stabilization will occur at 60 TW, 100 TW and 200 TW levels.

Even in the lowest forecast (Figure 2.16), integral consumption of primary energy sources is far greater than 2954 TW × yr, which characterizes the global organic fuel reserves (Hafele, 1981). These estimates point to the well-known fact that if the growth rates of organic fuel consumption, in particular oil and natural gas, are maintained,

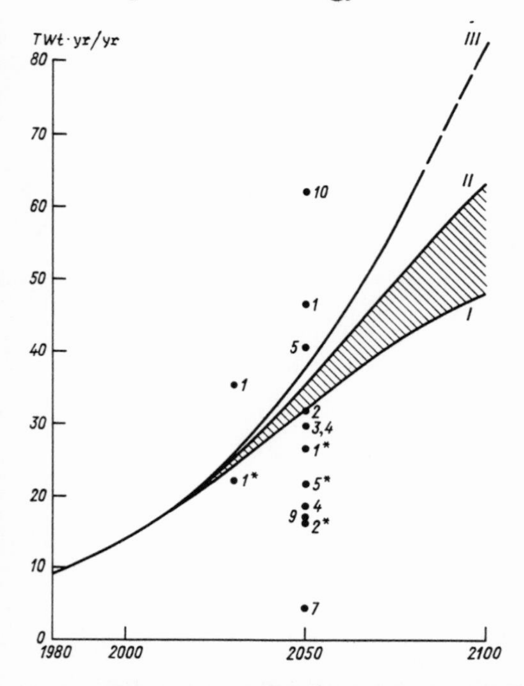

Figure 2.15. Diagnostic forecast of global energy consumption growth calculated by logistic function (ref. to eq. (2.2)). I-III, stabilization of per capita energy consumption at levels 6, 10 and 20kW × yr/yr, respectively; 1-11, forecasts of energy consumption in 2050 (see Figure 2.11).

mankind will face depletion and total extinction of these energy sources in the next decades (see Figure 2.16).

It is thus clear that the future energetics cannot be oriented only on traditional energy sources, including the renewable energy of rivers, tides, solar radiation, wind, waves, and currents, as well as geothermal energy. Analysis demonstrated (Hafele, 1981) that the technical (suitable for use) potential of these sources does not exceed 17.2 TW × yr/yr (Figure 2.17).

Such resources as water, wind, tidal, or wave power are insufficient. The theoretically unlimited resources of solar and geothermal energy are characterized by extremely low intensity of incoming energy. In addition, the use of geothermal, and particularly solar energy on the basis of modern

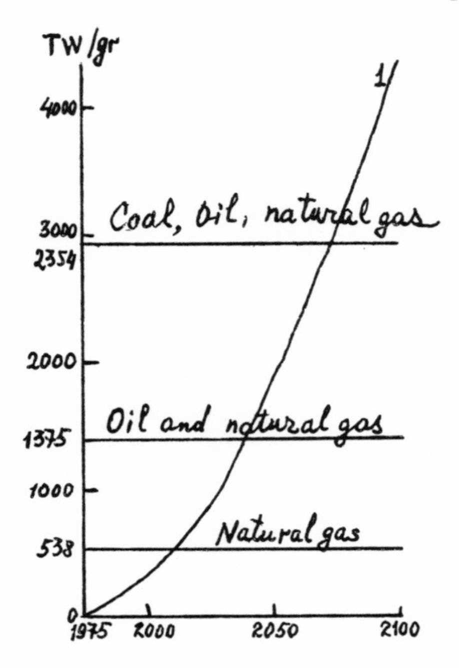

Figure 2.16. Integrated consumption of primary energy and resources of fossil fuel. 1, diagnostic forecast with minimum energy consumption (see Figure 2.15, curve 1).

technology will require considerable economic outlays, and the development of new technologies is time consuming.

Some alternative energy sources, namely nuclear reactions, are practically unlimited. Nuclear energy with thermal and fast neutron reactors, thermonuclear energy and energy from renewable sources must therefore be developed. Predictions of alternative energy source developments takes into account two tendencies. First, the share of alternative sources in the fuel and energy balance shall increase because of the natural depletion process and, consequently, rising cost of traditional sources, primarily oil. Secondly, factors restraining the maximum possible growth rates of alternative energy sources should come into play. They include lagging technologies, their high cost (this does not pertain only to nuclear energy), and finally, there are serious noneconomic factors that restrain growth, primarily of

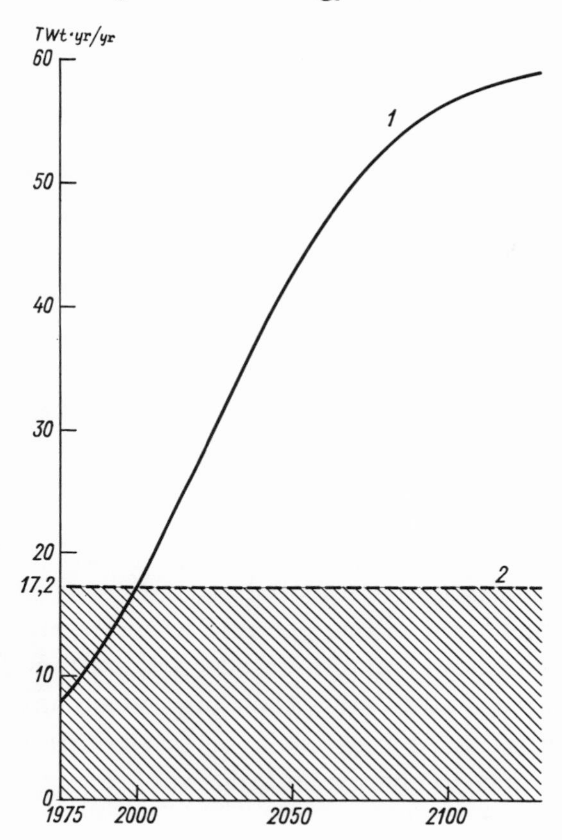

Figure 2.17. Predicted annual global primary energy consumption (1) (see curve II in Figure 2.15) and technological potential of renewable energy sources (2).

nuclear energy, and fast neutron nuclear energy; the latter is undoubtedly the most important sector of future energetics.

The majority of scientific projections for future global energy development thus assume that the current growth trends inherent to all of its sectors will persist into the next decade, after which the traditional energy sources will gradually be displaced from the global fuel and energy balance by alternative sources.

Due to the great inertia of energetics, several phases have to be isolated in developing a long-range strategy.

1. *The immediate phase* covers the period up to the year 2030. During this phase, the energy strategy should be based on the supposition that energy demand in this period can be met only by energy sources that can be widely and economically used given the extant level of scientific and technological development. We noted that these sources currently include fossil fuel based energy, hydraulic power, and nuclear energy based on thermal reactors. During this period there is a possibility that a few commercial power plants using alternative energy sources (solar energy, energy from nuclear synthesis, etc.) will be built. However, it is difficult to hope that at least during this term they will play a significant role in the energy balance.

2. *The beginning of the middle phase* in the year 2030 is characterized by possible extensive use of alternative energy sources. From this moment, it will be possible to accelerate the construction rates of power plants based on these new alternative energy sources and to start their large-scale application. These alternative sources could be fast-neutron nuclear energy, thermonuclear energy, and renewable energy resources (solar, wind power, etc.).

The exponential growth of energy consumption ceases in the middle phase. This corresponds to the turning point on the logistic curve at the moment $t = b/a$. In addition, in the middle phase, the use of traditional sources reaches its maximum, and their subsequent share in the global fuel and energy balance begins to decrease because alternative energy sources are used more widely. During this phase the fuel and energy balance should be optimized, taking into account socioeconomic factors, the availability of certain energy resources, and environmental impacts. The middle phase should be used for a smooth transition to the more distant phase of energy development.

3. *The distant phase* corresponds to the transition of global energetics to stabilization of energy consumption. By

the onset of this period, the following conditions should be achieved:

— the limits and restrictions should be identified for global energy consumption, i.e., the actual level of stabilization of per capita energy consumption should be determined;

— the global fuel and energy balance should be optimized, i.e., one or several of the few available variants of large-scale energy supply should be implemented.

The exact time for the onset of any phase (in years) can be estimated only approximately, and it will vary in different countries. It is obvious that in industrialized countries the distant phase will occur much earlier than in the rest of the world. Considering these stages, the energy quality characteristics and scales are emphasized, rather than prediction of an exact date for the beginning of a certain phase.

In summary, major forecasts used in our construction of the diagnostic scenario for energy development can be concisely formulated.

1. In the first half of the twenty-first century, the population growth is expected to slow down in all countries. By the year 2100, the Earth's population will stabilize at the 10 to 12 billion level.

2. By the end of the twenty-first century, the per capita global energy consumption will also stabilize.

3. In the visible future, the global fuel and energy balance cannot be one energy source, albeit the most attractive for its parameters (for example, nuclear energy). Organic fuels (coal, oil, and gas) will continue to play a noticeable role in society's life throughout the immediate and middle phases, i.e., to the end of the twenty-first century.

4. The typical trend of global energy is a lowering of the share of traditional energy sources (coal, oil, and gas) in the global fuel energy balance. Whereas it is currently about 90%, by the end of the next century it is expected to be no

more than 17 to 18% (Styrikovich et al., 1981). Organic fuel will be displaced by alternative sources, while continuing to play a noticeable role throughout the immediate and middle phases of global energy development (Styrikovich et al., 1981). Study of the projected energy development based on traditional energy sources is particularly important in terms of their climatic impacts on a global scale. Nuclear energy and fast return nuclear reactors, that apparently will provide most of the global fuel and energy balance by the end of the middle phase, do not have such global-scale "ecological impact" as traditional, primarily coal- and oil-based energetics (see, for example, Babayev et al., 1984).

Long-range global energy development considered in Styrikovich et al., 1981; Kuz'min et al., 1984; and Legasov et al., 1984, was chosen as the basic scenario to study the environmental consequences of anthropogenic impact. It is presented in Figure 2.18. This choice was made because this scenario has the greatest ecological safety compared to others that assume a higher growth in global energy consumption.

This scenario agrees with the above concept and is one of those scientifically substantiated forecasts that takes into account more comprehensively the trends toward transition to energy-conserving technologies. Alternative energy sources also have the maximum share in the global fuel and energy balance in this scenario. However, it is restricted by a number of technological and economic considerations. The entire energy consumption in this scenario will stabilize at the 60 TW $\times$ yr/yr level.

We will further assume that total energy consumption based on traditional sources (coal, oil, and gas) approximately corresponds to the basic scenario. However, more comprehensive consideration of ecological factors may make it necessary to optimize the structure of the fuel and energy balance, for example, alter it to reduce the use of coal with more intensive utilization of oil and natural gas.

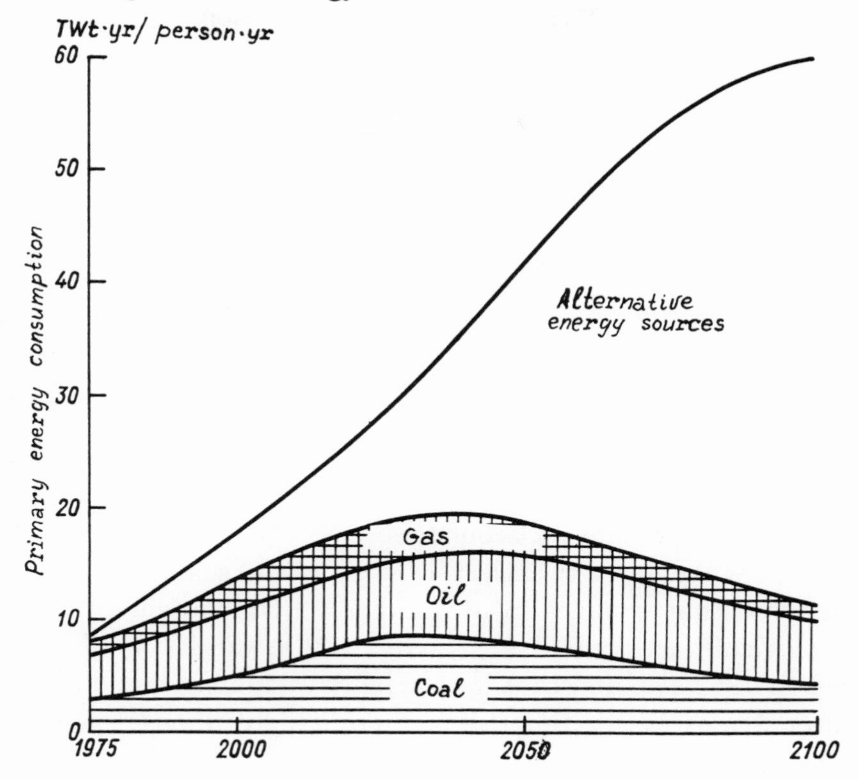

Figure 2.18. Development of the structure of global fuel and energy balance according to base variant.

The forecast of CO_2 emissions into the atmosphere up to the year 2100 based on the basic scenario is given in Legasov, Kuz'min, and Chernoplekov, 1984, and is used in Chapter 3 to calculate atmospheric CO_2 concentration.[1]

[1]More recent studies (Edmonds and Reilly, 1985; Edmonds et al., 1987) have corrected some deficiencies in the IIASA study. Legasov et al. (1984) and Edmonds, et al. (1987) arrived at about the same approximation of the total amount of fossil fuel burnt by the year 2075. Hence, the predicted atmospheric CO_2 concentrations have small diversions, but the source function $d\mu_f/dt$ according to Edmonds is increasing steadily by about 1.5% per year.

CHAPTER 3

ANTHROPOGENIC CHANGE IN CO_2 CONCENTRATION

3.1 Introduction

CO_2 is a small admixture to the major gases constituting the Earth's atmosphere. Its relative volumetric concentration P_a in the modern epoch is close to 300 ppm. However, since it is one of the gases creating the greenhouse effect in the atmosphere, and a necessary component for photosynthesis, even comparatively slight variations in its content can have important consequences.

Since the early twentieth century, some scientists have believed that changes in atmospheric CO_2 content in the geological past could lead to temperature variations of the Earth's surface, and that industrial development could also alter atmospheric composition and its greenhouse effect. Callendar (1938) hypothesized that warming of the 1930's was caused by man-induced increase in atmospheric CO_2 content. It was found out later that this warming was mainly the result of increased atmospheric transparency.

Data of monitoring atmospheric CO_2 content that was set up in March 1958 already clearly showed in 10 to 15 years that P_a was increasing steadily with time, and the rate of its annual increment was also rising due to more intensive consumption of fossil fuels (Budyko, 1982).

Using extrapolation of P_a monitoring data and estimates of greenhouse effect changes made by Manabe (1970), Budyko (1972) came to the conclusion that in the next decades, at least up to the mid-twenty-first century, increase in P_a will be the main factor changing the mean global temperature of the Earth's surface T_S. The authors believe that by the year 2000, the effect of this anthropogenic factor will exceed natural T_S fluctuations observed in the first half of the twentieth century, and by the mid-twenty-first century, the increment in mean global temperature ΔT_S will exceed 2° C.

Over 20 stations are currently monitoring P_a all over the globe. On 1 January 1980 the mean global value of P_a was 338 ppm (in 1958 it was about 315 ppm), and the 1984 mean was 343 ppm (see Komhyr et al., 1983; Bolin et al., 1986).

We know the amount M_f of CO_2 released into the atmosphere from fossil fuel combustion: comparatively reliable data on the rate of industrial emissions dM_f/dt have been obtained since 1950, the amount of fossil fuel burnt during he period of 1860 to 1950 has also been estimated. In the report "Climatic Effects of Increased Atmospheric Carbon Dioxide" (1982), Keeling pointed out that during the period of 1958 to 1980, on the average, about 55% of the CO_2 released from industrial sources remained in the atmosphere. This allowed us to estimate P_a values for the next decades based on projected intensity of industrial CO_2 emissions $dM_f/dt(t)$ into the atmosphere (Legasov, Kuz'min and Chernoplekov, 1984).

Since the Earth's climate is closely related to atmospheric CO_2 content, a number of international conferences and meetings were held to discuss various aspects of the interesting and complicated problem of the CO_2 global cycle and the factors determining CO_2 partial pressure in the atmosphere. The results of these meetings are presented in various publications (see, for example, Fate of Fossil Fuel CO_2 in the Oceans, 1977; Energy and Climate, 1977; Carbon Dioxide, Climate and Society, 1979; Problems of Atmospheric CO_2, 1980; Proceedings of the Carbon Dioxide and Climate Research Program Conference, 1980; Carbon Dioxide and Climate, Australian Research, 1980; Analysis and Interpretation of Atmospheric CO_2 Data, 1981; Changing Climate, 1983; The Carbon Cycle and Atmospheric CO_2–Natural Variations from Archean to Present, 1985; Bolin et al., 1986). Study of the planetary carbon cycle and its changes in the geological past due to natural causes was the basis for the calculation by many authors of anthropogenic variations in P_a. The results were generally the same: estimates of P_a by various models are sufficiently accurate to forecast climatic changes for the next century for a specific scenario of energy development, i.e., for a prescribed forecast of the variation in intensity of industrial CO_2 emissions into the atmosphere $dM_f/dt(t)$. A more comprehensive study of the global carbon cycle in particular, interaction of the ocean and benthic carbonate sediments should be made to compute longer range P_a changes. It has also been discovered that reliable forecasts of P_a values and estimates of their errors even for the next century require:

—estimation of the so-called biotic release of ΔM_b of CO_2 into the atmosphere in the last 100 to 120 years, which should have taken place due to anthropogenic impact on land vegetation and soils;

—obtaining globally averaged values of basic parameters determining the intensity of oceanic absorption of man-made CO_2;

—clarification of how P_a calculation results are influenced by various models of tracer transfer in the ocean.

Greater insight into these problems comes from recent experimental data on anthropogenic disorders in the cycle not only of the most widespread carbon isotope (^{12}C), but also the next most abundant (^{13}C) and radioactive isotopes (^{14}C), as well as through advances in modeling gas exchange between the atmosphere and ocean.

3.2. Carbon Cycle

When considering possible immediate variations in the atmospheric composition, the term "planetary carbon cycle" refers to transfer of this element within the so-called mobile resources reservoirs, i.e., within the atmosphere-ocean-land and ocean biomass system. Two stable isotopes with atomic weights 12 and 13 dominate in natural carbon of the mobile resources. The most abundant is isotope ^{12}C; the relative isotope ^{13}C content averages about 1.2%, but it differs somewhat in various carbon-containing compounds, because of isotope fractionation in the carbon cycle. Any changes in intensity of the exchange processes affect the ratio between ^{13}C and ^{12}C content in different reservoirs.

The radioactive isotope ^{14}C is always present in the mobile reservoirs of carbon, because there is a natural source of radioactive carbon in the upper atmospheric layers (for details see section 3.3). Release of CO_2 into the atmosphere as a result of burning fossil fuel that does not contain radioactive isotope disrupted the natural ^{14}C cycle. Nuclear weapon tests in the atmosphere have created a much more powerful source of ^{14}C than the natural one.

Figure 3.1 illustrates ^{12}C content (in 10^{12} kg) in different reservoirs. In the mid-nineteenth century, the atmosphere contained about $600 \cdot 10^{12}$ kg of C, mainly, in the form of CO_2. The biomass of terrestrial plants is about $560 \cdot 10^{12}$ kg of C, soil organic C content is about $2000 \cdot 10^{12}$ kg (Chapter 4 contains details about biotic components of the carbon cycle and substantiates these values of biomass).

About $37,000 \cdot 10^{12}$ kg of C are dissolved in oceanic waters in the form of inorganic compounds, consisting mainly of hydrocarbonate ion HCO_3^-, carbonate ion CO_3^{2-} and a dissolved CO_2.

The mass of marine organisms (about $3 \cdot 10^{12}$ kg of C) is small compared to the mass of terrestrial organisms (Suyetova, 1973). The content of aquatic humus, dissolved organic carbon, is close to that of organic carbon in soils, $1,500 \cdot 10^{12}$ kg of C. The total mass of organic carbon in the ocean is about 4% of the corresponding mass of inorganic carbon (Alekin, Lyakhin, 1984). The terrestrial M_b and ocean $(M_b)_w$ biomass contains about 10% of the carbon mass M in the reservoirs of the mobile resources.

The arrows in Figure 3.1 designate annual mean fluxes between reservoirs of the mobile resources. Productivity of

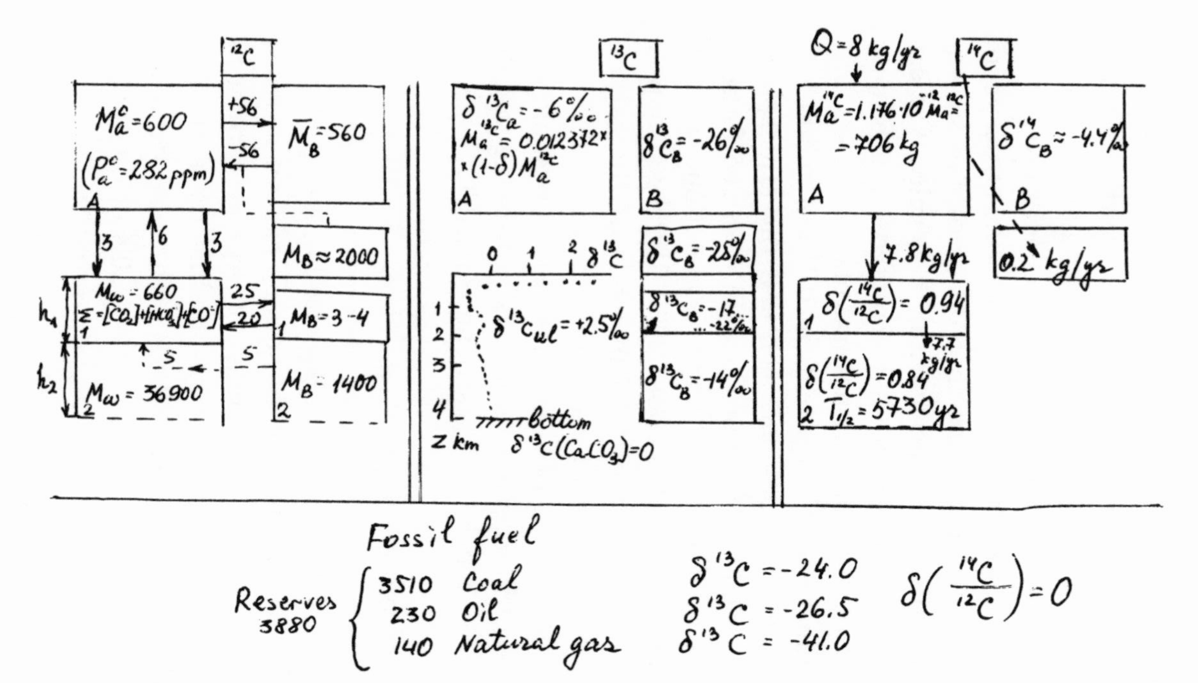

Figure 3.1. Natural distribution of isotopes ^{12}C, ^{13}C and ^{14}C in the atmosphere, ocean and biota. The ^{12}C content is in 10^{12} kg, ^{13}C is in absolute units (10^{12} kg) only for reservoir A; for other reservoirs deviations from the PDB-standard are given; ^{14}C mass is expressed in kg for reservoir A. Relative deviations from the NBS-standard are given for other reservoirs (the NBS-standard is taken to be 1.176×10^{-12} of ^{12}C mass in the corresponding pool). Magnitudes of fluxes are given in 10^{12} kg/yr.

terrestrial phytomass equals $56 \cdot 10^{12}$ kg C/yr and oceanic equals about $25 \cdot 10^{12}$ kg C/yr. Oceanic photosynthesis takes place only in the upper layer (reservoir 1 in Figure 3.1). Organic matter oxidized in the upper and deeper layers (reservoir 2) is distributed between them in an approximate 4 to 1 ratio (Skopintsev, 1977). The amount of carbon descending from photic to deeper layers is therefore about $5 \cdot 10^{12}$ kg yr. Finally, this carbon supplements the reserves of dissolved inorganic carbon in reservoir 2. Since the concentration of inorganic carbon Σ_2 dissolved in deeper layers exceeds the corresponding value Σ_1 in the surface layers, there is an ascending flux of inorganic carbon through the 2–1 interface, which compensates for the descending organic particles.

The surface water meridional temperature gradient causes annual mean CO_2 circulation between the ocean and the atmosphere: release of about $6 \cdot 10^{12}$ kg C/yr into the atmosphere in low latitudes and absorption from it of the same amount of CO_2 in high latitudes. These values are given in Figure 3.1. Seasonal variations in gas exchange intensity are not indicated in the diagram; they are small compared to the intensity of interlatitudinal circulation (Byutner, 1986). Absolute values of ^{13}C and ^{14}C isotopes are only given for the atmosphere; for other reservoirs, $\delta^{13}C$ deviations are indicated relative to the content in calcite fossil shells *Belemnitella americana* dating to the Cretaceous (so-called PDB-standard, in which the ratio between ^{13}C and ^{12}C masses is 0.012372). The $\delta^{13}C$ vertical profile in the ocean is taken from Broecker and Peng (1982).

In Figure 3.1, radioactive carbon content in the ocean is indicated relative to that of the atmosphere for preindustrial time. Values of $\delta^{14}C$ are not given in the diagram: they depend both on the fractionation coefficient and the characteristic residence time of ^{14}C in different reservoirs. We note that in contrast to stable ^{13}C cycle models, the

^{14}C fractionation coefficient can be ignored in modeling anthropogenic disorders in the radioactive carbon cycle. This is possible because changes in relative ^{14}C content caused by fossil fuel combustion and nuclear tests are major while the values of all ^{14}C fractionation coefficients exceed those for ^{13}C, but do not differ much from 1.

Since there is an external radioactive carbon source Q in the atmosphere, it is evident that there are ^{14}C fluxes into other reservoirs, compensating for radioactive decay of ^{14}C atoms. The values of these resulting fluxes (rather approximate) are given in the diagram.

Fluxes of carbon-containing compounds between reservoirs of the mobile resources are very intense, but in the natural state they compensate for each other with high accuracy. For example, measurements of seasonal variations in partial pressure P_a of atmospheric CO_2 showed that in the annual cycle of photosynthesis and destruction of organic matter, the mean annual values of corresponding carbon fluxes coincide with accuracy of about 1%. This was apparently true for CO_2 fluxes between the atmosphere and the ocean prior to anthropogenic carbon cycles perturbances.

In order to get some idea of the complete planetary carbon cycle, a comparison should be made of carbon content of the mobile resources and sedimentary rocks, and intensities of corresponding fluxes.

The total carbon content of the mobile resources M, equal to about $43.6 \cdot 10^{15}$ kg, as seen from Figure 3.1, is negligible compared to carbon accumulated in sedimentary rocks of the Earth's crust during hundreds of millions of years. According to Ronov (1976), about $71.3 \cdot 10^{18}$ kg of carbonate and $9.1 \cdot 10^{18}$ kg of incompletely oxidized organic carbon were deposited in sedimentary rocks of the Earth's curst during the Phanerozoic, i.e., throughout the last 570 million years.

During degassing of the Earth's interior, carbon entered the ocean and the atmosphere mainly in the form of CO_2

with a small admixture of CO and hydrocarbons. It was deposited in the form of carbonate rocks and organic compounds on the bottom of water bodies. In the course of the Earth's history, an enormous amount of mantle carbon passed through the atmosphere and ocean compared to that ever contained in the ocean, atmosphere, and biomass. The interior was not degassed intensively, however: in the last 507 million years, it averaged about $2 \cdot 10^{11}$ kg C/yr, and in the recent epoch, was probably several times lower (Budyko et al., 1985). CO_2 formed during fossil fuel combustion now enters the atmosphere. The intensity of this process dM_f/dt recently reached 5×10^{12} kg C/yr, two orders higher than the current rate of CO_2 emission into the atmosphere-ocean system due to degassing.

The reserves of coal, oil, and natural gas are about 0.5% of the incompletely oxidized organic carbon in sedimentary rocks of the Earth's crust, but their characteristic circulation over time is one or two centuries, which is very short compared to the geological time scale. Thus, the ocean-atmosphere system has currently been removed from equilibrium by brief but unusually strong perturbation. As Broecker (1974) said, mankind is conducting a unique geophysical experiment by burning fossil fuel reserves accumulated for hundreds of millions of years and by releasing CO_2 into the atmosphere.

We now have some monitoring data on anthropogenic changes in ^{12}C, ^{13}C and ^{14}C content in the mobile pools. All these data permit us to predict the variation in atmospheric CO_2 content in the near future, and to estimate global biomass losses in the last 100 years.

3.3. Radioactive Carbon

In 1955, Suess (1955) detected a decrease in atmospheric ^{14}C content as a result of measuring radioactive carbon in tree rings. The half-life $T_{1/2}$ of ^{14}C is 5730 years, therefore, it is not found in fossil fuel. It is produced in the upper atmosphere by interaction between neutrons of cosmic rays and atmospheric nitrogen:

$$^{14}N_7 +^1 n_0 \rightarrow^{14} C_6 +^1 P_1 \tag{3.1}$$

(here, the lower right index is the nuclear charge, the upper left one is the atomic mass). The resulting radioactive carbon oxidizes rapidly to $^{14}CO_2$ and mixes with atmospheric CO_2. According to different estimates, the average rate of ^{14}C production is 1.6–2 atom/(cm$^2\times$ sec), or about 8 kg/yr; it changes in a small range depending on solar activity and geomagnetic field intensity. Estimates of natural variability in atmospheric ^{14}C levels can be found in Stuiver and Quay, 1981. Measurements given in Figure 3.2 show that approximately in 1900–1910 the atmospheric ^{14}C level began to decrease noticeably. Suess' first measurements of this effect (1955) indicated a relative 1% decrease in ^{14}C content between the mid-nineteenth and the mid-twentieth centuries. Later measurements pinpointed this decrease to 2–3% by 1950, according to Lerman, Mook and Vogel (1970), and $2.0 \pm 0.1\%$ according to Stuiver and Quay (1981). A similar decrease has also been observed in the surface oceanic layer. The Suess effect in the mixed ocean layer was about 1.1% before the onset of nuclear tests, according to measurements of ^{14}C content in coral rings made by Druffel (1981).

Nuclear tests which started after 1952 released a large quantity of artificial radioactive carbon into the atmosphere (Figure 3.3a), consequently, completely masking the Suess effect. Comparison of data in Figure 3.3a with values of natural radioactive carbon fluxes shows that the rate of atmospheric influx of artificial radioactive carbon was four thousand times more than the intensity of its natural source. Perturbation thus almost doubled the atmospheric radioactive carbon level (Figure 3.3b). Monitoring data on radioactive carbon scattering obtained after the Nuclear Test Ban Treaty was signed in 1963 allowed us to determine the

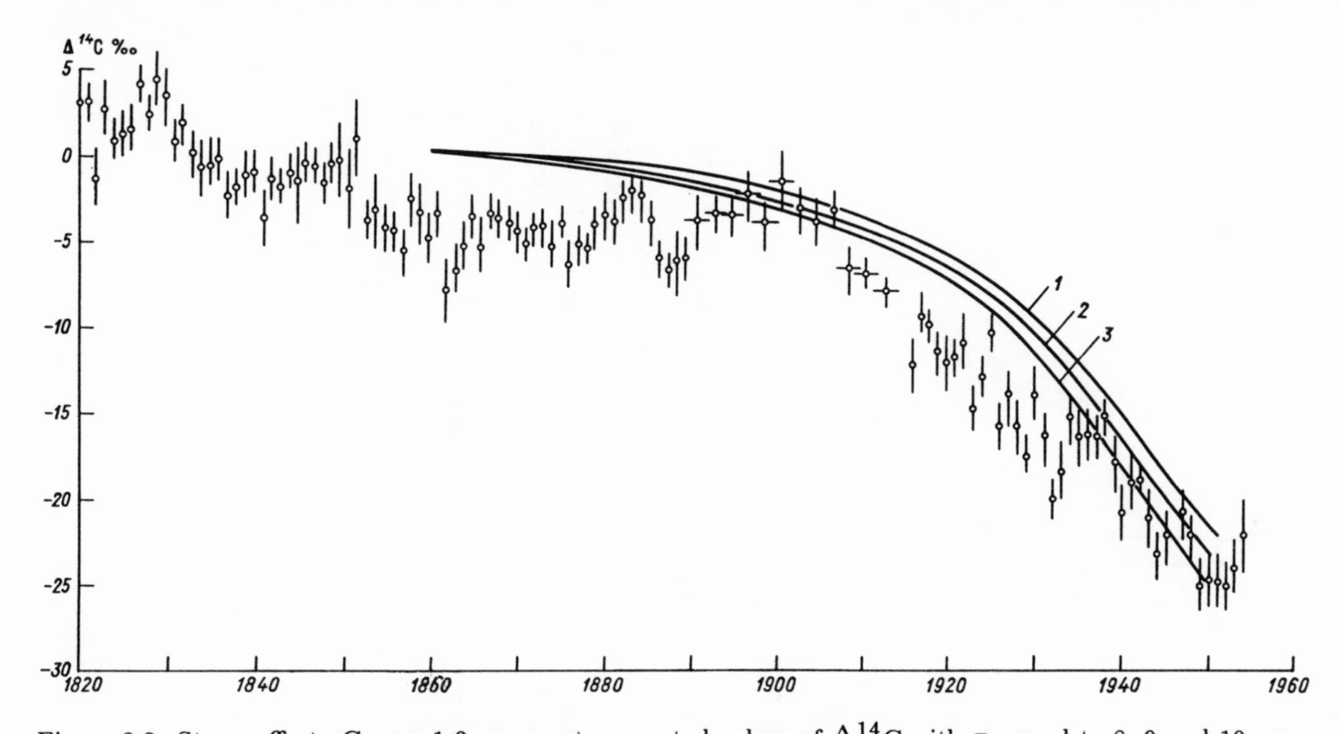

Figure 3.2. Stress-effect. Curves 1-3 represent computed values of $\Delta^{14}C$ with τ_p equal to 8, 9 and 10 years, respectively; the vertical bars indicate root-mean-square error of measurement data.

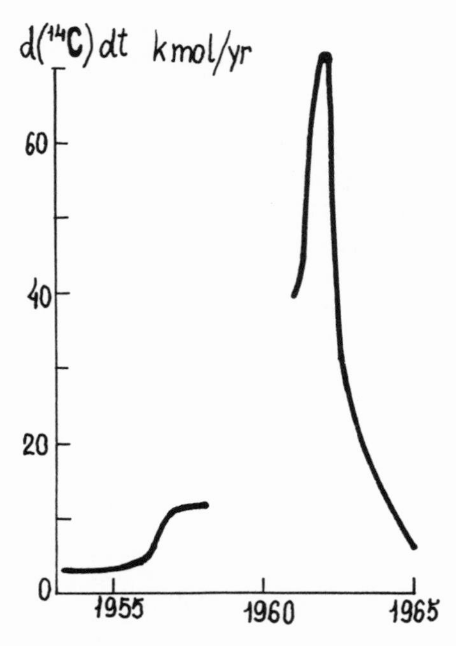

Figure 3.3a. Rate of radio-carbon influx into the atmosphere as a result of nuclear bomb tests.

characteristic relaxation time τ_r in the atmosphere (Turchinovich, 1983a and b). This parameter and Suess effect measurements were used to evaluate the amount of fossil fuel burnt from the mid-nineteenth to the mid-twentieth centuries. The variation in atmospheric ^{14}C content caused by release of CO_2 produced by fossil fuel combustion with the intensity dM_f/dt can be described by a simple relaxation equation:

$$\frac{d}{dt}\left(\Delta^{14}C\right)_a + \frac{1}{\tau_r}\left[\left(\Delta^{14}C\right)_a(t) - \left(\Delta^{14}C_a\right)^0\right] = -\frac{10^3}{M_a^0}\frac{dM_f}{dt}, \qquad (3.2)$$

where $\left(\Delta^{14}C\right)_a$ characterizes atmospheric radioactive carbon content pro mille versus the NBS-standard, M_a^0 is the

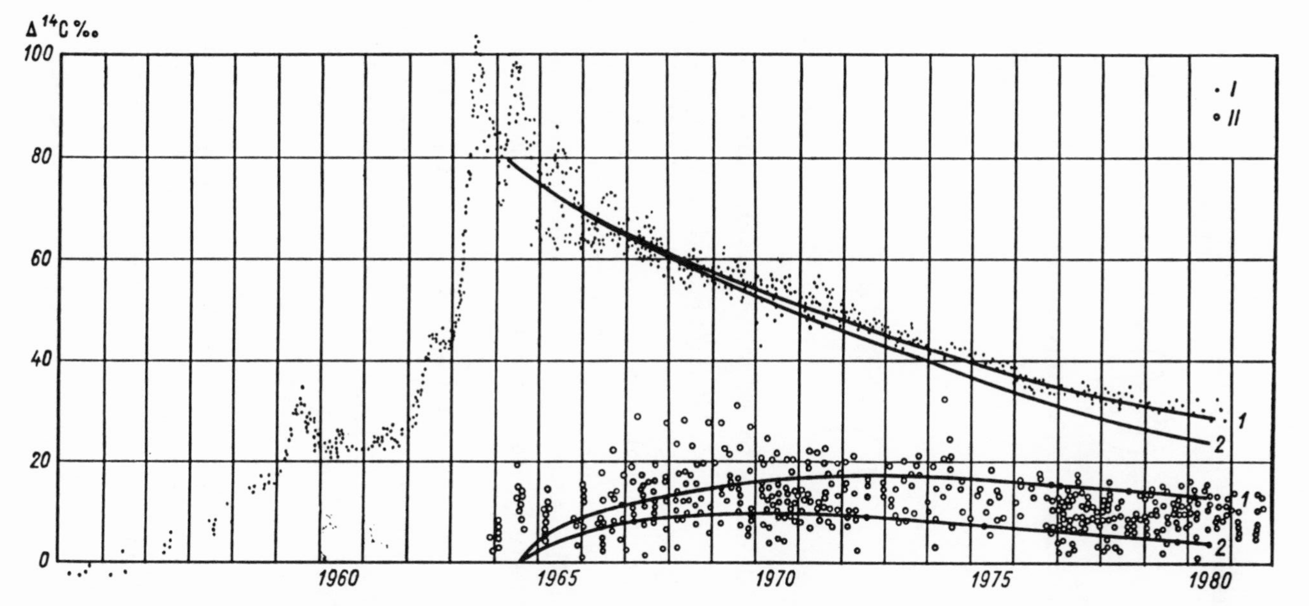

Figure 3.3b. Change in radio-carbon content in the atmosphere (I) and in the upper oceanic layer (II) from 1954 through 1981 from data of different authors (see Bolin et al., 1986) in percentages of its content under undisturbed conditions. 1-2, calculation results (Turchinovich, 1983b) with initial relative radiocarbon content in the upper quasihomogeneous layer and in deep layers equal to 0.94 and 0.84, respectively, $V_L = 40 \times 10^{-4}$ cm/sec, $P_a =$ 330 ppm; (1) $V_{md} = 2.5 \cdot 10^{-5}$ cm/sec, (2) $V_{md} = 5 \cdot 10^{-5}$ cm/sec.

atmospheric ^{12}C content before anthropogenic perturbations.

If the function dM_f/dt is represented in a simplified form

$$\frac{dM_f}{dt} = \left(\frac{dM_f}{dt}\right)_0 e^{\gamma t}, \tag{3.3}$$

we can obtain the relaxation equation solution (3.2) in an analytical form, namely:

$$(\Delta^{14}C)_a(t) - (\Delta^{14}C_a)^0 = -\left(\frac{dM_f}{dt}\right)_0 \frac{e^{\gamma t}[1 - e^{-(\gamma + \frac{1}{\tau_r})t}]}{(\gamma + \frac{1}{\tau_r})} \frac{10^3}{M_a^0}. \tag{3.4}$$

This solution allows us to estimate the amount of fossil fuel M_f burnt prior to 1950, according to measurements of the Suess effect. Beginning in 1950, the rate of industrial CO_2 release into the atmosphere was calculated by Marland and Rotty (1984). They analyzed data on consumption of coal, oil and natural gas throughout the world (Table 3.1). According to their study, dM_f/dt was $1.7 \cdot 10^{12}$ kg C/yr for 1950.

After substituting this value and 3.3 into 3.4, we find that for a 20% change in $(\Delta^{14}C)_a$ from 1860 to 1950, parameter γ should equal 0.024 and M_f amounts to $50 \cdot 10^{12}$ kg, while for a 30% change of $(\Delta^{14}C)_a$, values γ and M_f equal 0.026 and $70 \cdot 10^{12}$ kg, respectively (mean value of τ_r was assumed to equal 9 yrs).

Estimates of M_f by data on fossil fuel consumption during this time interval (about $65 \cdot 10^{12}$ kg) made by Keeling (1973) are in good agreement with the results from analyzing the Suess effect.

Estimates of M_f made prior to 1950 and data on radioactive carbon distribution between the atmosphere and ocean both in natural and perturbed states allowed us to

Table 3.1. The Amount of Anthropogenic CO_2 (ΔM_f) Emitted into the Atmosphere, According to Keeling (1973) and Marland and Rotty (1984)

Time Interval	ΔM_f, 10^{12} kg C	Time Interval	ΔM_f, 10^{12} kg C
1860–1870	1	1930–1940	7.5
1870–1880	2.3	1940–1950	18.0
1880–1890	3.2	1950–1960	21.2
1890–1900	4.2	1960–1970	32.8
1900–1910	6.4	1970–1980	48.4
1910–1920	10.6	1981–1983	15.4
1920–1930	10.6		
Total for 1860–1983		182*	

*Due to the uncertainty of data on ΔM_f prior to 1950, which sometimes is assumed to equal $73 \cdot 10^{12}$ kg C rather than $64 \cdot 10^{12}$ kg C as given here, the total value of M_f can be approximately $9 \cdot 10^{12}$ kg C higher.

pinpoint basic parameter values, that determine the intensity of gas exchange between the atmosphere and ocean (see Broecker, Peng, 1982; Turchinovich, 1983a and b).

It is evident that fossil fuel combustion is not the only source of anthropogenic CO_2 release into the atmosphere. The effect of human activities on the terrestrial biomass and soil could lead and has actually led to CO_2 release into the atmosphere from the biotic reservoir.

There are still discrepancies in the estimated intensity of the corresponding biogenic source $B(t)$ and the total amount of biomass that disappeared in the last century. Chapter 4 treats questions of direct estimates of $B(t)$ as well as which of the biotic reservoirs were most affected by human activities.

Section 3.4 presents the model for evaluating biogenic CO_2 source, developed by Turchinovich and Vager (1985).

It compares their results with those from direct estimates of $B(t)$.

Since the relative radioactive carbon content of the land plant biomass is close to that in the surface air layers, changes in the biota reservoir have practically no influence on the Suess effect. The intensity of the biogenic CO_2 source, $B(t)$, can be assessed by using combined modeling of anthropogenic disorders in ^{12}C and ^{13}C isotope cycles. The ^{13}C content averages 1.2% of ^{12}C concentration, but varies in different reservoirs because of isotope fractionation occurring during photosynthesis, atmosphere-ocean gas exchange, and chemical reactions in sea water (see Figure 3.1).

3.4. Estimate of Biogenic CO_2 Released into the Atmosphere from Data on Carbon Isotopes ^{12}C and ^{13}C

The biogenic source of atmospheric CO_2 in the last 120 years can be estimated by considering cycles of isotopes ^{12}C (major) and ^{13}C (the next in abundance) together.

Stable isotope ^{13}C content in the main reservoirs considered in the carbon cycle is usually measured by $\delta^{13}C$ expressed in pro mille (‰) with respect to the PDB-standard (see section 3.2). This value is related to isotope ^{13}C content with respect to isotope ^{12}C as follows:

$$\delta^{13}C = \frac{(^{13}C/^{12}C) - (^{13}C/^{12}C)_{PDB}}{(^{13}C/^{12}C)_{PDB}} 10^3.$$

Continental biomass and coal contain approximately equal quantities of ^{13}C isotope, and its relative content is approximately 20% lower than in the atmosphere. The $^{13}C/^{12}C$ ratio for the atmosphere and ocean changes both because of industrial CO_2 emission and size variations in the biota carbon reservoir. A lower carbon concentration in the continental biomass produces the same effect as industrial emission of CO_2, or smaller atmospheric $^{13}C/^{12}C$ ratio (Figure 3.4).

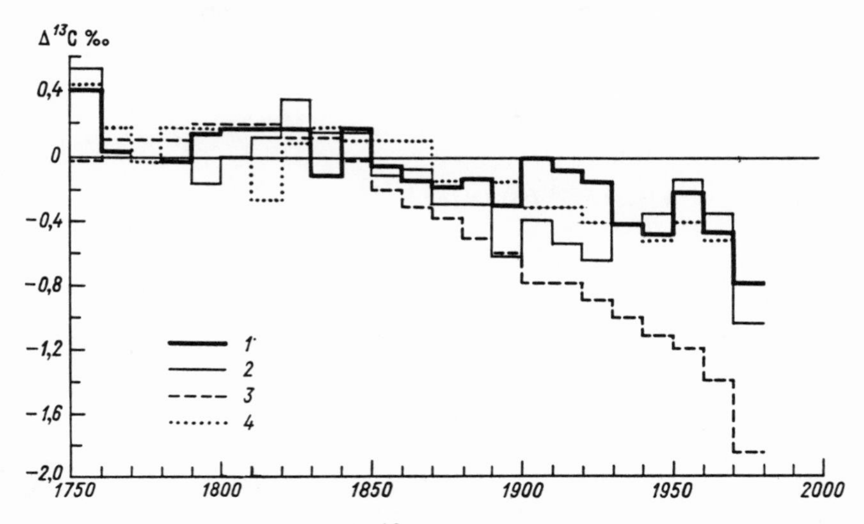

Figure 3.4. Mean values of $\Delta\ ^{13}C$ obtained from tree rings. 1, values of $\Delta\ ^{13}C$ normalized for ring thickness according to Stuiver et al. (1981); 2, $\Delta\ ^{13}C$ values only for open-site trees (Stuiver et al., 1981); 3, obtained by Freyer and Belacy (1981); 4, obtained by Peng et al. (1983).

Before atmospheric disturbances started, all fluxes in the stable isotope between the atmosphere and the continental biota, as well as between the atmosphere and the ocean were balanced; $\delta^{13}C$ on the average was constant, but varied in each of the reservoirs because of ^{12}C and ^{13}C fractionation during photosynthesis, dissolving of CO_2 in sea water (solubility coefficients K_0 for $^{12}CO_2$ and $^{13}CO_2$ are different) and chemical reactions in the system (Keeling, 1979)

$$[CO_2] \rightarrow [HCO_2^-] \rightarrow [C)_3^{2-}]$$
$$(x) \rightarrow (y) \rightarrow (z).$$

The model developed by Turchinovich and Vager (1985) describes the distribution of anthropogenic CO_2 between the atmosphere (a) and the ocean (w) with changes in land biota taken into account

$$\begin{cases} \dfrac{dM_a}{dt} = \dfrac{dM_f}{dt} - V_L S_w K_0 (P_a - P_w) + B(t), \\ h\dfrac{d\Sigma_w}{dt} = V_L K_0 (P_a - P_w) - V_{md}(\Sigma_w - \Sigma_0), \end{cases} \qquad (3.5a)$$

$$\begin{cases} \frac{d}{dt}(R_a M_a) = R_f \frac{dM_f}{dt} - V_L S_w K_0(\alpha_{ax} P_a R_a \\ \qquad\qquad - \alpha_{x\Sigma} P_w R_w) + R_b B(t), \\ h\frac{d}{dt}(\Sigma_w R_w) = V_L K_0(\alpha_{ax} P_a R_a - \alpha x \Sigma P_w R_w) \\ \qquad\qquad - V_{md}(\Sigma_w R_w - \Sigma_0 R_w^0). \end{cases} \tag{3.5b}$$

This equation system describes the atmosphere-ocean distribution of $^{12}CO_2$ (i.e., CO_2 composed essentially of $^{12}CO_2$) and that of $^{13}CO_2$. The following symbols are used here: $M_a(t)$ is the atmospheric CO_2 content by time t; Σ_w is the concentration of dissolved inorganic carbon in the mixed ocean layer; h is its thickness; S_w is the ice-free World Ocean's surface area ($3.37 \cdot 10^{14} m^2$); $P_a(t)$ is the partial pressure of atmospheric CO_2; $P_w(t)$ is the partial pressure of CO_2 dissolved in the upper oceanic layer; K_0 is the CO_2 solubility coefficient in sea water; dM_f/dt is the rate of CO_2 emission into the atmosphere from fossil fuel combustion; $B(t)$ is the intensity of biogenic release of CO_2 into the atmosphere from continental biota changes; $R_i = (^{13}C/^{12}C)_i$ ($i = a, w, f, b$) is the relative concentrations of ^{13}C isotope in atmospheric CO_2, in the upper quasihomogeneous oceanic layer, and in CO_2 emitted during fossil fuel combustion, and released from the biota, respectively; V_L is the effective atmospheric-ocean exchange rate of CO_2; V_{md} is the parameter that like V_L, has velocity dimensions and characterizes carbon exchange between the upper quasihomogeneous and deeper layers of the ocean; and α_{iK} is the fractionation coefficient.

For the period 1960 to 1980, the deep ocean can be considered as an infinitely large reservoir, absorbing anthropogenic CO_2 with constant concentration of dissolved inorganic carbon, i.e., $\Sigma = \Sigma_0 = \text{const}$.

The dependence of partial pressure of CO_2 dissolved in water (P_w) on the concentration of dissolved inorganic carbon ($\Sigma = x + y + z$) that contains CO_3^{2-} and HCO_3^- ions is calculated using hydrochemical relationships for the sea water carbonate system (for example, see Popov, Fyedorovg, Orlov, 1979). Byutner et al. (1981) demonstrated that when CO_2 exchange between the ocean and the atmosphere is computed, it is necessary to use accurate values of $P_w(\Sigma)$, calculated in terms of the contribution of borate alkalinity to the total alkalinity reserve. The results of corresponding computations for $^{12}CO_2$ are presented in Figure 3.5. This also applies to computing $^{13}CO_2$ exchange, but it is also necessary to take the fractionation coefficients into account correctly.

In the starting year of 1860, assuming that anthropogenic disturbances were small prior to that year, surface concentration of dissolved CO_2 should have been in equilibrium with atmospheric CO_2 concentration.

In turn, there was equilibrium between the components of the sea carbonate system, i.e., x, y and z, each of them should have been in thermodynamic equilibrium with the gas phase, i.e., with atmospheric CO_2.

Since the solubility coefficients for $^{12}CO_2$ and $^{13}CO_2$ differ, relative ^{13}C content in dissolved carbon dioxide at the air-water interface is 1.08% lower than in atmospheric CO_2. ^{13}C content in dissolved CO_2 in the upper quasihomogeneous oceanic layer is controlled by isotope equilibrium, that is reached during isotope substitution between all the components of the sea water carbonate system and is related to ^{13}C commutation in dissolved inorganic carbon by the ratio

$$R_w = R_x \left[\frac{x}{z} + \alpha_{xy} \frac{y}{z} + \alpha_{xz} \frac{z}{\Sigma} \right]. \tag{3.6}$$

Fractionation coefficient $\alpha_{x\Sigma}$ in equation system 3.5b that equals the R_x/R_w ratio is defined by expression 3.6, where α_{xy} and α_{xz} are the equilibrium fractionation coefficients for reactions of isotope substitution:

$$[^{13}CO_2] + [H^{12}CO_3^-] \overset{(\alpha xy}{\longleftrightarrow} [^{12}CO_2] + [H^{13}CO_3^-]$$

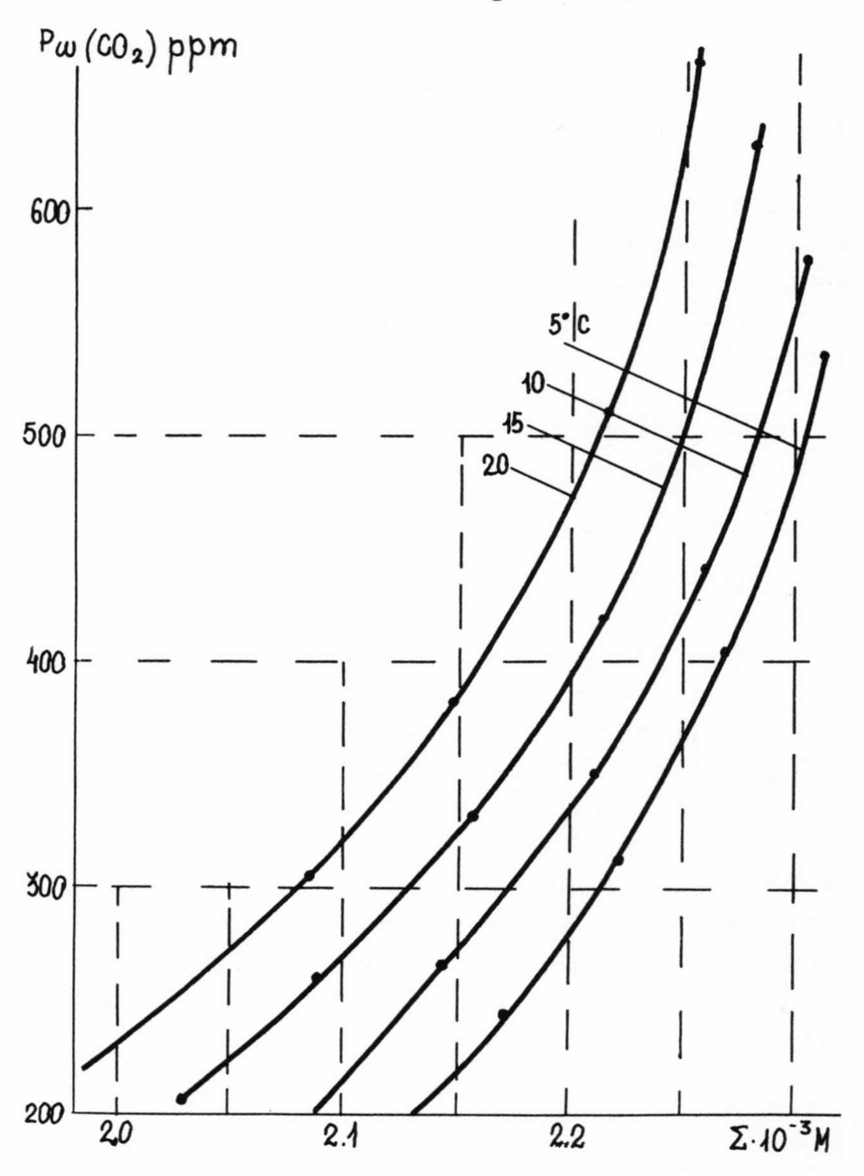

Figure 3.5. Relationship between partial pressure P_w of inorganic carbon dissolved in the upper oceanic layer and its concentration Σ at different temperatures.

and

$$[^{13}CO_2] + [^{12}CO_3^{2-}] \overset{(\alpha xz)}{\longleftrightarrow} [^{12}CO_2] + [^{13}CO_3^{2-}].$$

Values of $\alpha_{x\Sigma}$ depend not only on ambient temperature, but also on concentrations x, y and z in sea water.

Assuming that before 1860, isotope equilibrium existed between the atmosphere and the mixed ocean layer, the initial value of $\delta^{13}C_w^0$ can be determined from the given values of $\delta^{13}C_w^0$ and P_a^0 in the atmosphere. The equation describing isotope equilibrium between atmospheric CO_2 and dissolved inorganic carbon of sea water, is used for this purpose:

$$R_w^0 = R_a^0\left(\alpha_{ax}\frac{x}{\Sigma} + \alpha_{ay}\frac{y}{\Sigma} + \alpha_{az}\frac{z}{\Sigma}\right), \qquad (3.7)$$

where α_{ax}, α_{ay}, α_{az} are the equilibrium fractionation coefficients that characterize isotope substitution in the air-water gas exchange in the following chemical reactions in water (Viotkevich et al., 1977):

$$\begin{aligned} {}^{13}CO_2^{gas} + {}^{12}CO_2^{aq} &\overset{\alpha_{ax}}{\longleftrightarrow} {}^{12}CO_2^{gas} + {}^{13}CO_2^{aq} \\ {}^{13}CO_2^{gas} + H^{12}CO_3^- &\overset{\alpha_{ay}}{\longleftrightarrow} {}^{12}CO_2^{gas} + H^{13}CO_3^- \\ {}^{13}CO_2^{gas} + {}^{12}CO_3^{2-} &\overset{\alpha_{az}}{\longleftrightarrow} {}^{12}CO_2^{gas} + {}^{13}CO_3^{2-}. \end{aligned} \qquad (3.8)$$

The most pronounced fractionation effect is observed in the second reaction of this system, which results in an approximate 9% enrichment of oceanic waters with heavy isotope ^{13}C versus the atmospheric (Mook, Bommerson, Stoverman, 1974).

A number of studies (e.g., Keeling, Mook, Tans, 1979) define the coefficient α_{ax} from the kinetic effect of fractionation during absorption of CO_2 by alkaline solutions (α_{ax} = 0.986). Atmospheric CO_2 is absorbed by sea water during turbulent mass-exchange. Ariel', Byutner and Strokina (1981) showed that the intensity of this exchange should not depend on the molecular mass of the absorbed gas. In such a case, α_{ax} should equal its equilibrium value which at temperature 20°C is 0.9989. In order to compare the results of other authors, calculations were made by the model represented by equations 3.5a and b both with $\alpha_{ax} = 0.9989$ and 0.986.

The model represented by equation system 3.5a and b is based on atmosphere-ocean CO_2 exchange and the diffusion of inorganic carbon to the deeper layers through the lower boundary of the quasihomogeneous layer.

The biotic carbon cycle, including the photosynthesis, respiration and oxidization of organic matter of the continental biota, is considered closed and is not incorporated into the model, insofar as when atmospheric CO_2 concentration rises, the disturbances introduced into this cycle must be minor (see Chapter 4 for details). An essential factor influencing $^{12}CO_2$ and $^{13}CO_2$ atmospheric concentrations is the direct destruction of the continental biomass.

Primary parameters of the model are: effective rate of gas exchange between the atmosphere and the ocean, V_L; exchange rate, V_{md}, of total inorganic carbon between the upper quasihomogeneous layer and deeper ocean; initial partial pressure of atmospheric CO_2, P_a^0 (before anthropogenic perturbations), as well as ^{13}C isotope fractionation coefficients at the atmosphere-ocean interface.

We will briefly discuss analysis of the basic hydrodynamic parameters of the model V_L and V_{md}. They will be used further to forecast variations in partial pressure of atmospheric CO_2 for the next century.

Parameter V_L, that characterizes intensity of gas exchange via the atmosphere-ocean interface was obtained by Ariel' et al. (1979). Its dependence on wind speed for full-scale conditions was established based on analysis of experimental laboratory data on air-water gas exchange (Figure 3.6). The $V_L(u)$ relationship was subsequently extrapolated for the storm wind velocity interval ($u \geq 17$ m/sec) based on variations in a stormy sea surface studied by Bortkovskiy (1983). Ariel' and Strokina (1985, 1986) obtained both mean weighted values of V_L, by using probability distribution functions of wind speed at different points in the World

Ocean, and wind speed averaged for latitudinal belts and parameter V_L for different seasons (Figures 3.7–3.9). Figure 3.9 shows that the mean global value of gas exchange rate V_L equals about $40 \cdot 10^{-4}$ cm/sec. The value of V_L equal to $45 \cdot 10^{-4}$ cm/sec is used most often in models of anthropogenic carbon cycle disorders (Liss, 1973). Both values yield practically the same results (Zakharova, Byutner, 1985). The dependence of V_L on latitude and season for different oceans and on the average for the World Ocean has also been established.

Parameter V_{md} can be estimated from data on the distribution of various tracers in the World Ocean. For example, Revelle and Munk (1977) obtained a mean characteristic time of water exchange between the upper quasihomogeneous layer and deeper ocean of 12.5 years from data on oceanic distribution of tritium ^3H generated in the atmosphere as a result of nuclear tests. The authors assumed that the mean depth of the quasihomogeneous layer was 100 m and that $V_{md} \approx 2.5 \cdot 10^{-5}$ cm/sec.

Turchinovich (1983b) evaluates V_{md} by modeling dissemination of artificially produced ^{14}C from the atmosphere into the ocean, and compares the results with GEOSECS data on $\Delta^{14}C$ content changes in the upper quasihomogeneous layer of the Atlantic and Pacific oceans. He shows that the values of V_{md} should range between $2.5 \cdot 10^{-5}$ and $5 \cdot 10^{-5}$ cm/sec. Zakharova (1987) obtained $V_{md} \approx 6 \cdot 10^{-5}$ cm/sec by modeling F-11 and F-12 absorption by the oceans and comparing calculation results with experimental data of Gammon, Cline and Wisegawer (1982).

The value of V_{md} can be also calculated by using the model results obtained by Kagan and Ryabchenko (1981) for seasonal evolution of the upper quasihomogeneous oceanic layer caused by the effect of tangential wind stress and solar radiation absorption. If the annual mean flux of total inorganic carbon from the depth to the upper layer calculated by these authors (this flux exists when the ocean

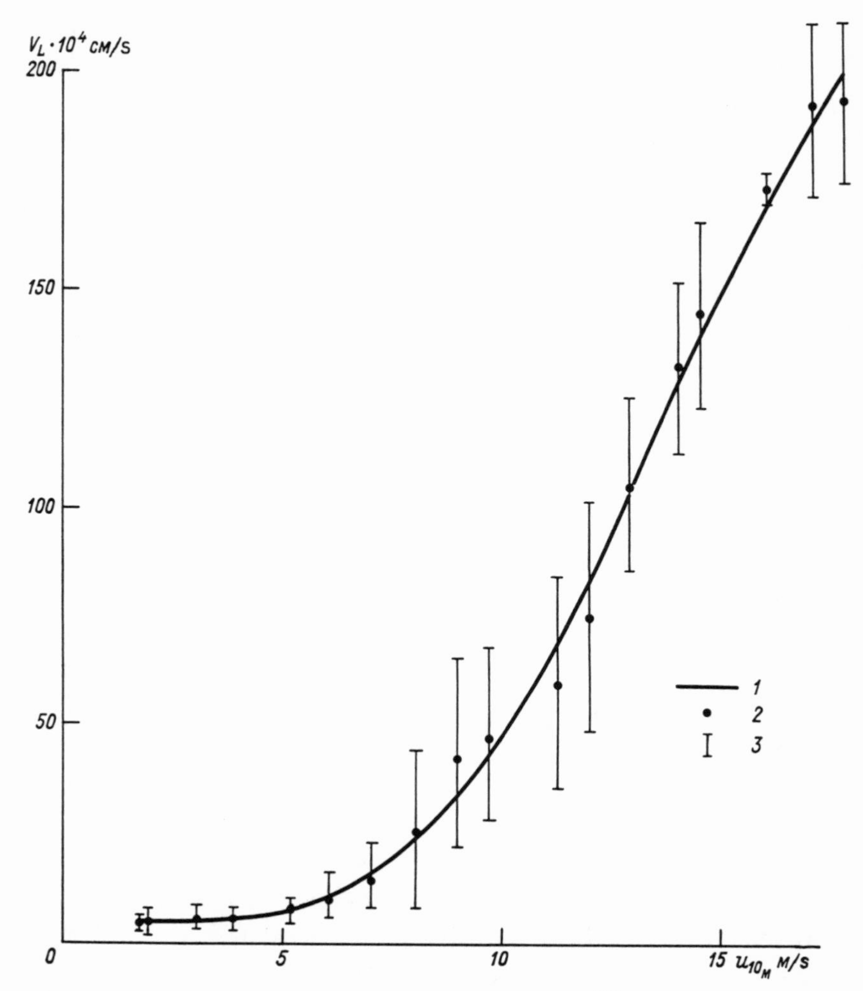

Figure 3.6. Dependence of gas exchange rate V_L on wind velocity above ocean (Ariel' et al., 1979). 1 and 2 are the running means V_L for averaging interval $\Delta u = 2$ m/sec; 3 is the standard deviation of measurement data by different authors.

is in an equilibrium state and it compensates for the flow of organic carbon-containing particles descending from the upper layer) is presented as $V_{md}(\Sigma - \Sigma_w)$, then we find that $V_{md} \approx 3 \cdot 10^{-5}$ cm/sec.

Tree rings record the change in isotopic composition of atmospheric carbon. Galimov (1925), Stuiver (1978) and

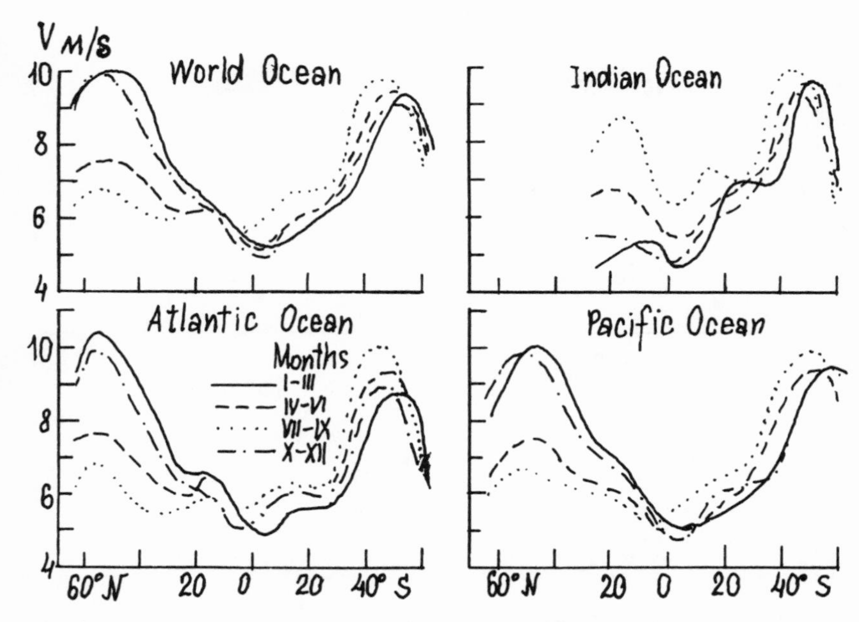

Figure 3.7. Latitudinal variations in mean seasonal values of wind speed above oceans (Ariel', Strokina, 1982, 1985).

Freyer (1975b) were the first in the mid-1970s to analyze the change in ^{13}C content in tree rings. They showed that the values of Δ^{13}C in tree rings depend not only on isotopic composition of atmospheric carbon, but also on air temperature, precipitation, local illumination and physiology of tree growth. Therefore, it was difficult to compare Δ^{13}C measured in tree rings with global changes in atmospheric $\Delta^{13}C_a$. Attempts are nevertheless still being made to obtain more reliable data on Δ^{13}C in tree rings. The most extensive data on Δ^{13}C in tree rings presented in Freyer and Belacy (1983) are given in Figure 3.4. These measurements were originally published in the proceedings from the Berne 1981 conference on CO_2 (Analysis and Interpretation of Atmospheric CO_2 Data, 1981). Peng et al. (1983) estimated 60% biogenic of the total CO_2 emission into the atmosphere for the past 150 years based on the data of Freyer and Belacy and the carbon cycle model of Oeschger et al. (1975).

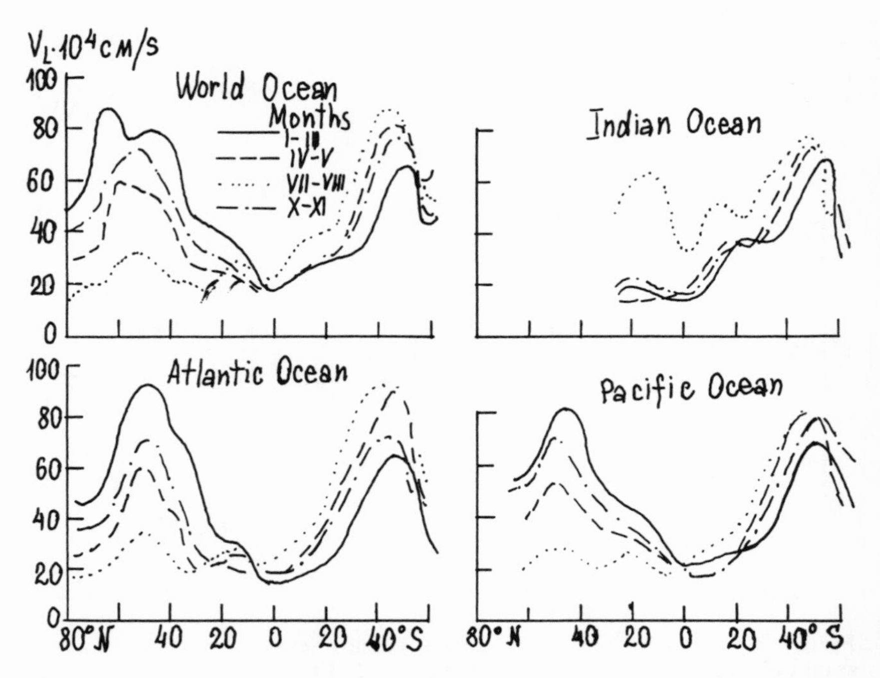

Figure 3.8. Latitudinal variations of mean seasonal rates of gas-exchange V_L between the ocean and the atmosphere (Ariel', Strokina, 1986).

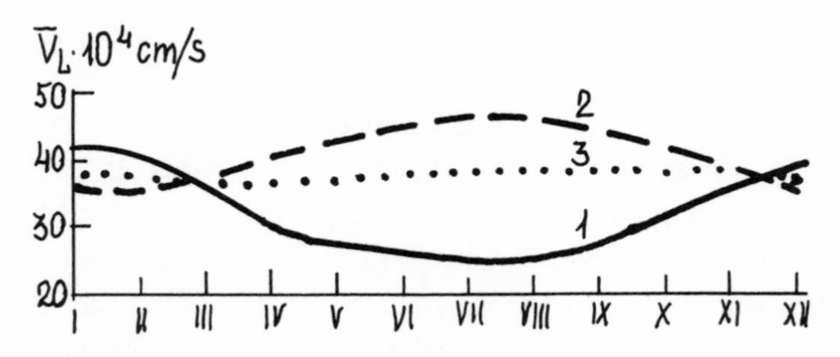

Figure 3.9. Mean values of $\overline{V_L}$ for the Northern (1), Southern (2) Hemispheres and the World Ocean (3) (Ariel', Strokina, 1986).

In addition to tree ring $\Delta^{13}C$ data, there is information derived from direct measurements of ^{13}C in the atmosphere, made by Goodman (1980) and also by Mook, Keeling and Herron (1981). First measurements of $\delta^{13}C_a$ were made in 1956. Comparison of Table 3.2 and Figure 3.4 data show

Table 3.2. The Results of Measuring Atmospheric $\delta^{13}C_a$.

	1956		1978		1980	
	P_a ppm	$\delta^{13}C_a$ (‰)	P_a ppm	$\delta^{13}C_a$ (‰)	P_a ppm	$\delta^{13}C_a$ (‰)
Northern Hemisphere	314.1	-6.69				
Cape Grim, Australia				-7.8 (for 1977)		-0.03 ‰/yr 1977–1980 linear trend
Fanning Island (4°N, 159°W)			335.14	-7.66 (-7.35)	338.28	-7.77 (-7.46)
South Pole			332.74	-7.67 (-7.36)	335.86	-7.80 (-7.49)
Mauna-Loa, Hawaii			335.20	-7.65 (-7.34)		

Note: Values of $\delta^{13}C_a$ given in parentheses take into account atmospheric N_2O content with the same molecular weight as CO_2.

good agreement between direct measurements of $\Delta^{13}C_a$ and tree ring data.

The $\delta^{13}C$ content also changes in the upper quasihomogeneous oceanic layer. Lengthy observations of $\delta^{13}C$ from coral rings, published in Nosaki et al. (1978), yielded a total value of $\Delta^{13}C$ = -0.5% for the 1850–1978 period. Equations

3.5a and b permit calculation of simultaneous changes in $^{12}CO_2$ and $^{13}CO_2$ in the atmosphere and the ocean.

The rates of $^{12}CO_2$ industrial emissions up to 1980 were set according to data from Keeling (1973) and Marland and Rotty (1984). The initial P_a^0 values in 1860 varied within 270–290 ppm. The initial concentration of total inorganic carbon was calculated from the initial value $P_w^0 = P_a^0$ and the relationships in Figure 3.5, i.e., based on chemical equilibrium between the atmosphere and ocean surface using mean values of temperature and salinity in the World Ocean surface.

The following dimensionless parameters were introduced into the computations:

$$\tau = \frac{V_{md}}{h}t; \; \tilde{M}_a = \frac{M_a}{M_a^0}; \qquad \tilde{M}_w = \frac{\Sigma_w S_w h}{M_a^0};$$
$$\tilde{M}_f = \frac{M_f}{M_a^0}; \qquad \gamma = \frac{V_L}{V_{md}} \frac{K_0 P_w}{\Sigma_w}; \quad \tilde{B} = \frac{h}{V_{md}} \frac{B}{M_a^0}. \tag{3.9}$$

Equation system 3.5a and 3.5b was resolved for derivatives. After a transform, the following normal system of differential equations was obtained:

$$\frac{d\tilde{M}_a}{d\tau} = \frac{d\tilde{M}_f}{d\tau} - \gamma \tilde{M}_w \left(\frac{P_a}{P_w} - 1 \right) + \tilde{B} = F_1,$$
$$\frac{d\tilde{M}_w}{d\tau} = \gamma \tilde{M}_w \left(\frac{P_a}{P_w} - 1 \right) - (\tilde{M}_w - \tilde{M}_w^0) = F_2,$$
$$\frac{d\tilde{R}_a}{d\tau} = \frac{1}{\tilde{M}_a} \left[\tilde{R}_f \frac{d\tilde{M}_f}{d\tau} - \gamma \tilde{M}_w \left(\alpha_{ax} \tilde{R}_a \frac{P_a}{P_w} - \alpha_{x\Sigma} \tilde{R}_w \right) + \tilde{R}_b \tilde{B} \right] - \frac{\tilde{R}_a}{\tilde{M}_a} F_1 = F_3,$$

$$\frac{dR_w}{d\tau} = \frac{1}{\tilde{M}_w} \left[\gamma \tilde{M}_w \left(\alpha_{ax} \tilde{R}_a \frac{P_a}{P_w} - \alpha_{x\Sigma} \tilde{R}_w \right) - (\tilde{M}_w \tilde{R}_w - \tilde{M}_w^0 \tilde{R}_w^0) \right] - \frac{\tilde{R}_w}{\tilde{M}_w} F_2 = F_4. \tag{3.10}$$

Cauchy's problem for system 3.10 with conditions

$$\tilde{M}_a = \tilde{M}_a^0; \qquad \tilde{M}_w = \tilde{M}_w^0; \qquad \tilde{R}_a = \tilde{R}_a^0; \qquad \tilde{R}_w = \tilde{R}_w^0$$

and $\tau = 0$ was solved numerically by the Runge-Cutt procedure.

The influence of parameters P_a^0, $\delta^{13}C_a^0$, V_{md}, $\delta^{13}C_f$ and α_{ax} on $P_a(\tau)$ and $\Delta^{13}C_a(\tau)$ was studied while solving equation system 3.10. Sensitivity of the findings to selection of these parameter values is shown in Figures 3.10 and 3.11.

The change in initial atmospheric CO_2 concentration (for 1860) from 270 up to 290 ppm did not affect $\Delta^{13}C(\tau)$, but considerably influenced the temporal variations in atmospheric CO_2 concentration, i.e., function $P_a(\tau)$.

A 0.5% change in initial value of $\delta^{13}C_a^0$ leads to a 0.08% variation in $\Delta^{13}C_a$ value, while atmospheric CO_2 remains the same (curves 3 and 4 in Figure 3.10).

The fractionation coefficient α_{ax} and isotopic CO_2 composition (R_{ff}) released during burning of fossil fuel (coal, oil, gas) affect the $\Delta^{13}C_a$ value, but should influence atmospheric CO_2 content (Figure 3.10, curves 1 and 2, 5 and 6).

A lower rate of dissolved inorganic carbon diffusion from the upper quasihomogeneous layer into deeper oceanic layers leads to more rapid growth of P_a and greater changes in $\Delta^{13}C_a$ (see Figure 3.11). This trend becomes more pronounced when an additional source of CO_2 is introduced into the atmosphere, i.e., with $B \neq 0$. Figure 3.12 illustrates model results with $V_{md} = 5 \times 10^{-5}$ cm/sec.

By comparing model results for ^{12}C and ^{13}C isotopes with data on atmospheric CO_2 monitoring and variations in relative ^{13}C atmospheric content, we conclude atmospheric CO_2 content increases not only because of CO_2 emissions from burning fossil fuel, but also because of changes in the continental biomass reservoir. Inclusion of the indefinite values for basic model parameters does not alter this conclusion.

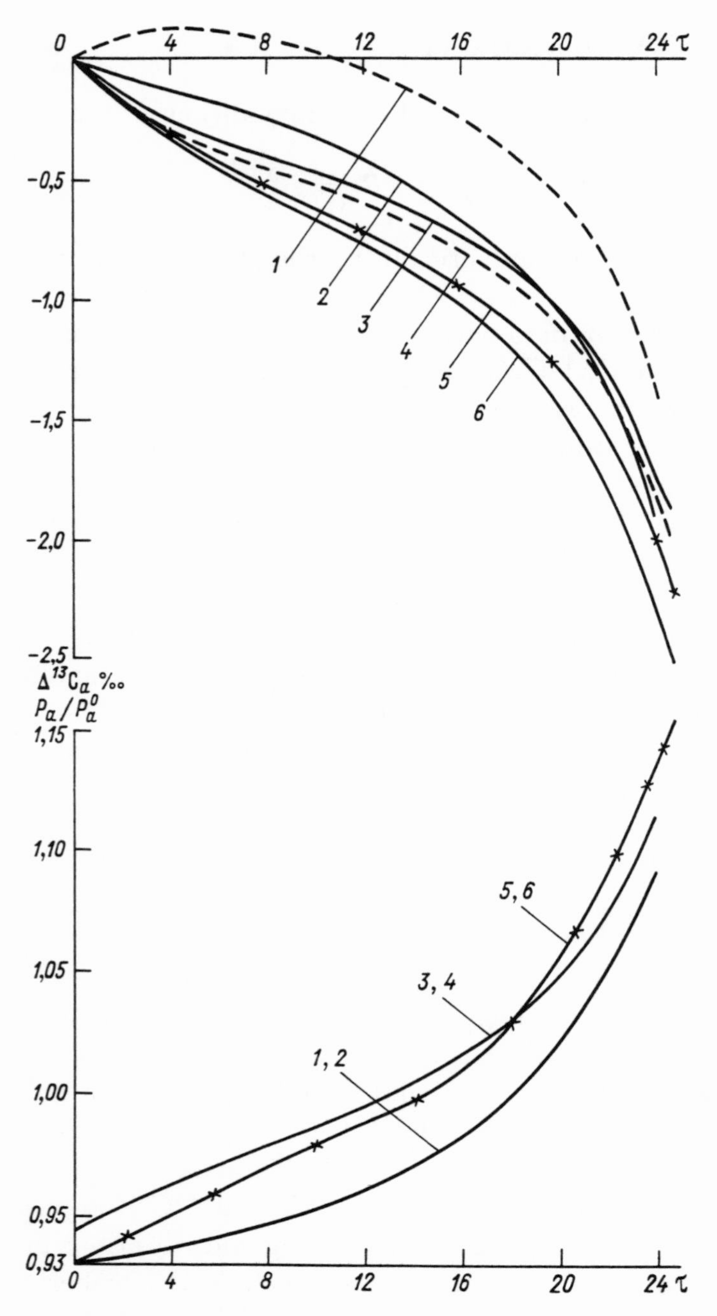

Figure 3.10. Influence of values of fractionation coefficient α_{ax}, isotope composition of fossil fuel $\delta\ ^{13}C_{ff}$ and initial content of ^{13}C in the atmosphere on $\Delta(\delta\ ^{13}C_a)$ (abbreviated as $\Delta\ ^{13}C_a$) and on P_a. (1) $\alpha_{am} =$ 0.986; (2) $\alpha_{ax} =$0.9989; (3) $\delta\ ^{13}C_a^{\circ} =$ -6‰; (4) $\delta\ ^{13}C_a =$ 5.5‰; (5) $\delta\ ^{13}C_{ff} = -25$‰; (6) $\delta\ ^{13}C_{ff} = -28$‰.

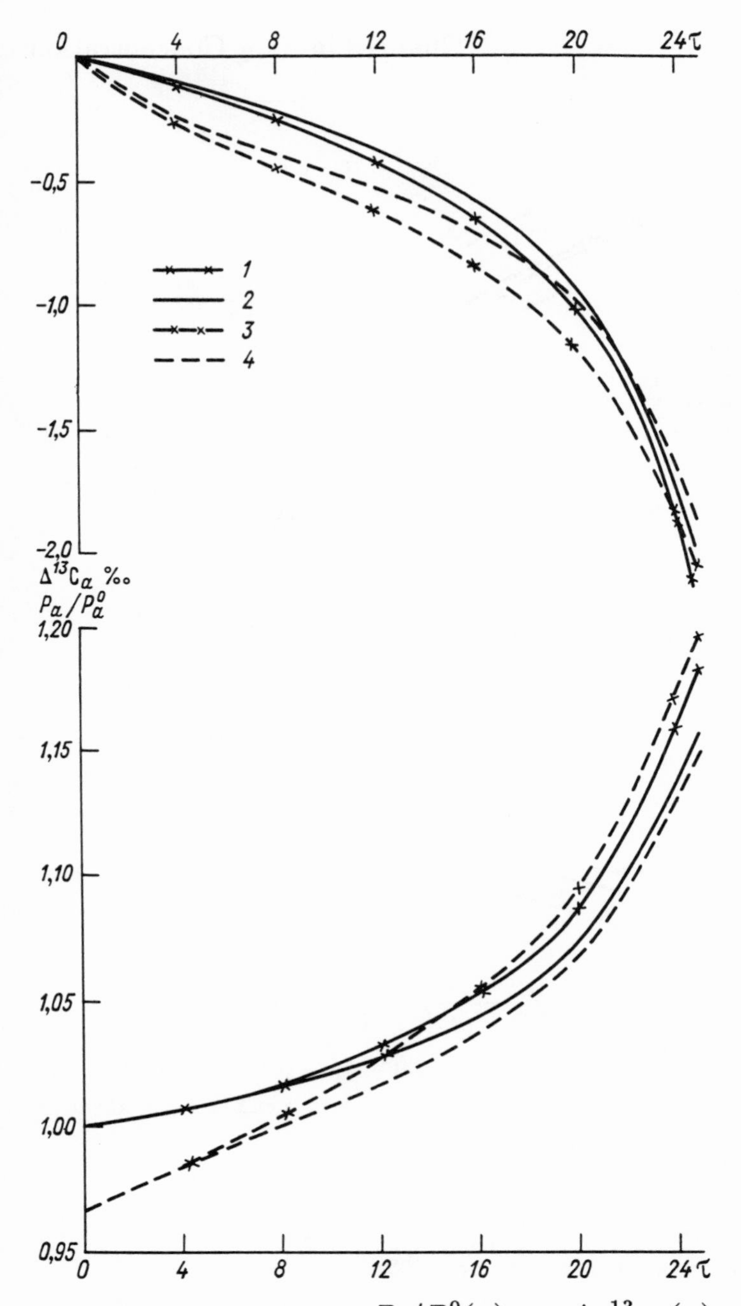

Figure 3.11. Dependence of $P_a/P_a^0(\tau)$ and $\Delta\ ^{13}\mathrm{C}_a(\tau)$ on different values of V_{md} and B.

Curve	1	2	3	4
V_{md}, 10^{-5} cm/sec	2.5	5.0	2.5	5.0
B, kg/yr	0	0	$0.5 \cdot 10^{12}$	$0.5 \cdot 10^{12}$

Dimensionless parameter $\tau = V_{md}t/h$ (see (3.9)).

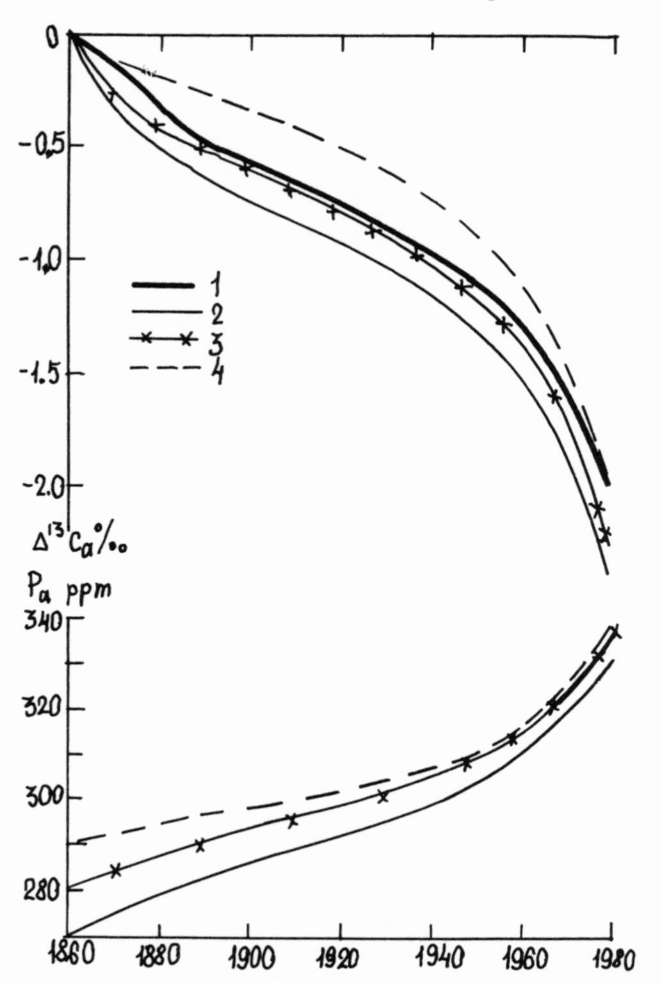

Figure 3.12. Model calculation results for $\Delta\ ^{13}C_a$ and P_a with $V_{md} = 5.0 \cdot 10^{-5}$ cm/sec. 1, Empirical data on $\Delta\ ^{13}C_a$ and monitoring data on P_a.

Curve	2	3	4
P_a^0, ppm	270	280	290
$\int Bdt, 10^{12}$ kgC	120	90	30

$\delta\ ^{13}C_a^0$, values are assumed to be –5.5%

The most probable value of initial atmospheric CO_2 concentration P_a^0 from these calculations is close to 280 ppm. When P_a^0 equals 270 ppm, the calculation results do not agree with empirical data on ^{12}C and ^{13}C: $P_a(\tau)$ during the period 1958–1980 is less than the observed, while the ^{13}C calculations yield higher $\Delta^{13}C_a$, then those from tree rings. An increase in the value of B leads to even greater discrepancy between calculated and observed $\Delta^{13}C_a$ values.

For the initial value of $P_a^0 = 290$ ppm, $P_a(\tau)$ values close to monitoring data are obtained for low values of B (0.25 Gt/yr), but then the overall pattern of $\Delta^{13}C_a$ variation disagrees with empirical data.

In the case of $V_{md} = 2.5 \cdot 10^{-5}$ cm/sec, model results agree satisfactorily with data on $P_a(\tau)$ and $\Delta^{13}C_a$ with total biogenic emission of 60–75 Gt for 120 years. This corresponds to mean intensity of biogenic CO_2 emission of 0.5–0.625 Gt/yr.

When the effective rate of carbon exchange between the upper quasihomogeneous and deeper ocean layers is greater ($V_{md} = 5 \cdot 10^{-5}$ cm/sec), total biogenic CO_2 emission into the atmosphere over 120 years proves to be about 90 Gt, i.e., averages 0.75 Gt/yr (see Figure 3.12). The appearance of function $B(\tau)$ (Figure 3.13) significantly influences the relationship of $P_a(\tau)$ and $\Delta^{13}C_a(\tau)$. By comparing model results with data on $\Delta^{13}C_a$, published in Freyer and Belacy (1983), we can conclude that the $B(t)$ function should be bell-shaped with peak $B_{\max} = 0.75$–0.8 Gt/yr in 1918–1920, while during the monitoring years, i.e., beginning in 1958, B is close to zero. Similar conclusions were drawn by Siegenthaler, Heiman and Oeschger (1978) and Pearman (1980b), however, their estimated biogenic CO_2 influx into the atmosphere was less reliable since the values of P_a^0 and function $\Delta^{13}C(t)$ were less definite then than now.

The aforementioned analysis shows that for the most probable values of V_{md} equal to $2.5 \cdot 10^{-5}$ cm/sec and 5.0×10^{-5} cm/sec, the integral biogenic release is 60–90 Gt of C between 1860 and 1980 with initial P_a^0 value close to 280 ppm (Vager, Turchinovich, 1987).

More reliable data on the altered ^{13}C content in the atmosphere and the ocean are needed to pinpoint the value of $\int B dt$ (or $B(t)$).

The $P_a(\tau)$ curve in Figure 3.13 for total biogenic emission of 75 Gt for different $B(\tau)$ and $dM_f/d\tau$ functions derived from Table 3.1 data, can now be compared to curve $P_a(t)$ (Figure 3.14) obtained by Neftel et al. (1985) after analyzing the composition of air bubbles in young glaciers of the West Antarctic.

Comparison of the model and experimental data shows that although it is simple, the model satisfactorily describes gas-exchange in the ocean-atmosphere system. The total biogenic emission is actually about 75×10^{12} kg of C and in any case does not exceed 100×10^{12} kg of C, but the onset of biogenic CO_2 release should be dated 2 to 3 decades prior to 1860 (if the data of Neftel et al. are correct). Comparison of function $B(t)$ with data from Table 3.1 on CO_2 release into the atmosphere due to burning fossil fuel indicates that biogenic emission was predominant, apparently up to the end of the nineteenth century after which intensive consumption of coal, oil and natural gas made the biogenic source less important.

After 1950, its contribution to $P_a(t)$ change was negligible. This conclusion is drawn in practically all the studies which compare the model results for $P_a(t)$ function with data of continuous observations of atmospheric CO_2 changes.

3.5. Forecasting CO_2 Concentration Changes

The problem of the distribution of industrial CO_2 emissions between the atmosphere, ocean and continental biota has been stated and modeled by many specialists. Earlier studies (1975–1979) assumed that utilization of all the accessible fossil fuel reserves $(M_f)_\infty$ would occur in the next 200 to 300 years. Function dM_f/dt was distributed differently in time, and the temporal variation of atmospheric CO_2 content $M_a(t)$ (see, e.g., Energy and Climate, 1977) was calculated for each specified dM_f/dt. The resulting functions $M_a(t)$ had a peak several decades ahead of the

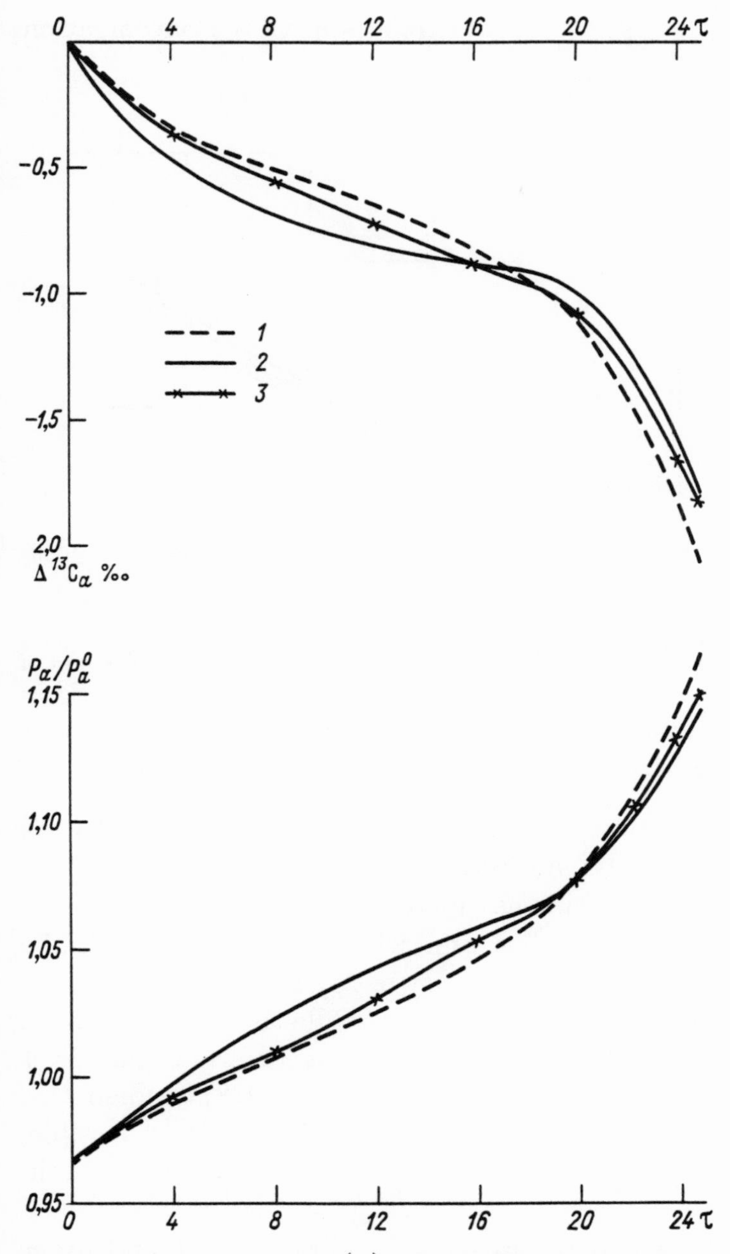

Figure 3.13. Influence of $B(\tau)$ dependence on the shape of functions $\Delta\ ^{13}C_a(\tau)$ and $P_a(\tau)$ with fixed value of $\int Bdt$ equal to $75 \cdot 10^{12}$ kgC, $P_a^0 = 290$ ppm. (1) $B = \text{const} = 0.625 \cdot 10^{12}$ kgC/yr; (2) $B = 10^{12}$ kgC/yr with $0 \leq \tau \leq 10.2$ and linearly decreases to zero with $\tau \geq 20.5$; (3) $B = 0.75 \cdot 10^{12}$ kgC/yr with $0 \leq \tau \leq 16.4$ and linearly decreases to zero with $\tau \geq 16.4$.

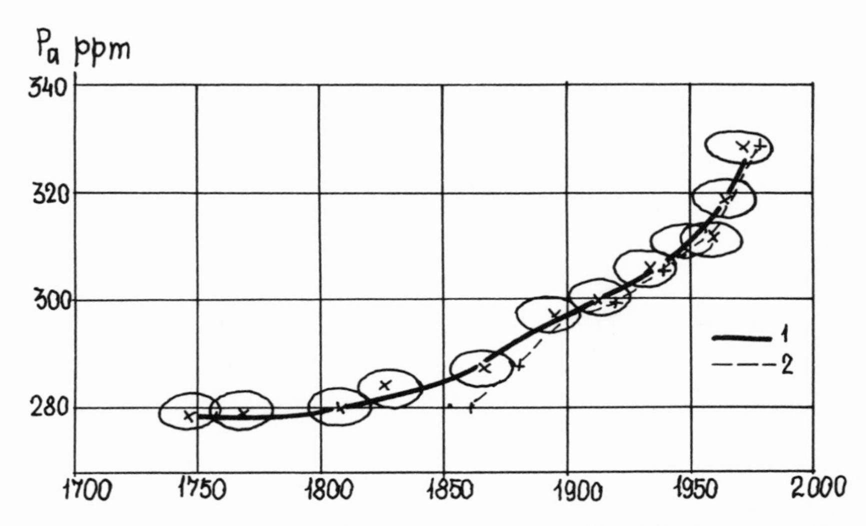

Figure 3.14. Changes in relative concentration P_a with time according to data from (Neftel et al., 1985). 1, average values of P_a (large and small semiaxes of ellipses represent errors in dating air bubbles and of measured P_a; 2, model calculation results of $P_a(t)$ with $P_a = P_a^0 =$ 280 ppm in 1860; B(t) = 0 with $t \leq 1860$, B(t) = $0.75 \cdot 10^{12}$ kgC/yr with $1860 \leq t \leq 1940$ and decreases linearly to zero by 1980.

dM_f/dt maximum. After reaching its peak, function $M_a(t)$ diminished slowly; its asymptotics at $t \to \infty$ depend on whether possible biomass increase M_b was limited in the model or not. If such a limitation was not imposed, the result was absurd: the majority of M_f turned into biomass, which thus increased 5 to 7 times, while the CO_2 content in the atmosphere and upper quasihomogeneous oceanic layer returned to the initial values. Olson (1978) studied this variant. Since unlimited growth of biomass is impossible, the majority of authors incorporated this circumstance in some way.

Another trend began recently in the USSR and other countries. Equations describing biota's reaction to anthropogenic disturbances are excluded from the model, and changes in the biomass reservoir are represented by $B(t)$. This function can be defined for the past (see section 3.3), while for the future, it is extrapolated within reasonable limits, and the corresponding uncertainty is incorporated into the estimated errors of forecasting $P_a(t)$ or $M_a(t)$.

This statement of the problem permits limited parametrization modeling of ocean-atmosphere gas exchange; its experimental and theoretical study has made marked progress in recent years.

Calculated distribution of industrial CO_2 emission M_f between the atmosphere and ocean depends on the way CO_2 transport into the deep ocean is described, especially in case of high rates of perturbations dM_f/dt disturbing the equilibrium of the ocean-atmosphere system. There are currently two basic ways to describe diffusion of anthropogenic CO_2 into the deep layers of the ocean. The first one assumes that the diffusing CO_2 flux is proportional to the difference in mean concentrations between the upper quasi-homogeneous and uniformly mixed deep layers of the ocean. The coefficient of proportionality is determined by integral diffusion resistance of a very stratified barrier layer at the interface (seasonal thermocline). Therefore, in contrast to standard box models, we will call the carbon cycle model based on the first method the model with integral coefficients. This method is based on the results of the work by Kagen and his colleagues (1979, 1981, 1982), who studied basic laws of seasonal and latitudinal variations in the mixing intensity between the upper and deep oceanic layers. Many studies on forecasting $P_a(t)$ use a model that presents the deep oceanic layers as a uniform turbulized medium with constant diffusion coefficient K, while ignoring the barrier between the surface and deep layers. The most detailed description of the fundamentals of this diffusion model is presented in the work by Oeschger et al. (1975), and its applications and later modifications can be found in Seigenthaler, Heimann, Oeschger, and Wenk (1978, 1981, 1984).

The deep oceanic diffusion model incorporates a transient diffusion equation with constant coefficient

$$\left(\frac{\partial}{\partial t} - K\frac{\partial_2}{\partial z^2}\right)(\Sigma_2 - \Sigma_2^0) = 0, \qquad (3.12)$$

where Σ_2 is inorganic carbon concentration in deep oceanic layers. Time-dependent mean concentration $\Sigma_w(t)$ in the upper quasihomogeneous oceanic layer serves as the boundary condition for this equation.

Siegenthaler-Oeschger's method is widely used to compute diffusion of various tracers in the ocean and, in particular, distribution of industrial CO_2 emissions between the atmosphere and ocean.

Most of the equations in the carbon cycle model with integral coefficients and the diffusion model are analogous. These are first order equations that describe changing carbon content M_i in the atmosphere (a), upper (1) and deep oceanic layers (2) under the influence of the perturbing function $M_f(t)$ which determines the amount of industrial CO_2 released into the atmosphere by moment t. The equation system in which integral coefficients are used, looks like

$$M_f(t) = \Delta M_a(t) + \Delta M_1(t) + \Delta M_2(t) + \int_0^t B dt, \qquad (3.13)$$

$$\begin{aligned}\frac{dM_1}{dt} &= S_w h_1 \frac{d\Sigma_1}{dt} \\ &= S_w V_L K_0 (P_a - P_w) - S_w V_{md}(\Sigma_1 - \Sigma_2),\end{aligned} \qquad (3.14)$$

$$\frac{dM_2}{dt} = S_w h_2 \frac{d\Sigma_2}{dt} = S_w V_{md}(\Sigma_1 - \Sigma_2). \qquad (3.25)$$

Equation 3.13 is the balance of emitted CO_2. Eq. 3.14 describes the altered total content $M_1 = S_w h_1 \Sigma_1$ of dissolved inorganic carbon in the upper quasihomogeneous oceanic layer, where h_1 is the thickness of this layer and S_w is the ice-free surface area of the World Ocean. The main component of the sea water carbonate system is the hydrocarbonate ion. Dissolved gas concentration $x = K_0 P_w$ is a small percentage of y, but it does determine the intensity of gas-exchange with the atmosphere (term $S_w V_L K_0(P_a - P_w)$) in 3.14, where V_L is the effective rate of gas exchange through the air-water interface, $K_0 P_a$ is CO_2 concentration at the atmosphere-ocean interface, and $K_0 P_w$ is the concentration in the upper oceanic layer. Parameter V_{md}, with the same velocity dimensions V_L, characterizes the mean annual intensity of total inorganic carbon exchange between the upper quasihomogeneous and deeper oceanic layers during seasonal variations in their interface properties boundary (see section 3.2).

Unlike Eq. 3.12, Eq. 3.15 conforms to the idea of the deep ocean as a well-mixed reservoir with characteristic mixing time τ_2 equal to $\frac{h_2}{V_{md}}$.

Eqs. 3.14 and 3.15 are different in the diffusion model of the carbon cycle:

$$\frac{dM_1}{dt} = S_w h_1 \frac{d\Sigma_1}{dt} = S_w V_L K_0(P_a - P_w) - F, \qquad (3.14a)$$

$$\frac{dM_2}{dt} = F, \qquad (3.15a)$$

where F is the total flux of inorganic carbon from the upper quasihomogeneous into the deep oceanic layers.

Both models incorporate a nonlinear function of correlation between partial pressure of CO_2 dissolved in the ocean P_w and total concentration of dissolved inorganic carbon Σ. Investigation of the sensitivity of computational results showed that precise hydrochemical relationships between P_w and Σ have to be used, taking into account the borate alkalinity contribution (Zakharova, Byutner, 1985). Selection of the mean annual surface layer temperature slightly influences the computational results, all things being equal. The main dynamic parameters which determine intensity of anthropogenic CO_2 absorption by the ocean are V_L and

V_{md} for the model with integral coefficients of exchange and V_L and K for the diffusion model.

The study by Byutner and Zakharova (1983) compared parameters V_{md} and K yielding similar results for the projected function $P_a(t)$ to the year 2050. By establishing this correlation, we can define V_{md} from the values of K obtained by analyzing monitoring data for different tracers in the ocean. In estimates of forecasting errors for P_a (CO_2) one can apparently assume V_{md} equals $(4 \pm 1.5) \cdot 10^{-5}$ cm/sec. Another common parameter V_L for both variants of the model can be considered more precise. The method for determining it is presented in section 3.3 (for details see Byutner, 1986).

As already mentioned, we now have more accurate values for the source of both past and future industrial CO_2. Figure 3.15 illustrates the atmospheric CO_2 partial pressure up to the year 2100, computed using the source function dM_f/dt based on the last scenario of future energy development with gradual transition to alternative energy sources (Legasov, Kuz'min, Chernoplekov, 1984). The computation used the model with integral coefficients for $V_L = 39 \cdot 10^{-4}$ cm/sec and for two values of V_{md}, $5 \cdot 10^{-4}$ cm/sec (curve 1) and $2.5 \cdot 10^{-5}$ cm/sec (curve 3). Curve 1 corresponds to maximum absorbability of the ocean at which there is still satisfactory agreement with experimental data on changes in ^{12}C, ^{13}C and ^{14}C content in mobile carbon reservoirs, for the last 120 years. Forecasting errors with a specified source function dM_f/dt (curve 5 in the same figure) were assessed for this case. The results are represented by the cross-hatched area bordered by curves 2a and 2b. The uncertainty of possible changes in continental biomass reservoir size contributes most to the error. It was assumed that these changes could be up $\pm 120 \cdot 10^{12}$ kg by the year 2100. The resulting calculation error agrees with the corresponding value in the U. S. National Academy of Sciences report (Changing Climate, 1983, p. 20).

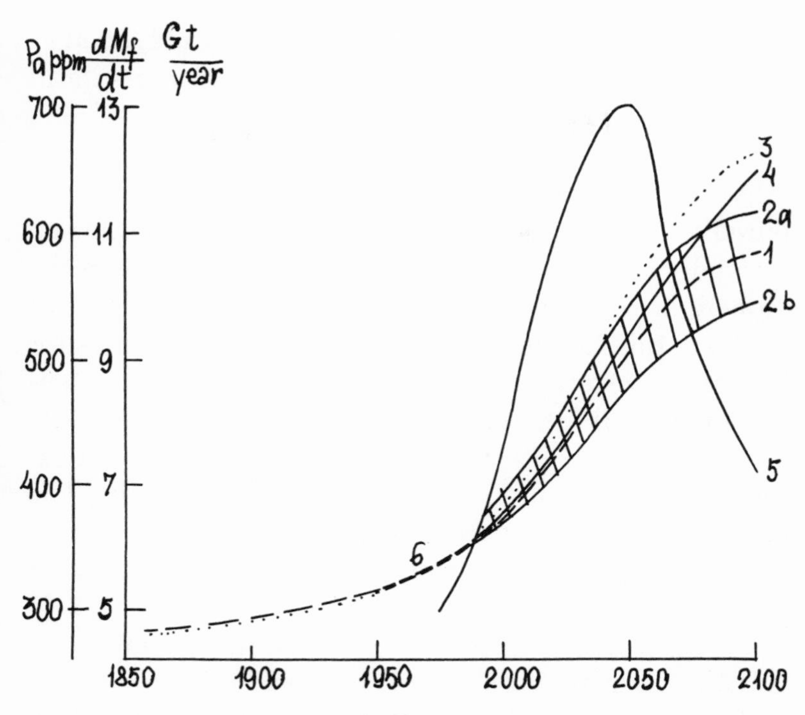

Figure 3.15. The forecast of P_a-values up to 2100 by the scenario proposed in (Legasov, Kuz'min, Chernoplekov, 1984) using a model of integral exchange coefficients. 1, most probable dependence of P_a on time with $V_{md} = 5 \cdot 10^{-5}$ cm/sec and the volume of biotic reservoirs unchanged since 1980; 2a and 2b, with $\int B dt = 120 \cdot 10^{12}$ kg and 120 $\cdot\ 10^{12}$ kg, respectively; 3, the same as 1, but with $V_{md} = 2.5 \cdot 10^{-5}$ cm/sec; 4, estimate made by Legasov, Kuz'min and Chernoplekov (1984); 5, source function dM_f/dt used in computations; 6, monitoring data.

A secondary factor in the error is the uncertain parameter V_{md} (about ±20 ppm in absolute magnitude); the uncertain V_L parameter is a minor contribution.

Since the predicted source function dM_f/dt can change, it was necessary to study the factors governing the remaining percentage of industrial CO_2 emissions in the atmosphere (in % of total emission M_f for the present moment t, as well as for individual time intervals, i.e., $\Delta M_a/\Delta M_f \cdot$ 100%. Monitoring of atmospheric CO_2 concentration indicates that the function $\Delta M_a/\Delta M_f$ has increased between

1958 and now. Figure 3.16 presents values of $\Delta M_a/\Delta M_f \cdot 100\%$ calculated for ten-year running mean increments in $\Delta P_a(CO_2)$ based on monitoring data and corresponding mean values of ΔM_f from Table 3.1. Figure 3.16b shows that the first four points (1) are in good agreement with computed values, while the last one (2) indicates either an unexpectedly large portion of emitted CO_2 remaining in the atmosphere or some inaccuracy in the initial data. We can assume that this inaccuracy stems from Table 3.1 data for the last three years. According to these data, the intensity of global fossil fuel consumption did not increase in the last three years, despite the fact that the rates of industrial development seem not to have changed markedly. The higher rate of fossil fuel consumption mentioned in Chapter 2 is about 1% per year and cannot ensure a stable dM_f/dt function.

It should be pointed out that the increment increase in the $\Delta M_a/\Delta M_f$ ratio with time is obtained only when precise hydrochemical relationships are used between dissolved CO_2 concentration and HCO_3^- and CO_3^{2-} ions. When these relationships are incorporated into the model in a simplified form, the $\Delta M_a/\Delta M_f$ ratio becomes constant with the exponential emission function typical of the present time (Oeschger et al., 1975).

The portion of industrial CO_2 emissions remaining in the atmosphere also depends on the type of source function dM_f/dt. The slower the growth rate of dM_f/dt, the larger the portion of CO_2 absorbed by the ocean, and hence, the smaller the portion remaining in the atmosphere.

For the next decades, this percentage for the most likely emission function (curve 5 in Figure 3.15) will be approximately constant because of the effect of two opposite factors: a) non-linearity of hydrochemical relationships and b) decrease in $(d/dt)(dM_f/dt)$ function with time (the latter indicates the deviation of dM_f/dt from an exponential increase). Curve 4 in Figure 3.15, resulting from simple

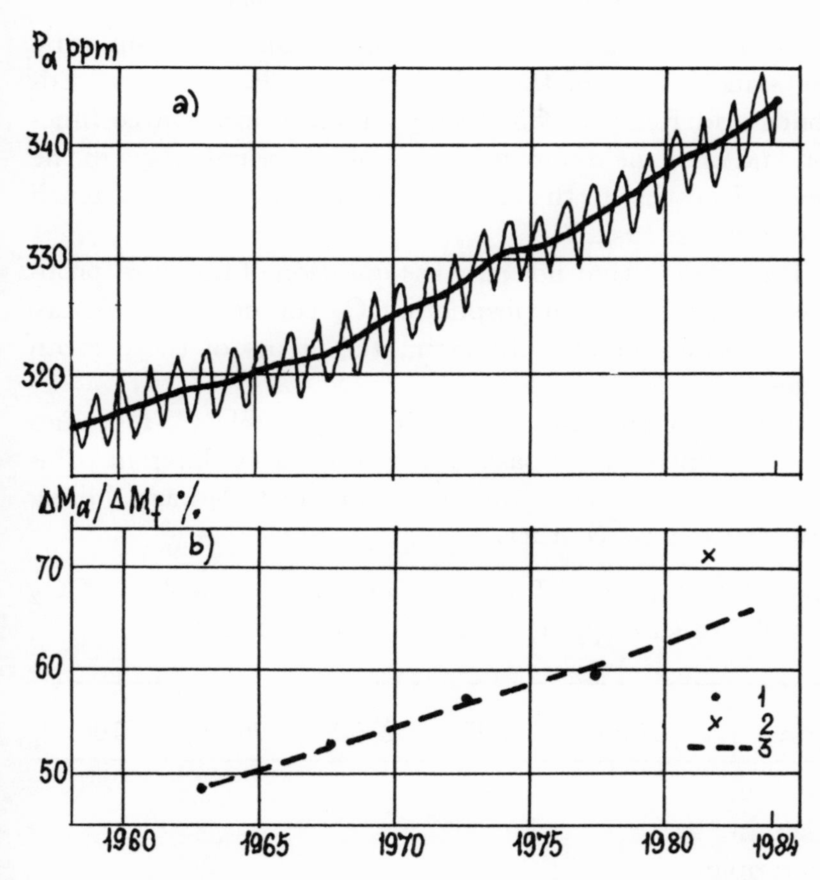

Figure 3.16. Monitoring data on P_a from Mauna-Loa station (Hawaii) up to January 1, 1984 according to Bolin et al. (1986) (a) and the remaining atmospheric $\Delta M_a/\Delta M_f$ of industrial CO_2 emissions (b). 1 and 2, empirical data; 3, estimates by the model with integral exchange coefficients.

estimation of the $P_a(t)$ function by extrapolation for the twenty-first century of the mean ratio $\Delta M_a/\Delta M_f$ that was calculated by empirical data for 1958-1980 was therefore close to the results of more precise computations by carbon cycle models.

We also note that P_a values obtained for at least the next 50 years (up to the year 2030) were close to those of Borisenkov and Altunin (1983), despite the fact that these authors used another carbon cycle model and source function dM_f/dt based on principles a bit different from ours. The same is true of the calculations of P_a values by Edmonds and Reilly (1985) based on their own source function and on some different models of tracer pathways in the ocean. To estimate the errors of such predictions is a much more difficult task.

It is evident that not only the question of the most probable mean values of atmospheric CO_2 content in the future is important but also the estimated errors of these mean values.

Table 3.3 indicates the limits of possible $P_a(t)$ values $P_a(t)$ obtained on the basis of two absolutely different principles for estimating calculation errors for CO_2 partial pressure for the next century.

Table 3.3. Limits of Possible Values of Atmospheric CO_2 Partial Pressure (ppm) for the Period 2000–2100

Source	2000	2025	2050	2075	2100
Legasov, Kuz'min, Chernoplekov, 1984	375–385	430–445	490–550	530–610	540–640
Bolin et al. 1986	360–380	380–470	400–580	410–720	420–900

Note: The first line is the limits of calculation errors, the second one is errors of possible P_a (CO_2) values adopted at the Villach conference (Bolin et al., 1986).

The upper line indicates limits that correspond to Figure 3.15 and were obtained by estimating the total error of $P_a(t)$ values, resulting from indefinite values of the basic factors i.e., V_{md} and ΔM_b that affect the calculation result. This is obviously a small error, and the calculation accuracy is acceptable for forecasting corresponding changes in mean global temperature (see Chapters 9, 11). This error also diminishes as we gain more knowledge about the dissemination of various tracers in the ocean and changes in the carbon isotope composition in difficult reservoirs.

The limits of possible $P_a(t)$ values given in the lower line, are considerably broader than ours. This is primarily because they include uncertainty associated with different published forecasts of the source function dM_f/dt. In principle, the limits of $P_a(t)$ values can be expanded even more in this manner. The book "Changing Climate" (1983) for example, provides a set of source functions on page 92 that show $P_a(t)$ values for the year 2100 ranging from 370 to 1400 ppm. However, this is an incorrect approach to estimating forecasting errors of the $P_a(t)$ function, since source functions close to the upper or lower limit of the aforementioned graph are simply impossible: it is difficult to assume that intensity of CO_2 industrial emissions into the atmosphere will start to decline right now, nor that fossil fuel will remain the basis of the power industry until the year 2100. Since it is impossible to assess the forecasting error of the source function dM_f/dt by standard method, it is better to start from common sense and the results of investigating "transitional" periods (see Chapter 2), than to ascribe an arbitrary error to this function. Calculations of function dM_f/dt based on the assumption that in the next 150–200 years all available fossil fuel reserves, i.e., about $5000 \cdot 10^{12}$ kg, will be burnt, evidently led to exaggerated values of this function (see Energy and Climate, 1977). The function used here corresponds to the total reserve of fossil fuels of about $1600 \cdot 10^{12}$ kg that will be burnt by the year 2100, i.e., no more than 1/3 of the currently available fossil fuel reserves; the consumption rate should peak by the year 2030 at 2.6 times above the modern value. It will decline since alternative energy sources will become the basis of energy development. The projected function dM_f/dt up to 2100 is uncertain, of course, but it probably does not exceed the error limits in the upper line of Table 3.3.

Analysis of empirical data and the results of modeling air-ocean gas exchange allows us to draw the following conclusions.

1. Different methods of modeling transfer processes in the ocean yield similar values of $\Delta M_w/(\Delta M_f + \Delta M_b)$, i.e., the relative amount of anthropogenic CO_2 absorbed by the ocean.

2. The value of $\Delta M_w/(\Delta M_f + \Delta M_b)$ depends significantly on the type of function that links concentrations of different compounds containing dissolved inorganic carbon in the surface oceanic layer, and depends little on temperature. The calculated anthropogenic CO_2 distribution between the atmosphere and the ocean therefore can use the mean global surface water temperature, but it is necessary to calculate accurately the hydrochemical relationships between concentrations of dissolved CO_2 and total inorganic carbon $\Sigma = [CO_2] + [HCO_3^-] + [CO_3^{2-}]$.

3. Analysis of modeling the stable isotope ^{13}C cycle shows that terrestrial biomass served as an atmospheric CO_2 source from 1860 to 1980; its total change ΔM_b was 60–90 Gt C, i.e., 30–45% of the carbon in the fossil fuel consumed during the same period (see Table 3.1). The resulting ΔM_b value generally agrees with estimates made independently by other methods (see Chapter 4).

4. Assessments of the continental biomass decline ΔM_b caused by human activity in the previous 120 years made it possible to evaluate the limits of its possible change for a similar future interval. In this case, it is more rational to calculate atmospheric CO_2 partial pressure before the year 2100 by modeling the CO_2 exchange between the atmosphere and the ocean, excluding the equations that describe functioning of biota on a planetary scale, whose parameters are still obscure. This method yields more reliable mean values of function $P_a(t)$ and a better estimate of possible forecasting error.

CHAPTER 4

THE INFLUENCE OF BIOTIC FACTORS ON THE CO_2 CYCLE

4.1. Introduction

The enormous amount of atmospheric oxygen, produced mainly through photoautotrophic assimilation of plants and microorganisms, as well as lithospheric reserves of incompletely oxidized organic carbon indicate that the interrelated processes of photosynthesis and organic destruction proceed at somewhat different rates, and that the carbon cycle in the Earth's outer crust is not completely closed. Biota accumulates CO_2 naturally, i.e., the algebraic sum of biospheric fluxes has a negative sign, and the so-called net discharge of carbon from the atmosphere occurs. Its rate is certainly low (Budyko, 1984; Budyko et al., 1985) and can be revealed only for geological time scales.

As noted in previous chapters, the atmospheric CO_2 increment that affects the climate by intensifying the Earth's greenhouse effect depends on many different factors. Prediction of anthropogenic climatic changes therefore required a detailed study of all the carbon cycle components, and the biota is one of the most difficult aspects to investigate. The first question was whether the higher atmospheric CO_2 content was associated only with CO_2 release from fossil fuel burning or also with anthropogenic perturbations of ecosystems. It was also difficult to clarify the reaction of plants to the higher CO_2 concentration.

Some specialists were surprised that CO_2 concentration was higher in the atmosphere, for it has long been known that the intensity of photosynthesis of the majority of plants depends almost linearly on CO_2 content within the range 0–0.3%. The CO_2 increment produced by fossil fuel combustion could supposedly be compensated for by higher photosynthetic intensity and greater biomass.

In the early 1970s, when higher atmospheric CO_2 was compared with content fossil fuel emissions, it was found that about half the anthropogenic CO_2 stays in the atmosphere, and it was assumed that the terrestrial biota is a net discharge of anthropogenic carbon from the atmosphere. However, then other opinions were expressed. From 1976 up to the mid-1980s, quite different estimates were made of the annual total carbon fluxes between the terrestrial ecosystems and the atmosphere (Figure 4.1). Even now the role of biota is uncertain: is it total discharge or total source of atmospheric carbon.

The discrepancy in the available estimates is explained primarily by inaccurate information on both CO_2 fluxes into and from the atmosphere, and the size of many carbon reservoirs. Only the atmospheric carbon content can be reliably determined, because its mass is known and systematic measurements of atmospheric CO_2 concentrations and emissions from fossil fuel burning ($\sim 5 \cdot 10^{12}$ kg C/yr) have been made since the late 1950s.

By early 1984, the atmosphere contained $728 \cdot 10^{12}$ kg C as carbon dioxide. The size of other reservoirs was less definite and the range of uncertainty often unknown. In some cases available estimates differed tremendously. Thus, research in the 1960s and 1970s by Bogorov (1974) established that the oceanic phytomass (phytoplankton and phytobenthos) is $1.7 \cdot 10^{12}$ kg (raw weight), while the biomass of zooplankton, zoobenthos and nekton is $32.5 \cdot 10^{12}$ kg, i.e., the zoomass was 20 times greater than the phytomass. However, other studies (Bowen, 1966; Whittaker and Likens, 1973, 1975; Cauwet, 1978) gave the opposite ratio with phytomass averaging eight times more than zoomass. Although this is an ecologically important problem, it is not significant in evaluating the global carbon balance components and is only mentioned here to show how contradictory the

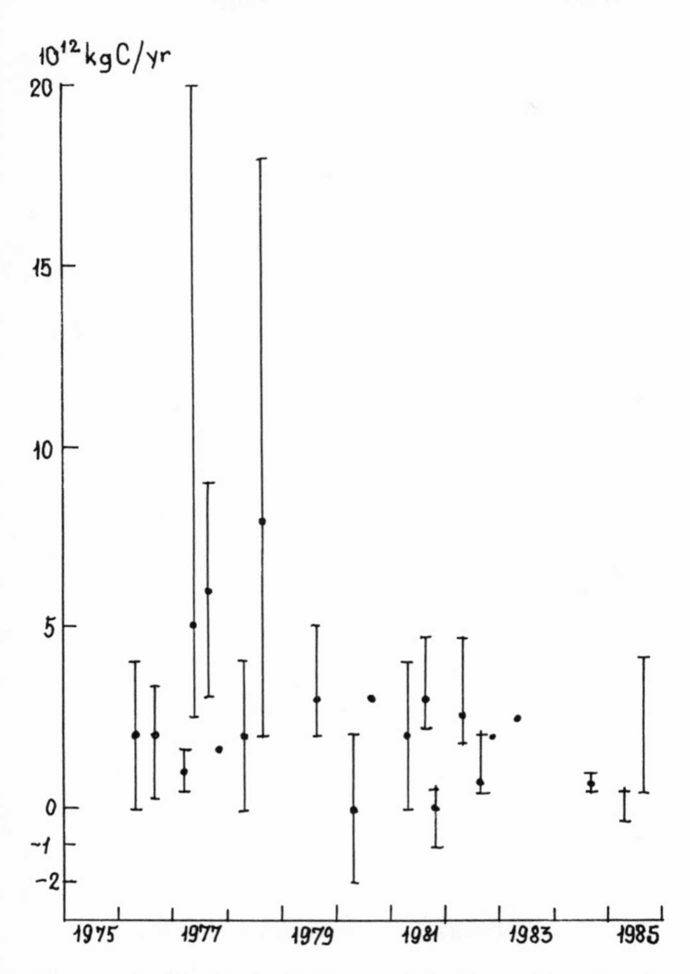

Figure 4.1. Estimates of net carbon fluxes into the atmosphere during anthropogenic disturbances of land biota obtained by different specialists from 1976 to 1985. From Clark et al., 1982, with supplements.

available estimates are. The total oceanic biomass assessments (phytomass and zoomass) differ little and are about $3 \cdot 10^{12}$ kg C. This value was used to compute a summary table of organic carbon content in the main world reservoirs (Table 4.1).

Despite the inaccuracy of some data, we can establish that most of the organic carbon is concentrated in oceanic

Table 4.1. The Main World Reservoirs of Organic Carbon

Reservoir	10^{12} kg C	Source
Biosphere		
terrestrial phytomass	560	Watts, 1982; Olson, 1982; Ajtay, 1979
terrestrial zoomass	0.5–5	Bazilevich et al., 1970; Whittaker, 1975
soil carbon (humus, peat, detritus)	2000	Kobak and Kondrasheva, 1986
oceanic biomass (phytomass plus zoomass)	3	Bowen, 1966; Cauwet, 1978; Whittaker, 1975
Dissolved organic carbon plus particulate matter	1400–2000	Skopintsev, 1971; Romankevich, 1977
Lithosphere		
fossil fuel	7660	Sundquist, 1985
- coal	6800	Sundquist, 1985
- oil	210	Sundquist, 1985
- gas	170	Sundquist, 1985
- other types	480	Sundquist, 1985
Sedimentary rocks of the ocean and continents	12,400,000	Budyko et al., 1985

and continental sedimentary rocks ($12.4 \cdot 10^{18}$ kg, or 99.9% of the total carbon content of the planet). The biosphere and the accessible fossil fuel contain no more than 0.04%

and 0.06% of the total reserve, respectively. Organic carbon is distributed almost equally in the biosphere between the land ≡ (57.4%) and the ocean (42.6%), where it is almost all dissolved (so-called aquatic humus). There is significantly more carbon in the continental soil (44.8%) and only slightly more than 12% in the terrestrial biota.

The immense reservoir of organic carbon in sedimentary rocks is almost invariable and its role appears only in the geological time scale. The biotic reservoirs are unstable, which is reflected in the seasonal variations of atmospheric CO_2. Studies of global CO_2 balance (Müller, 1960; Kobak, 1964) in the 1960s found that the greatest carbon dioxide fluxes from and into the atmosphere are associated with biological processes, i.e., CO_2 absorption via photosynthesis and its emission through the destruction of organic matter ($\sim 60 \cdot 10^{12}$ kg C/yr).

Terrestrial biota and soils were the first to feel the anthropogenic effect, and further study of the carbon cycle needed reliable assessments of their carbon content.

4.2. Change in Carbon Content of Terrestrial Phytomass

Global estimates of terrestrial phytomass. The first attempts to determine the storage of carbon in terrestrial biota were made in the early twentieth century, in particular, by V. I. Vernadskiy (1926); however, reliable results appeared only in the 1960s (Rodin and Bazilevich, 1965; Bowen, 1966; Duvigneaud and Tung, 1968). The summary by Rodin and Bazilevich estimated terrestrial biota in terms of dry organic matter as $2400 \cdot 10^{12}$ kg. This is rather the potential terrestrial phytomass, since the authors tried to include anthropogenic changes in vegetation, which hypothetically resulted in destruction of 10% of the forests. The estimate by Dobrodeyev and Suyetova (1976) of $2600 \cdot 10^{12}$ kg of dry organic matter also refers to the so-called undisturbed or restored vegetation. At about the same time, Whittaker and Likens (1973, 1975) obtained a value of $827 \cdot 10^{12}$ kg C, or $1841 \cdot 10^{12}$ kg of dry organic matter. There are no grounds to use these data to characterize the 1950 carbon reserve in the terrestrial biota.

New evaluations were made later, at the end of the 1970s and early 1980s by Ajtay et al. (1979), Olson (1982), Watts (1982), Brown and Lugo (1982, 1984). They agree fairly well with each other and are generally much lower than earlier estimates (Table 4.2).

Table 4.2. Estimated Carbon Content in Terrestrial Phytomass

Source	10^{12} kg C	10^{12} kg of dry organic matter
Kovda, 1973	1350	3000
Bazilevich, Rodin and Rozov, 1970	1080	2400
Whittaker and Likens, 1975	827	—
Dobrodeyev and Suyetova, 1976	—	2600
Ajtay et al., 1979	560	—
Olson, 1982	560	—
Brown and Lugo, 1982, 1984	470	—

Some publications interpret the lower estimates as dwindling carbon content in the terrestrial biota in response to human activities, mainly to forest clearing. But it is hard to imagine that the terrestrial biota decreased by $260 \cdot 10^{12}$ kg C in 10 years, as follows from a comparison of the data of Whittaker with Ajtay et al. and Olson (Whittaker and Likens, 1975; Ajtay et al., 1979; Olson, 1982). This fact is probably explained by the inaccurate estimates made in the 1970s and earlier. Olson (1982) and Ajtay et al. (1979) already had a more reliable basis (satellite data, aerial photographs and land valuations) which significantly improved their estimates, primarily for the tropical regions. Aerial photographs showed that many areas, uncultivated forests that were previously considered dense, were either partially dense or sparse. Olson (1982) consequently assumed that the carbon content of the tropical forests was $230 \cdot 10^{12}$ kg, half the value of $460 \cdot 10^{12}$ kg of Whittaker and Likens (1975).

It is quite clear that these data are not conclusive. The latest information obtained by the Food and Agricultural Organization (FAO) and the United Nations Environmental Program (UNEP) within the Environmental Global Monitoring System is valid only for 35 of 76 countries in tropical Africa, America, and Asia comprising 40% of dense tropical forest area, and is not very reliable for 23 countries with about 20% dense and 29% sparse forests. Further investigations will possibly correct these assessments. We can now only rely on the data of Ajtay et al. and Olson, who make the identical estimate of $560 \cdot 10^{12}$ kg for global carbon content, but different value of carbon content in some biomass.

The generalized data in Table 4.3 give not only carbon distribution among the main biomass, but also the calculated carbon residence time in them (τ) as the quotient of biomass divided by productivity. With mean global τ of 9.3 years, this value has a high range from 1.3 years in annuals to 24 years in various arboreal communities of the boreal zone.

Estimation of variations in forest biomass according to inventory statistics. As already mentioned, there have been repeated attempts to use full-scale research to estimate anthropogenic changes in the terrestrial biota pool (forest clearing, agricultural use of forests and so on). The focus was on variation in the forest biomass, since forests comprise at least 85% of all carbon in terrestrial ecosystems and a significant decrease in the forest biomass could appreciably affect the components of the global carbon cycle.

Table 4.3. Carbon Content, Productivity and Residence Time (τ) in Living Biomass of the Main Terrestrial Biome (from Ajtay et al., 1979, and Olson, 1982)

Biome	Area, 10^{12} m^2	C Storage, 10^{12} kg	Primary net Production, 10^{12} kg C/yr	τ, years
Tropical rainforests	11.2	178.6	9.9	18.0
Tropical seasonal forests	5.1	44.6	3.2	14.0
Temperate forests	7.8	76.6	4.7	16.3
Boreal forests (taiga)	10.6	111.6	3.4	24.2
Woodland and shrubs	8.6	40.5	4.6	11.9
Savanna	23.5	57.6	14.2	4.0
Steppes	9.6	10.0	3.5	2.8
Tundra and alpine meadows	11.6	9.5	1.3	7.3
Desert and semidesert	39.4	6.5	1.4	4.6
Cultivated areas	16.0	12.5	9.4	1.3
Swamps (marshes and peat bogs)	3.2	11.5	4.0	2.9
Lakes and streams	2.6	0.5	0.4	1.2
Total	149.2	560.0	60.0	9.3

Information on forested areas and their usable timber in different time periods shows the dynamics of the carbon content in forest ecosystems. Of course, the forestry statistics are incomplete, except for the last three decades (and even less for the tropics).

The current forested area of the Earth is $41 \cdot 10^{12}$ m^2, including boreal ($10 \cdot 10^{12}$ m^2), temperate ($13 \cdot 10^{12}$ m^2) and tropical forests ($18 - 19 \cdot 10^{12}$ m^2).

Comparison of the data on timber resources and temperate forested zone shows that from 1950 to 1970 the forested area increased at a rate of about $70 \cdot 10^7$ m^2 a year, while the total commercial timber stands increased by $42.5 \cdot 10^9 \times$ m^3. Whereas in the 1950s, the temperate forest timber resources were about $15 \cdot 10^{10} \times$ m^3, in the 1970s, it was about $20 \cdot 10^{10} \times$ m^3 (Bukshtynov, 1959; Bukshtynov et al., 1981). According to these data, the 1950s temperate forest biomass is assessed at $116 \cdot 10^{12}$ kg of organic matter, or $58 \cdot 10^{12}$ kg of carbon, and for the 1970s, at $\sim 150 \cdot 10^{12}$ of organic matter, or $74 \cdot 10^{12}$ kg of carbon. The 20-year (1950–1970) increase in total temperate forest biomass was $16 \cdot 10^{12}$ kg C, i.e., the rate of carbon accumulation from the atmosphere was $\sim$ (0.8–0.9) $\cdot 10^{12}$ kg C/yr. Armentano and Hett (1980) estimate the total carbon absorption from the atmosphere by the temperate forests at (1.0–1.2) $\cdot 10^{12}$ kg C/yr, which is somewhat high because the authors used inaccurate data on the dynamics of the USSR forest pool previously noted (Kobak et al., 1985; Kobak and Kondrasheva, 1985).

According to Miller's data (1980), the boreal ecosystem biomass is also increasing and carbon is accumulated at a rate of 0.03 to $0.3 \cdot 10^{12}$ kg/yr. We recall that a lot of the carbon in these ecosystems is concentrated in dead organic soil matter. This means that until the temperature of this region undergoes significant changes, organic decay will not actively influence carbon release into the atmosphere. Even the expected increase in the surface air temperature will not significantly accelerate decay. Thus, the northern ecosystem in the visible future will discharge atmospheric carbon (Kobak and Kondrasheva, 1985).

Consequently, for the last 30 years, the boreal and temperate (subboreal) forests have been functioning as a net discharge of carbon at an absorption rate of about $1.0 \cdot 10^{12}$ kg/yr, with a total accumulation of about $30 \cdot 10^{12}$ kg.

It is much more difficult to calculate the changes in biomass and area of temperate forests in the latter nineteenth to early twentieth century. However, comprehensive information on certain countries and individual regions (primarily, European Russia, U. S. southeastern and northeastern regions) proves that extensive forest clearing to satisfy the increased demand for wood and agricultural land made the temperate forests noticeably smaller. As a result, from 1860 to 1900 these forests acted as a net source of carbon for the atmosphere, releasing $(0.6–1.3) \cdot 10^{12}$ kg C/yr (Stuiver, 1978; Loucks, 1980; Moore et al., 1980), averaging 0.95 $\cdot 10^{12}$ kg C/yr, or a total of $38 \cdot 10^{12}$ kg C.

At the beginning of the twentieth century, and particularly after World War II, deforestation of the temperate zone declined because fossil fuel became more important as a fuel source than timber. At about the same time, many countries started reforestation, which undoubtedly affected the total carbon balance of forest ecosystems. By 1925, temperate forests had gradually been converted from net source to net discharge of atmospheric carbon. Specialists have different ideas about the role of forests from 1900 to 1950, however. Armentano and Ralston (1980) think that at that period the temperate forest did not discharge much carbon ($\sim 1.0 \cdot 10^{12}$ kg C/yr). Stuiver (1978) believes there was a balanced carbon gain and loss in temperate forests. On the whole, the resulting carbon atmospheric influx was close to zero, or the net carbon release into the atmosphere could be no more than $0.5 \cdot 10^{12}$ kg/yr, or $25 \cdot 10^{12}$ kg for 50 years, taking account of the maximum estimates by Moore et al. (1980), and that in some countries, for instance, in

European Russia, there was extensive forest clearing in the early twentieth century.

The most difficult and debatable question is the role in the carbon cycle of tropical forests, since they contain about half of all carbon of the Earth's forests. We are fairly confident that from 1860 to 1950 human interference in the tropical ecosystem was insignificant, therefore, the biomass did not change much and there was zero carbon flux into the atmosphere from forest ecosystems.

Intense cultivation of tropical forests started in the 1950s. Whereas during the entire history of the Earth, the tropical forest area decreased by $300 \cdot 10^{10}$ m^2, about two-thirds of this decline occurred in the last three decades (UNESCO, 1978). However, because of inaccurate information, it is difficult to represent quantitatively the tropical biomass changes that occurred in the 1950s through the 1970s. Indirect evidence shows that from 1950 to 1970, an area of $10 \cdot 10^{10}$ m^2 was cleared annually. The biomass loss can be estimated from the following data: the carbon storage in plants is 12 kg/m^2 and in the upper 30-cm tropical soil layer it is 4–6 kg/m^2 (Brown and Lugo, 1980), the oxidation rate of plant biomass, of which only one-third turns into the long-preserved carbon products and about two-thirds are oxidized rather quickly (Loucks, 1980), the humas mineralization rate, with 90–95% of organic matter oxidized within a year (Andreas, 1980). This loss was $28 \cdot 10^{12}$ kg C for 20 years, or $1.4 \cdot 10^{12}$ kg C per year, of which $0.9 \cdot 10^{12}$ kg C/yr is plant material and $0.5 \cdot 10^{12}$ kg C/yr is humus.

The information on the change in the tropical forest resources over the past decade is more reliable. Comparison of the inventory data of 1973 and 1980 (Bukshtynov et al., 1981; Lanly, 1982) on wood resources has shown that over that period the forest area decreased by $113 \cdot 10^{10}$ m^2, or by $16 \cdot 10^{10}$ m^2 per year, which agrees with the data of Sommer (1976) and Myers (1980), who give the rate of tropical forest disappearance as 0.73–0.99% of their total area, i.e.,

$15.8 \cdot 10^{10}$ m^2 were cut annually. Thus, annual cutting of $16 \cdot 10^{10}$ m^2 of forest area releases $1.3 \cdot 10^{12}$ kg C/yr into the atmosphere through oxidation of plant biomass.

The FAO and UNEP investigations in tropical forests in 1979 to 1981 demonstrate that from 1976 to 1980, no more than $10 \cdot 10^{10}$ m^2 forested area were cleared annually, $6.9 \cdot 10^{10}$ m^2 being in dense forests (Lanly, 1982; Hadly and Lanly, 1983). From 1981 to 1985 the rate of clearing was almost the same, $11.3 \cdot 10^{10}$ m^2/yr, or 0.58% of the total area (Table 4.4), which means that the disappearance of primary tropical forests was not as fast as Myers believed (1980). The estimates of Myers $(16–20) \cdot 10^{10}$ m^2/yr and these of the FAO-UNEP that are based on the same initial information vary because of different scientific and commercial criteria and approaches to the problem of tropical forest clearing. It is quite natural that forest clearing of large areas has raised serious concern among all specialists, because it actually threatens with extinction many plant and animal species, loss of a unique gene bank as well as soil erosion and other grave consequences. Myers was consequently justified in trying to take account of the slightest anthropogenic variations in forest area. However, the solution to agriculture and forestry problems in the tropical belt, were more than one-third of the entire world population (1.6 billion people) lives requires specifics on the extent of forest clearing, the nature of transformations in these areas, etc. The FAO-UNEP results obtained within the Global Environmental Monitoring System in 1979-1981 mainly indicate intensity of tropical deforestation.

The nature of forest transformation has not been properly understood. Huguet, Director of the Technical Centre of Tropical Wood Research (1982), thinks that an area of about $5.1 \cdot 10^{10}$ m^2 (of $11 \cdot 10^{10}$ m^2 cleared annually) can be reconverted either naturally or artificially into forests. Rapid reforestation occurs in many clearings, if they are not too large (Lugo et al., 1980; Roche, 1984; Gilbert, 1984).

The estimates, therefore, show that more than half of the cleared forest areas cannot be reforested.

Table 4.4. Deforestation Rate in the Tropics in 1981–1985 from Lanly (1982), Hadly and Lanly (1983)

Continent	Total area 10^{10} m^2	Area cleared 10^{10} m^2	% of the total area
America	896	5.6	0.63
Africa	703	3.7	0.53
Asia	336	2.0	0.60
Total	1935	11.3	0.58

According to our calculations based on the FAO-UNEP estimates of forest cutting ($11 \cdot 10^{10}$ m^2/yr), the carbon flux should be decreased to $0.9 \cdot 10^{12}$ kg/yr.

Change in terrestrial biota productivity. The use of standard biological methods to study global productivity of land plants does not permit an analysis of whether it varies because of anthropogenic perturbations of the biota and atmosphere. Indirect methods to help solve this problem are consequently very important, for example, analysis of the seasonal amplitudes in partial CO_2 pressure measured at global monitoring stations for several years in a row.

Seasonal variations in atmospheric CO_2 concentration are associated with a phase shift in the annual photosynthesis-destruction cycle of continental biota. With a time lag between discharge (photosynthesis) and source (organic destruction) of the atmospheric CO_2, the amplitude of p_a (CO_2) seasonal changes would have been 26 ppm (with global primary net production of $56 \cdot 10^{12}$ kg C). If these processes were completely synchronized, the p_a (CO_2) amplitude should have equaled zero. Slight changes in the phase shift between the CO_2 source and discharge can modify the amplitude. Its magnitude and latitudinal dependence are determined by atmospheric circulation, the allocation of the main photosynthesizing biomes and the intensity of the seasonal succession. It is evident that the p_a (CO_2) seasonal dynamics at a certain latitude depends mainly on the seasonal variations in net production of the ecosystem (i.e., on the difference between primary net production and carbon loss in respiration of heterotrophic organisms) and not on the primary net production of photosynthesizing plants. In equatorial latitudes, despite the maximum primary production of photosynthetic plants, p_a (CO_2) seasonal variations are very small.

However, the higher primary net production (and hence, the higher level of organic matter destruction) under definite conditions can elevate the p_a (CO_2) seasonal amplitude.

The problem of its changing absolute value has been repeatedly discussed for the last two decades in different studies, but no unanimous opinion has been reached. We made a statistical analysis of the data obtained at the Mauna Loa stations and weather ship P, that were also examined by Bacastow et al. (1981). It was found that the standard deviation from the straight line in Figure 4.2 is $\pm$ 0.22% per year (Buytner et al., 1985). The p_a (CO_2) trend dispersion is such that the probability of its positive values is about 85%.

Since there are still no statististics for the p_a (CO_2) amplitude trend, we tried to use data on anthropogenic changes in terrestrial plant productivity to estimate the possible variations in p_a (CO_2) amplitude. The phase shift between photosynthesis and destruction is determined by seasonal variations in surface and low Earth air temperatures, and does not depend on anthropogenic perturbations, whereas the continental biota productivity can vary because of these factors. When phase shift is fixed, altered seasonal course should be linked to modified productivity. We have consid-

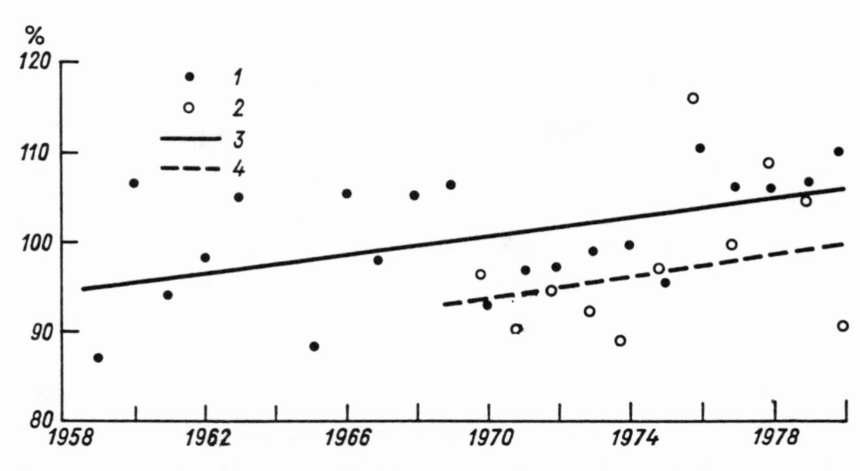

Figure 4.2. Relative changes (%) in seasonal amplitudes of CO_2 concentration by data from Mauna-Loa station (1) and weather ship P (2). 3, mean dependence from data of Mauna-Loa; 4, from weather ship data P.

ered the main factors influencing the changed global production, i.e., more intensive farming, greater de- and reforestation as well as higher photosynthesis due to greater atmospheric CO_2 concentration.

We have information that grain production in the main supplying countries intensified $0.8 \cdot 10^{12}$ to $1.3 \cdot 10^{12}$ kg per year from 1960 to 1977. It is easier to assess improved productivity from grain production date, since grain is about 25% of the green mass (the mean for grains), the carbon content of the green mass is 45%, while grains constitute 60% of all agricultural crops. It was established that productivity of agrophytocenoses increased by $1.8 \cdot 10^{12}$ kg C/yr (from $2.4 \cdot 10^{12}$ to $4.2 \cdot 10^{12}$ kg C/yr), from 1960 to 1980, or by 3% of global land plant productivity. This should have increased the amplitude of p_a (CO_2) seasonal fluctuations by 0.16%/year.

In addition to the previously mentioned increase in area and biomass of the temperate forest ecosystems from 1950 to 1980, productivity could have risen by $0.68 \cdot 10^{12}$ kg C/yr. It should have diminished in the tropical forests in the same time span by $3.5 \cdot 10^{12}$ kg C/yr because of intensive clearing. The total decrease in the forest yield can be estimated at $2.8 \cdot 10^{12}$ kg C/yr ($2 \cdot 10^{12}$ kg C/yr for the last 20 years), which should have reduced p_a (CO_2) by 0.18% a year. The total anthropogenic impact on terrestrial ecosystems thus did not evidently influence the p_a (CO_2) amplitude.

The 85% probable rise in p_a (CO_2) amplitude might result from the biota response to higher atmospheric CO_2 concentration, but this effect on plants is not conclusive yet. Ecological and physiological investigations show, however, that higher CO_2 concentrations lead to more intensive photosynthesis and larger leaf surface in many species, and altered areas of plant community distribution, species composition and increased primary net production (Lieth et al., 1980).

The CO_2 concentration increased by 7.5% from 1959 to 1980. Koval et al. (1983), Oeschger et al. (1980) and others have found that this factor could stimulate a 2.5% increase in agricultural yield. If the reaction of global production were hypothetically the same, seasonal changes in p_a (CO_2) would rise, but by no more than 0.2% per year.

In all probability, the trend of p_a (CO_2) amplitudes based on the Mauna Loa station and whether ship P data has not been proved because of the brief observations and the difficulties in determining the amplitudes while p_a (CO_2) increased from year to year. The growth rate of annual mean p_a (CO_2) is not a monotonous function of time; it fluctuates considerably from year to year and is not even constant within one year. Additional errors inevitably arise in calculating p_a (CO_2) seasonal amplitudes. Since the trend of the p_a (CO_2) amplitudes is not reliable, it is difficult to use this method to determine the changes in global production of terrestrial phytomass. However, it will undoubtedly be possible in the near future, when sufficient data have been gathered at the global CO_2 monitoring stations.

4.3. Change in the Amount of Soil Carbon

Soil carbon pool. Soil is an enormous reservoir of organic carbon on our planet. There is much more carbon in it than the biomass of terrestrial ecosystems. it is on a par with the quantity of dissolved organic carbon in the ocean and its reserves in fossil fuels.

All organic matter produced by living organisms in the process of their functioning enters the soil in the long run and forms organic matter there. Through various transformations, some of that organic matter forms simple compounds, while the rest turns into more stable elements of soil organic matter such as humus, peat, sapropel and so on.

Most of soil organic matter is humus and peat. Two opposite processes constantly occur in soil, namely, humification that supplements the humus reserves on Earth, and mineralization of humus compounds to final decomposition products that diminishes the amount of humus. This annual influx of organic matter and its loss through mineralization determines the accumulation of organic matter in soil. Varied organic carbon content in different soil types of the world results from the contrasts in hydrological and thermal conditions in geographical zones and the nature of vegetation.

The worldwide estimates of soil organic carbon lie in a wide range from $700 \cdot 10^{12}$ kg to $3,000 \cdot 10^{12}$ kg (Bohn, 1976, 1978). This great variance occurs because scientists used different data on the organic carbon in various soil types and the area they occupy.

Our estimate of the total soil organic carbon, 2100×10^{12} kg, is based on information from a detailed soil map of the world published in 1975 by Kovda and Lobova, and the mean values of organic carbon concentration obtained by different specialists in field studies. This figure practically coincides with that of Ajtay (Ajtay et al., 1979) and Zinke (1984), but is $1,000 \cdot 10^{12}$ kg smaller than the maximum estimates by Bohn (1976) and (500–600) · 10^{12} kg greater than the frequently used data of Kononova (1976), Post et al. (1982) and Büringh (1984). It is quite probable that the last two authors underestimated the organic carbon amount because they did not make full allowance for the role of such organic soils as peat and others. It should also be mentioned that comparison between Kononova's and our data is not entirely justified, since she emphasized the carbon content in humus and not its total amount.

The greatest reserves of organic carbon (41.5%) are concentrated in organic soils (22.1% of the land) (peat and bogs, frozen taiga and Arctic soils). The soils of deserts (23.6% in area) and arid and semi-arid regions (19.5% in area), that occupy approximately the same area as organic soils, contain much less carbon: 8.4% and 13.9%, respectively. Humid tropical and subtropical zone soils (25.8% in area) hold 22.9% of the total carbon reserve, forest soils (4.5% in area) hold 5.2%, and steppe soils (5.9% in area) have 8.1% of the total carbon.

We have found that the soil carbon content increases latitudinally from the equator toward the poles, with the peak in the boreal zone and a slight decrease in the polar region (Figure 4.3). We have also obtained the distribution of soil organic carbon in different global zones (Table 4.5). The greatest amount of organic carbon (35%) is concentrated in the boreal zone, the tropical belt (27.5%) comes next. Its area is rather large, which is why on the whole the carbon concentration in this zone is comparatively high, although its soil concentrations are lower than other areas. The carbon content in the subtropical and subboreal zones is almost the same (15.7% and 15.2%, respectively), while in the polar region, organic carbon reserves are the lowest, since both the area and the soil carbon concentration are small.

The available estimates of changes in carbon content of the soil pool are also contradictory. Information on the

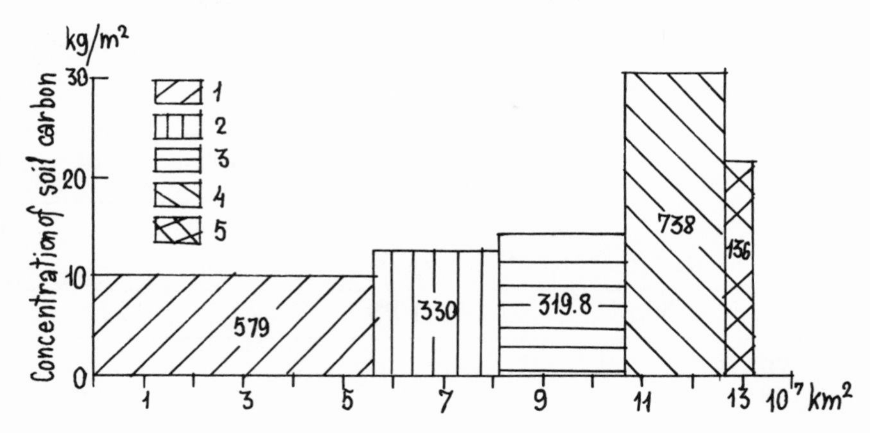

Figure 4.3. Distribution of organic carbon content (10^{12} kg) of soils over different belts. 1, tropical belt; 2, subtropical; 3, subboreal; 4, boreal; 5, polar.

rate of mineralization and accumulation of soil organic carbon is quite fragmentary. The annual rate of humification is assessed as from $0.6 \cdot 10^{12}$ kg C/yr (Olson, 1982) to $(1.0\text{–}2.5) \cdot 10^{12}$ kg C/yr (Kononova and Alexandrova, 1974; Gorshkov et al., 1980). It was therefore necessary to estimate the global rate of formation and mineralization of humus in ecosystems undisturbed by man.

We have already mentioned that humification, which leads to accumulation of organic carbon, and humic mineralization are the two processes that constantly take place in soil. Accumulation of soil organic carbon results mostly from humification of leaves and dead roots. Mineralization occurs at the same time, also by a complex set of microorganisms. The process of mineralization undoubtedly is shorter than humification, which includes catabolism producing lignin and tannin (in molecular form), as well as resynthesis and anabolism. The mass of fallen vegetation can be defined with greater accuracy than, for instance, the mass of the forest floor, allowing us to observe fairly well the seasonal changes in fallen vegetation in different

Table 4.5. The Distribution of Soil Organic Carbon by Zones and Regions

	Area		Carbon Content		
Belt	10^{12} m^2	% of total	kg/m^2	10^{12} kg	
Tropical	56.3	42.1	10.3	579.1	27.5
Region					
humid	25.9	19.4	13.5	349.5	16.6
semi-arid and arid	17.4	13.0	8.9	154.9	7.3
semi-desert and desert	13.0	9.7	5.7	74.7	3.6
Subtropical	25.7	19.3	12.9	330.5	15.7
Region					
humid	6.5	4.9	20.3	133.3	6.4
semi-arid and arid	8.6	6.4	16.0	137.7	6.5
semi-desert and desert	10.6	8.0	5.6	59.5	2.8
Subboreal	21.9	16.5	14.6	319.8	15.2
Region					
forest	6.1	4.5	18.1	108.6	5.2
steppe	7.9	6.0	21.3	169.0	8.0
semi-desert and desert	7.9	6.0	5.3	42.2	2.0
Boreal	23.7	17.8	31.1	738.6	35.1
Region					
taiga-forest	15.4	11.5	35.8	549.3	26.1
frozen taiga	8.3	6.3	22.6	189.3	9.0
Polar	5.7	4.3	23.9	136.4	6.5
Total	133.3	100.0	15.8	2104.4	100.0

latitudinal zones (from several dozen kilograms per hectare in the Arctic tundra and deserts to several dozen tons in the tropical grasslands). Leaves in the forest usually fall on the soil surface throughout the year, but in communities with growing and dormant periods (metabolism with a pause), more leaves fall at the beginning and end of the pause. For example, according to Karpachevskiy (1981) in the oak and spruce forests of the southern taiga subzone, the number of needles shed decreases from 30% in April to 5% in June and remains at this level to the end of August. Then, in September-October, it increases to 45% of the annual, and in mid-October drops to 2–3%. Dropping of leaves and branches is similar. Build-up of the forest floor is primarily related to the quantity of litter and its decomposition rate. Despite the complexity of the analysis, it has been found experimentally that from 2/3 to 3/4 of the detritus in the soil is mineralized in a year (Kononova, 1976; Kononova and Alexandrova, 1974; Rodin and Basilevich, 1965). The remaining 20–30% of the litter decay during subsequent years, and part of it becomes humus.

Build-up of the floor is similar in different plant communities. Accumulation occurs initially as the community grows in age and density (the accumulation rate is determined by the ditrital composition and decay conditions). Then equilibrium is established between the detrital influx and floor decomposition, i.e., the mass of incoming detritus equals that of the floor that decomposes during a year. The entire amount of fallen vegetation can therefore be used to characterize the changes in the forest floor.

As early as 1950, Kitredge proposed the so-called forest floor–detritus coefficient (the ratio of forest floor and detritus masses, τ_{det}) to determine the mineralization rate of the floor. τ_{det} also shows the residence time of carbon in the floor. Later, in constructing ecosystem destruction models, Olson (1963) used the inverse ratio ($k = detritus/floor$) and found that this index is distinctly different in forest

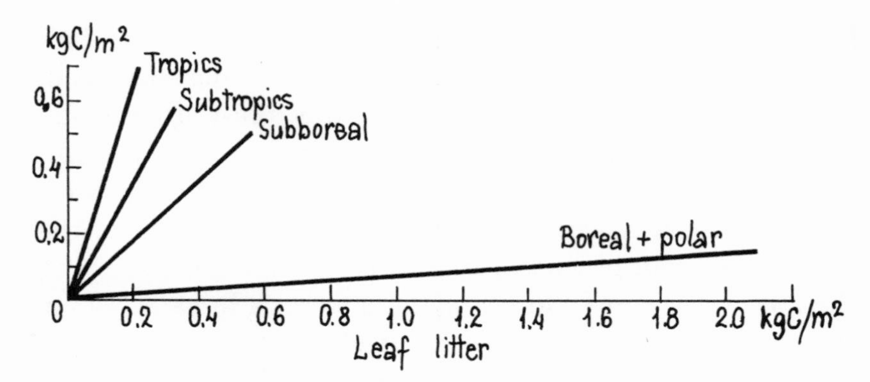

Figure 4.4. Ratio of carbon content in fallen leaves and floor in different belts.

ecosystems at different latitudes. We have already stated that one of the goals of our study was to reveal the global characteristics of soil organic matter mobility. We have therefore used data on the mass of detritus and forest floor (Rodin and Basilevich, 1965; Hampicke and Bach, 1980) and calculated the floor-detritus coefficients (τ_{det}) for different climatic belts (Table 4.6). We ignored possible anthropogenic changes in the soil pool, and assumed that all ecosystems are balanced, i.e., influx of detritus equals changes in the forest floor. The calculations show that when the mean τ_{det} is 2 years, this parameter varies significantly in different climatic zones, from 4 months in the tropics to 25 years in the polar zone (Figure 4.4).

It was then necessary to obtain global estimates of the net carbon releases into the biospheric soil pool. The calculations were based on data on the annual detrital amount in ecosystems of different climatic belts (Rodin and Basilevich, 1965), since it is the detritus (and root material) that are the primary source of organic matter, that is transformed into humus, and on the value of net humus output, which according to Kononova (1976), is 2.5% of the detritus. Our estimates of the soil organic carbon pool in different zones and organic carbon concentrations in different soil types were also used.

Table 4.6. Content and Residence Time of Carbon in Detritus of Different Climatic Belts

Belt	Soil carbon concentration (a)	Carbon mass (b)	Detritus (a)	Detritus (b)	Litter fall (c)	Litter fall (b)	Years
Tropical	10.3	579.1	0.12	6.8	0.39	22.3	0.30
Subtropical	12.9	330.5	0.20	5.1	0.37	9.6	0.50
Subboreal	14.6	319.8	0.24	5.3	0.22	4.9	1.10
Boreal	31.1	738.6	1.70	57.0	0.18	4.2	13.6
Polar	23.9	136.4	2.40	10.0	0.07	0.4	25.0
Total		2104.4		84.2		41.4	2.0

(a) kg/m^2. (b) 10^{12} kg. (c) $kg/(m^2 \times yr)$.

The amount of fresh humus substances does not exceed 10% of the initial detrital mass (Kononova, 1976, 1984; Kononova and Alexandrova, 1974), averaging about 6% (Gil'manov and Bazilevich, 1983; Oberländer and Roth, 1968), however, not all of it remains in the soil and enters the stable soil carbon pool. Some of it (more than half) is involved in new biological processes, therefore it is called labile or biologically active humus (Gerasimov and Chichagova, 1971; Zavelskiy, 1975). Some is deposited in the soil, forming the so-called stable reservoir ("net output" of humus substances from the carbon cycle) or net carbon release into the biosphere. The labile compounds constitute one-third and stable compounds two-thirds of the total soil organic carbon.

The stable humus substances are closely associated with mineral soils, and are the oldest elements of these soils according to radioactive carbon dating. In chernozems, they are represented by humic acids, the oldest elements among the examined samples. Humic coal and humin (an intermediate group of humus substances between humic acids and coal) are the youngest products in this soil type. In podzolic soils, the oldest fraction is either humin or grey humic acids. In all the fractions of podzolic soils, the organic carbon turnover is much faster than in the youngest labile chernozem fraction. Even the youngest chernozem fraction is older than the most stable podzolic fraction (Gerasimov and Chichagova, 1971).

In estimating the humus accumulation rate, we have relied on detrital data while ignoring the role of root material. It is quite possible that our estimates are consequently somewhat lower, particularly for grassy communities.

Table 4.7 summarizes the calculation results of carbon fluxes (from fresh humus and net carbon release into the stable reservoir) into soils of different climatic belts. The calculations were based on detrital data in different plant communities and coefficients experimentally obtained by Oberländer and Kononova, that helped to estimate the detrital portion that forms fresh humus (6%) and the "net output" (2.5%).

Our calculations showed that the annual carbon flux into the Earth's soils from fresh humus is $2.5 \cdot 10^{12}$ kg C, or 0.0011 (0.11%) of the total soil organic carbon reserve. As already mentioned, most of the fresh humus substances (about $1.5 \cdot 10^{12}$ kg C) undergo further biological reprocessing and the stable reservoir receives about $1.0 \cdot 10^{12}$ kg of carbon per year. This is more than 30 times greater than the annual carbon influx into the geochemical cycle. The Pliocene estimates of organic carbon dispersed in the Earth's sedimentary rocks (Ronov, 1976) allow us to obtain an average rate of buried or incompletely oxidized organic carbon of about $0.03 \cdot 10^{12}$ kg C/yr.

The amount of humus formed under different climatic conditions varies, and this is due both to the rate of humification and the sizes of the climatic zones. The rate of the formation and accumulation of humus is about the same in

Table 4.7. Net Carbon Discharge into Stable Soil Pool in Different Climatic Belts

	Soil carbon, 10^{12} kg					Formation of humus substances, 10^{12} kg/yr		
Belt	Total (humus + forest floor)	Detritus (litter)	Labile	Stable	Litter fall 10^{12} kg/yr	Labile	Stable	Total
Tropical	579.1	6.8	190.8	381.6	22.3	0.78	0.56	1.34
Subtropical	330.5	5.1	108.5	216.9	9.6	0.33	0.24	0.57
Subboreal	319.8	5.3	104.9	209.3	4.9	0.17	0.12	0.29
Boreal	738.6	57.0	227.2	454.4	4.2	0.14	0.11	0.25
Polar	136.4	10.0	42.1	84.2	0.4	0.02	0.01	0.03
Total	2104.4	84.2	673.5	1346.4	41.4	1.44	1.04	2.48

the tropical and subtropical belts (the humus increment is 0.02 kg/m^2 per year and the "net output" is 0.01 kg/m^2 per year). These figures are much higher than in other belts: in the subboreal 1.8 and 2 times, in the boreal, 2.2 and 2.3 times, and in the Arctic, 4.6 and 5.9 times higher, respectively. The highest rate of humification occurs in red, yellow and cinnamonic soils (the annual increment is 0.05 $kg/m,^2$), in meadow, flood-plain, reddish-black soils and brunizems (0.03 kg/m^2), and in red, red-cinnamonic soils and chernozems (0.022–0.24 kg/m^2).

The tropics, especially the humid regions, have the most humus in the world, and the polar belt has the least (Table 4.7). Almost the same amount of humus is formed in the humid and arid regions of the subtropics, although the rate of formation and accumulation vary. In the subboreal zone, the accumulation rate is half that of the subtropical. Humification rates in the forest and steppe regions differ little, and are four times lower only in semi-desert and desert regions.

Anthropogenic effect on the soil pool. The low net carbon output generally shows that in most ecosystems, undisturbed by man, accumulation and mineralization of organic matter in soils are almost balanced and the organic carbon content is relatively stable. Carbon can accumulate in the soil only if there is excess moisture inhibiting breakdown of fallen vegetation.

Anthropogenic perturbation and primarily, cultivation of virgin lands and low farming efficiency, disturb the soil carbon balance. Soils quickly lose a considerable amount of native organic compounds through enhanced microbiological activity and intense ditrital reprocessing. The results of many years of microbiological experiments in deep chernozems carried out by Grinchenko and Chesnyak (1966) demonstrate that plowing leads not only to a higher population of many microorganism groups, but also the appearance of new groups.

Soil degradation means breakdown of structure and density, higher rate of humus mineralization because of low influx of organic matter from plant residues and fertilizers, the loss of nutrients, and higher soil toxicity since toxic aluminum, manganese and iron can accumulate. Erosion and secondary salinization of irrigated areas become more threatening. Although these processes are generally alike, they can proceed at varied rates in different zones (Levin, 1983).

Carbon loss depends on the intensity of cultivation, the crop yield and the amount and quality of fertilizers.

The general trend of breakdown in organic compounds and soil structure has been repeatedly noted by many researchers. There is information that from 1882 to 1952, 15% of all the world's cultivated lands became unsuitable for agricultural usage, and 38.5% lost half of their humus layer (Lysak, 1980). Water and wind erosion have currently spread to about two billion hectares, which represents 15% of the agricultural lands (Word and Dubeau, 1975).

It is very difficult to estimate the global carbon loss from the world soil pool, because of extremely inadequate systematic long-term monitoring of the content of certain soil elements, including carbon, in cultivated soils. The information about areas that have been lost to agriculture over a long time period can only be utilized partially to characterize carbon loss by soils and influx into the atmosphere, since the term "unsuitable land" includes areas that have become deserts or salinized.

Table 4.8 shows that the global estimates of soil carbon losses lie within a wide range, and one of the causes is different initial characteristics of C_{org} content of the soil pool. Thus, Revelle and Munk (1977), based on a global estimate of the soil carbon pool of $700 \cdot 10^{12}$ kg C, calculated the carbon loss at $0.3 \cdot 10^{12}$ kg C per year (i.e., about 15% of the atmospheric carbon increment due to the total anthropogenic CO_2 influx). Bohn (1976, 1978), estimating

Table 4.8. Global Soil Carbon Loss over the Last Hundred Years

Reference	Carbon loss, 10^{12} kg/yr	Reference	Carbon loss, 10^{12} kg/yr
Kurakova, 1975	2.2	Bohn, 1978	1–2
Kovda, 1977	2.3–2.9	Ryabchikov, 1980	1.2–1.3
Bolin, 1977	1–5	Gorshkov et al., 1980	1–3
Revelle and Munk, 1977	0.3	Olson, 1982	0.9

the world total C_{org} content at $3{,}000 \cdot 10^{12}$ kg C, found that mineral soils annually lose $(1\text{–}2) \cdot 10^{12}$ kg of organic carbon.

Information on land subject to erosion and deflation, and on the native organic matter content of soils (19.2 kg/m^2 of humus or 22.5 kg/m^2 of carbon) allowed specialists to estimate the soil carbon decrease. The annual reduction in arable areas is from $(6\text{–}7) \cdot 10^{10}$ m^2 to $(20\text{–}25) \cdot 10^{10}$ m^2 (Ryabchikov, 1980), whereas the annual carbon content drops from $1.2 \cdot 10^{12}$ kg to $(2.3\text{–}2.9) \cdot 10^{12}$ kg. According to Gorshkov et al. (1980), a hundred years of farming using poor technology have led to the total loss or depletion of $20 \cdot 10^{12}$ m^2 of agricultural lands ($524 \cdot 10^{12}$ kg of humus, or about $300 \cdot 10^{12}$ kg of carbon have been lost, of that $384 \cdot 10^{12}$ kg through erosion and deflation and $104 \cdot 10^{12}$ kg through soil depletion by lengthy cultivation, i.e., $220 \cdot 10^{12}$ kg C and $80 \cdot 10^{12}$ kg C, respectively).

The discrepancy in global estimates of soil carbon loss gave the impetus to new calculations using a different approach. Our calculations were based on the data referring to the mass of the agricultural area, CO_2 concentration in different soil types, and observations of the decreasing carbon content in soils cultivated for a long time. The mass

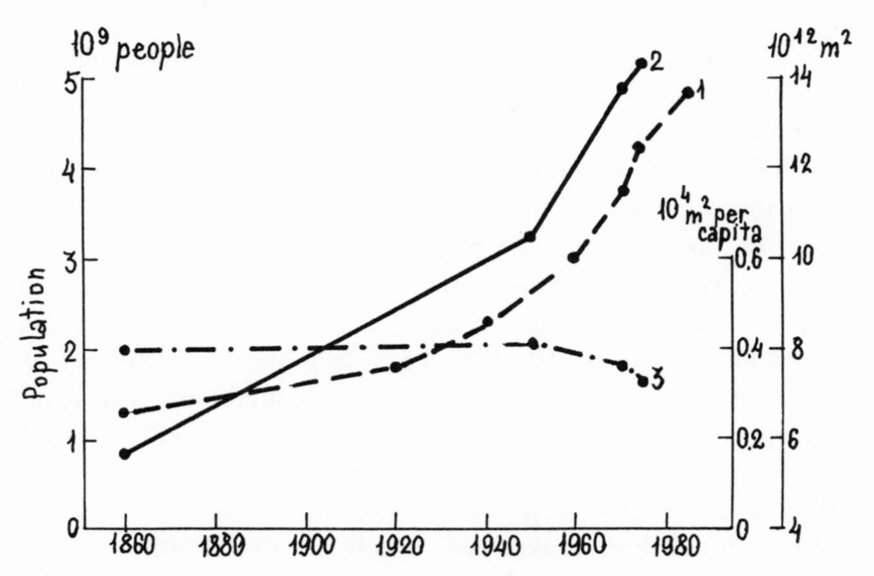

Figure 4.5. Population growth and the dynamics of cultivated areas in the world. 1, population growth (Population of the countries of the world, 1984); 2, increase in area (10^{12} m^2) of cultivated land (Revelle and Munk, 1977; Rozov and Stroganova, 1979); 3, dynamics of per capita cultivated areas (10^4 m^2 per capita).

of the world soil pool was first pinpointed at $2000 \cdot 10^{12}$ kg C (Kobak and Kondrasheva, 1986), and the soil carbon reserve was defined as 10.9 kg/m^2 in the mineral and 47.4 kg/m^2 in the organic soils, averaging 15.8 kg/m^2.

Population growth has generally required more cultivated land (Figure 4.5). The per capita crop area has tended to decline since the mid-1950s.

The data of the late 1970s (Rozov and Stroganova, 1979) have shown that 10.8% ($1.44 \cdot 10^{13}$ m^2) of ice-free land has been plowed, i.e., the global average land use coefficient (K) 0.11. The major crop-producing areas are in the subboreal and subtropical belts. The leached and common chernozems of the southern regions, as well as chestnut and brown soils have been exploited most of all. To estimate the C_{org} loss, we used only soils with $K > 0.01$, that is typical of 11 soil types encompassing about 85% ($1.22 \cdot 10^{13}$ m^2) of the cultivated lands of the world.

The scales of organic matter mineralization were determined by nomographs constructed by Gonchar-Zaykin and Zhyravlev (1979), using the model dynamics of soil humus content. It was found that after ten years of grain cultivation, the total carbon losses are $\sim 31.5 \cdot 10^{12}$ kg (we have ignored the organic compound compensation that occurs because stubble enters the soil, depending on the crop yield). Most of the loss ($21.5 \cdot 10^{12}$ kg C) occurs in the tropical belt, while $10 \cdot 10^{12}$ kg C is released in the extratropical zone. Taking the aforementioned compensation into account, C_{org} losses decrease to $30.5 \cdot 10^{12}$–$30.2 \cdot 10^{12}$ kg C (with a harvest of 0.25 kg/m^2 and 0.4 kg/m^2, respectively) in a ten-year period, i.e., at an annual rate of 2% of the initial humus content. These results should be considered tentative, mostly because they were based on the assumption that virgin lands are breaking down. These are rather hypothetical figures, representing the peak organic carbon loss by soils if all were plowed simultaneously.

The calculated tropical/extratropical C_{org} loss ratio, however, is very important. The tropical soils lose two-thirds, and this should be emphasized in the estimated biogenic carbon flux into the atmosphere. The calculations also are in satisfactory agreement with the monitoring data on decreasing soil humus content within the first ten years of soil cultivation.

The actual humus reduction over a long period of time for farming is about 0.4–0.5% a year of the initial content and is exponential (Grinchenko and Chesnyak, 1966; Grinchenko et al., 1979). The C_{org} loss stabilizes at no more than 0.01%/yr in some, e.g. chestnut, soils within 40–50 years.

The majority of soils in the subtropical and subboreal zones have been farmed for a long time. Most of the developed countries have not expanded the crop areas in the last three decades, so that the current C_{org} losses are no more than $0.2 \cdot 10^{12}$ kg C/yr.

In recent years agricultural areas have been enlarged mainly in Asia, Africa and Latin America. Most of the soil carbon released into the atmosphere comes from transformation of tropical forests into farms.

It is well known that organic matter is mineralized quickly in tropical soils, however it is estimated qualitatively rather than quantitatively. It has been reported (Andreae, 1980) that as much as 95% of the soil organic matter at a depth of 30–40 cm is mineralized throughout the year, others contend that the carbon content of cultivated soils decreases somewhat more slowly.

In our calculations we used the maximum rates of organic matter oxidation and did not consider all the cultivated lands in the tropics, but only the changed areas. The most recent estimates by Lanly (1982), Hadly and Lanly (1983) and Huguet (1982) show that only $6.2 \cdot 10^{10}$ m^2 of forest area are turned into agricultural lands at a forest cutting rate of $11.3 \cdot 10^{10}$ m^2 a year, and on the remaining $\sim 5.10^{10}$ m^2 the forest may grow anew. In this case, the net carbon release through the destruction of soil organic matter in the tropics is $(0.3\text{–}0.4) \cdot 10^{12}$ kg C/yr and the total soil flux is $(0.5\text{–}0.6) \cdot 10^{12}$ kg C/yr.

Considering that for the last 30 years, temperate and boreal forests have been acting as a net sink for atmospheric carbon at an absorption rate of $\sim 1.0 \cdot 10^{12}$ kg C/yr, the carbon flux from extratropical soils has been $\sim 0.2 \cdot 10^{12}$ kg C/yr, while the tropical forests have been a clear net source at a rate of $\sim 1.5 \cdot 10^{12}$ kg C/yr, the total net production of carbon from the biota is estimated at $0.7 \cdot 10^{12}$ kg C/yr or $21 \cdot 10^{12}$ kg for the period from 1950 to 1980. This is in good agreement with Loucks' estimates (1980). The biogenic flux from 1860 to 1983 was $\sim 60 \cdot 10^{12}$ kg C ($85 \cdot 10^{12}$ kg C at the peak).

It is quite possible that this figure ($(60\text{–}85) \cdot 10^{12}$ kg C) is somewhat high, because in our calculation we used the maximum value of the rate of annual clearing in the tropical forests ($\sim$ 16 mln. hectare a year).

As already mentioned in the previous chapter, attempts have recently been made to use other methods to estimate the biogenic flux. Comparison of the calculated results from the carbon cycle model incorporating atmosphere-ocean interaction and terrestrial biota functioning (Byutner and Zakharova, 1983) with the data on atmospheric CO_2 concentration (from current observations and from those of the late nineteenth century obtained by indirect methods) shows that the atmospheric carbon influx from biomass changes has been $(60–100) \cdot 10^{12}$ kg for 120 years (from 1860 to 1980).

A similar value ($(60–90) \cdot 10^{12}$ kg C) for the same time interval was obtained by using experimental data on the change in ^{13}C isotope content in the atmosphere and in upper quasihomogeneous layer of the ocean, and by modelling the carbon cycle for two stable carbon isotopes, namely ^{12}C and ^{13}C. The changes in atmospheric ^{13}C can be detected both through direct measurements of ^{13}C content (of which only a few have been made since the second half of the 1950s) and through observations of ^{13}C dynamics in tree rings. This method was introduced in the late 1970s (Stuiver, 1978) and required considerable refinement. Its current modification includes effects of isotope fractionation at the ocean-atmosphere boundary and changes in carbon isotope composition in the upper oceanic layer (Turchinovich and Vager, 1985).

Three independent techniques have thus established that through anthropogenic perturbations of the terrestrial biota over the last 120 years, the atmosphere should have received $(60–90) \cdot 10^{12}$ kg C, and one-third of it resulted from changes in the soil carbon pool. There is good reason to believe that this value reflects the actual biotic changes. It is much smaller than CO_2 emissions from fossil fuel combustion ($193 \cdot 10^{12}$ kg C from 1860 to 1983) and is 31–45% of the industrial discharge.

Repeated attempts have been made in the last decade to assess the carbon flux produced by anthropogenic perturbations of the biosphere. Some of the results are given in Table 4.9 and show that the data spread is rather large. Our estimates are lower than the majority of those given in the table, which can be discussed further. We note, however, that recent estimates are clearly tending to be lower than those obtained earlier. For instance, in 1985 Peng suggested a value of the biogenic carbon flux almost 2.5 times smaller than that published in 1983. Woodwell (1978), whose estimates were the greatest (as high as (18–20) $\cdot 10^{12}$ kg C/yr), has now decreased them to (0.5–4.2) $\cdot 10^{12}$ kg C/yr (Houghton et al., 1985).

We note in conclusion that organic carbon turnover is an integral and the most important part of the planetary carbon cycle. The cyclicity of carbon transformations is associated primarily with photoautotrophic absorption of CO_2 by plants, as a result of which about 8% of the carbon in the atmosphere is annually extracted from it ($\sim 60 \cdot 10^{12}$ kg C/yr). This is the greatest carbon flux escaping from the atmosphere. About the same amount of carbon returns to the atmosphere through the destruction of organic matter. Only a minute incompletely oxidized portion is metamorphized and buried in the lithosphere. Residence time of carbon varies in different biospheric reservoirs. It is about 9 years (from 1.3 to 24) in the terrestrial phytomass and 2 years (from 0.3 to 25) in the upper soil layer (detritus). The soil carbon pool generally changes little and carbon stays there for more than 1,000 years.

Omitting reserves of incompletely oxidized organic carbon in sedimentary rocks, the terrestrial phytomass and soil hold a lot of the atmospheric carbon ($2,600 \cdot 10^{12}$ kg). Anthropogenic changes in this reservoir could probably

Table 4.9. Net Carbon Influx into the Atmosphere from Anthropogenic Perturbations of Terrestrial Ecosystems (from different sources)

	Carbon flux, 10^{12} kg		
Source	For the whole period	Per year, average	Time interval years
Tans, 1976	150	1.5	1850–1950
Bolin, 1977	40–100	0.2–0.6	1800–1975
Bolin, 1978	40–120	0.2–0.7	1800–1975
Revelle and Munk, 1977	70–80	0.6–0.7	1860–1970
Siegenthaler et al., 1978*	133–195	1.2–1.7	1860–1974
Stuiver, 1978*	120	1.2	1850–1950
Wagener, 1978*	170	1.2	1800–1935
Freyer, 1979	70	0.6	1860–1974
Hampicke, 1979	180	1.5	1860–1980
Chan and Olson, 1980	151	1.4	1860–1970
Moore et al., 1981	148	1.3	1860–1970
Woodwell, 1983	180	1.5	1860–1980
Peng et al., 1983*	240	1.9	1850–1975
Peng, 1985*	144	0.8	1800–1980

*These are the data based on ^{13}C changes in tree rings.

affect the atmospheric CO_2 content, since they are obviously closely associated with the atmospheric gas composition.

The world forest inventory data (and regional information) allow us to track the dynamics of the forest tracts and timber reserves for the last 120 years. Since forests comprise more than 85% of the entire phytomass, we can thereby understand how the whole terrestrial biota pool changes. It has been found that in the last 30 years the temperate and boreal forest biomass increased, which made it possible to calculate the rate of carbon accumulation from the atmosphere ($\sim 1 \cdot 10^{12}$ kg C per year). Before that time, when these forests were intensively felled, they represented a net source of carbon, so that in the 90-year period from 1860 to 1950, the atmosphere should have received from 40 to $60 \cdot 10^{12}$ kg C.

Man had virtually no impact on the tropical forests before 1950, but in the last 30 years, annual clearing of 10–16 mln hectares of the tropical area released net carbon from the biota of about $45 \cdot 10^{12}$ kg C ($31 \cdot 10^{12}$ kg C through oxidation of felled biomass and $13.5 \cdot 10^{12}$ through oxidation of soil organic matter during conversion of forests into agricultural land). The soil carbon flux caused by anthropogenic factors could hardly be more than our value, because in some cases the use of cleared areas for agricultural purposes causes an increase, not a decrease in humus content. This is particularly typical of forests converted into hay fields (Karpachevskiy, 1981).

The total carbon released from anthropogenic disturbances of terrestrial biota is estimated as $(60\text{–}90) \cdot 10^{12}$ kg C for the period from 1860 to 1983, i.e., 30–45% of the carbon flux from fossil fuel burning during the same period. For the last two or three decades, the biotic carbon flux has not exceeded 15% of the industrial emissions and 12% of the total.

Similar results have been obtained by other independent methods, and there is reason to believe that the future biotic contribution to the total rise in atmospheric CO_2 content will be much smaller than that of the industrial emissions.

The CO_2 release into the atmosphere through oxidation of felled biomass and soil organic matter is only one of the anthropogenic effects on the biota. Many complex ecological problems arise that are of great concern to specialists.

The most important is decrease in the biological diversification on our planet. The number of populations and of whole biomes is currently declining not only in the extratropical zone, but on a large scale in the tropics as well. Felling of vast forests leads to catastrophically fast extinction of many species and loss of a unique gene pool.

CHAPTER 5

TRACE GASES IN THE ATMOSPHERE

5.1. Introduction

Scientists have recently focused not only on the greenhouse effect caused by higher atmospheric CO_2 content, but also on the greenhouse effect due to higher atmospheric content of other radiative trace gases: ozone, methane, nitric oxides, and chlorofluorocarbons (Karol' et al., 1983; Ramanathan et al., 1985; Karol' et al., 1986b). The mean atmospheric concentration of these gases, except for water vapor, is about a thousand times less than that of CO_2. They participate in numerous photochemical reactions with many other gases and aerosols. A lot of them actively absorb infrared radiation from the Earth's surface and the atmosphere. Ozone and NO_2 also intensively absorb solar ultraviolet radiation, and therefore, contribute much to the formation of thermal conditions in the stratosphere (Karol' et al., 1986b; Ramanathan et al., 1986). Current estimates indicate that in the next decade, the contribution of trace gases ("minor gases") and CO_2 to the greenhouse effect will be comparable. At the same time, information about the atmospheric cycles of many trace gases and their concentration distributions in different parts of the atmosphere is still inadequate compared to current knowledge about the CO_2 cycle (Problems of Atmospheric CO_2, 1980).

This chapter, that updates earlier works (Karol', 1986a; Karol' et al., 1986b), surveys current ideas on radiation properties, atmospheric cycles, distribution of trace gases in the atmosphere, and of extant trends of their anthropogenic changes. Expected changes in atmospheric gas composition and greenhouse effect induced by trace gases in terms of thermal inertia of oceans are assessed.

Table 5.1 contains basic radiative trace gases and their average concentrations in the troposphere and stratosphere

Table 5.1. Characteristic Global Trace Gas Concentrations in the Northern Hemisphere Atmosphere and Changes in Radiation Surface Air Temperature When the Trace Gas Content Changes

	Mixture Ratio				
Gas	Tropo-Sphere	Strato-Sphere	Mixture Ratio Change	ΔT_R °C	Source
		Microconstituents (ppm)			
O_3	10^{-2}– 10^{-1}	1–10	$2(O_3)^{a,b}$	0.7	Karol' et al., 1986b
				0.9	Report of the Meeting of Experts, 1982
			$(O_3)/2^{b,c}$	-0.7	Karol' et al., 1986b
CH_4	1.5–1.7	<1.5	$1.5 \rightarrow 3.0$	0.4	Ibid.
				0.6	Report of the Meeting of Experts, 1982
N_2O	0.3	< 0.3	$0.3 \rightarrow 0.6$	0.7	Karol' et al., 1986b
H_2O		2–10	$2 \rightarrow 4^c$	0.4	Ibid.
			$3 \rightarrow 6^c$	0.6	Report of the Meeting of Experts, 1982
		Nanconstituents (ppb)			
SO_2	0.05–0.2	~ 0.01	$2 \rightarrow 4$	0.02	Report of the Meeting of Experts, 1982
NH_3	1–20	~ 1(?)	$6 \rightarrow 12$	0.09	Ibid.
HNO_3	0.1–1	0.5–5	$2(HNO_3)^b$	0.06	Ibid.
$CFCl_3$	0.1–0.3	< 0.1	$0 \rightarrow 1$	0.15	Ibid.
CF_2Cl_2	0.2–0.4	< 0.2	$0 \rightarrow 1$	0.13	Ibid.
CF_4	0.05–0.07		$0 \rightarrow 1$	0.07	Ibid.

Table 5.1 (Continued)

Gas	Mixture Ratio Tropo- Sphere	Strato- Sphere	Mixture Ratio Change	ΔT_R °C	Source
CCl_4	0.1–0.2	< 0.1	$0 \rightarrow 1$	0.10	Ibid.
CH_3Cl	0.5–0.8	< 0.5	$0 \rightarrow 1$	0.01	Ibid.
CHF_2Cl	0.03–0.05	< 0.03	$0 \rightarrow 1$	0.04	Ibid.
CH_3CCl_3	0.1–0.2	< 0.1	$0 \rightarrow 1$	0.02	Ibid.

Notes: *a*, change in the troposphere; *b*, similar change in the vertical concentration profile; *c*, change in the stratosphere.

(Report of the Meeting of Experts, 1982; Ramanathan et al., 1985, 1986; Karol' et al., 1986b). Trace gases with absorption bands in "transparency windows" for infrared (IR) radiation (3–5 and 8–13 μm ranges), formed by water vapor in the troposphere, where absorption band width increases under the influence of higher air pressure than stratospheric are especially important. Visible solar radiation is absorbed poorly by ozone and NO_2 in the stratosphere and almost not at all by trace gases in the troposphere. Near infrared solar radiation (0.7–4.0 μm), infrared radiation of the Earth's surface, of clouds, and of the atmosphere is intensively absorbed in the troposphere, mainly, by clouds, by CO_2 and water vapor, and re-emitted back to Earth and into outer space. Some of the longwave radiation is absorbed and re-emitted by various minor gases only in transparency windows (especially in the 18–13 μm range), and this flux produces about 10% of the total atmospheric greenhouse effect that is about 30°C at the Earth's surface (Karol', Rozanov, Timofeyev, 1983; Ramanathan et al., 1986). The mean number of clouds in different layers, their properties, as well as tropospheric water vapor content have a strong influence on the greenhouse effect. It is a matter of general experience that these factors have great spatial and temporal variation, so it is difficult to determine their influence on climatic characteristics and to reveal their long-term trends. However, this high variability depends on many reasons and probably can affect the expected global climatic changes comparatively little in the near future. This is confirmed by quantative estimates of correlations between latitudinal and seasonal temperature variations and global cloud cover (Mokhov, 1984; Ramanathan et al., 1986). At the same time, an increase in CO_2 and trace gases in the atmosphere is a permanent though small factor unilaterally altering radiation fluxes and radiation heating in the atmosphere.

These changes primarily determine the altered thermal conditions at different stratospheric levels outside polar latitudes that have a poor vertical mass-exchange. In the troposphere, where mass-exchange is intensive, temperature conditions depend strongly on dynamic processes, primarily on large-scale horizontal transfer and vertical convection. Changes in the vertical temperature profile in the troposphere, more precisely — in the troposphere — Earth's surface system, depend on variations in radiation heating. Such changes are usually revealed by alterations of effective radiation fluxes at the tropopause level. Some authors, who only considered the radiation and heat balance of the Earth's surface severely underestimated the dependence of the greenhouse effect on the higher atmospheric CO_2 content (Ramanathan et al., 1986).

Earth's CO_2 absorption bands (the primary one is centered on $\lambda = 15 \mu m$) are "saturated" by longwave outgoing radiation even in the lower tropospheric layers, from which energy is re-emitted to the Earth and to the upper atmosphere where the radiation is absorbed by clouds, and by H_2O, CO_2 and other trace gases, thus producing additional radiation heating. A radiation heating increase ΔF of the troposphere — underlying surface system is proportional to a logarithmic increase in ΔCO_2. Absorption bands for low contents of trace gases in the atmosphere which are not "saturated" by radiation of the Earth's surface and of lower atmosphere for such optically "thin" absorbing media $\Delta F \sim \Delta$ C/C affect atmospheric radiation conditions relatively more effectively (Report of the Meeting of Experts, 1982; Ramanathan et al., 1986).

Radiation effects induced by increased atmospheric CH_4 and N_2O are analogous to those of CO_2 and H_2O; they cause warming of the lower atmosphere and cooling of the stratosphere. This is due to the increased amount of longwave radiation emitted into the outer space and decreased portion of longwave radiation flux of the Earth's surface and lower atmosphere, reaching the stratosphere. In contrast to these gases, radiation heating caused by higher content of chlorofluorocarbons peaks not in the lower troposphere, but in the tropical tropopause. Radiation heating of the Earth-troposphere system from the addition of one F-11 ($CFCl_3$) or F-12 (CF_2Cl_2) molecule exceeds that from addition of 10^4 CO_2 molecules because the stronger absorption bands of these chlorofluoro-hydrocarbons occur at the "transparency window" $\lambda = 8$–$13\ \mu m$ and at the peak intensity of the Earth's surface longwave radiation (Ramanathan et al., 1986).

The mean annual energy fluxes of solar and longwave radiation of the Earth-atmosphere system, absorbed by the stratospheric ozone layer are 12 and 8 W/m^2 respectively. Of this radiation energy flux 16 W/m^2 are spent to heat the stratosphere; 2.5 W/m^2 are emitted into space, and 1.5 W/m^2 into the troposphere. As the stratospheric ozone content decreases, the upper layer is particularly cooled, since ozone absorption of solar radiation is maximum there and surface radiation fluxes are affected in two ways: more solar, but less longwave radiation from the cooler stratosphere is transported downwards. Therefore, the total thermal effect in the lower atmospheric layer will depend strongly on vertical distribution of atmospheric ozone.

The troposphere accounts for 10–15% of the total atmospheric ozone content, but tropospheric ozone absorption in the 9.6 μm band almost coincides with that by the entire layer of stratospheric ozone because the band is wider as pressure rises. Therefore, radiation warming of the Earth-troposphere system is considerably more sensitive to altered ozone content in the stratosphere than in the troposphere, where it is analogous to the greenhouse effect induced by CH_4 and N_2O (Alexandrov et al., 1982; Ramanathan et al., 1986). In contrast to radiation effects of higher atmospheric H_2O and CO_2, which have a number of overlapping absorption bands, radiation effects produced by trace gases can be considered, with a small error, to be additive and independent. The complicated and nonlinear photochemical interaction between these gases, especially in the presence of ozone (Ramanathan et al., 1986) may be ignored.

Radiation effects of atmospheric aerosols are different and are mostly of local and regional nature. Volcanic aerosols that penetrate into the stratosphere mainly during major eruptions have a global, but mild effect. Aerosols scatter and absorb solar radiation, thereby decreasing its influx into the Earth-troposphere system, and absorb its longwave radiation fluxes. All this results in significant radiation heating of the stratospheric layer containing these aerosols and usually in weak radiation cooling of the lower atmosphere (Volcanos..., 1986).

The direct influence of changes in radiation heating on temperature of different atmospheric layers is modified considerably by many feedbacks that play the basic role in climatic formation of different scales. For example, water vapor content in the lower tropospheric heated air increases so that air relative humidity changes little. This leads to additional absorption of longwave radiation, that significantly intensifies heating of this layer and the entire greenhouse effect, thus forming positive feedback. Other climatic feedbacks will be indicated below.

5.2. Radiation-Photochemical Models

Minor gases present in low concentrations in the atmosphere have a relatively weak effect on radiation fluxes. The altered ozone content indicated in Table 5.1 thus changes the infrared radiation flux and the rate of radiation cooling by less than 10%, and less than 1% by other trace gas content alteration. More precise methods are therefore needed to determine the influence of altered trace gas content in the atmosphere than the calculations used in general atmospheric circulation models (Report of the Meeting of Experts, 1982; Karol', Rozanov, Timofeyev, 1983). However, calculations made by more precise methods require too much computer time and therefore are impractical for detailed three-dimensional models, even for the best modern computers. Besides, the influence of the altered minor gas content on the rate of radiation heating and cooling is too small to be detected by the normal precision of computing of the main meteorological elements in the general atmospheric circulation models and are not manifest even in model "perturbations" of their fields because of the interference from simulated nonradiation atmospheric processes.

Models of radiation equilibrium with a convective adjustment (radiation-convective models) are primarily used to assess the climatic effect of changes in trace gas content. The temperature of different atmospheric layers in these models is determined by the equilibrium between the radiation inflow and outflow. This radiation equilibrium temperature T_* in the lower troposphere decreases with altitude more rapidly than adiabatic and observed temperatures. This stratification in the real atmosphere is unstable and leads to the formation of convection and to sharp increase of vertical heat transfer from the Earth's surface. Therefore, in atmospheric layers with $\gamma_* = -\partial T_*/\partial z > \gamma$, the temperature gradient is assumed to be the mean climatic $\gamma = 6.5°$ C/km. This makes the model temperature profile close to that observed throughout the troposphere and stratosphere beyond polar latitudes (Report of the Meeting of Experts, 1982; Karol' et al., 1986b).

Dozens of radiative convective models have been built by now. They take into account many additional factors and relations and serve as an effective tool for climatic theory research (Karol' et al., 1986b). Table 5.1 also presents estimates of the change in surface air "radiation" temperature ΔT_R when different trace gas concentrations vary in the troposphere and stratosphere. These estimates are based on radiative convective models with $\gamma = 6.5°$ C/km, and on the assumption that relative humidity and all cloud cover characteristics are preserved in the troposphere. The major drawback of these models is that they omit changes in the hydrologic cycle and dynamic processes, and none of the attempts to eliminate such drawbacks have been successful.

It is evident from Table 5.1 that doubling of the atmospheric content of trace gases with mixing ratio about 10^{-6} raises the temperature (ΔT_R) by several tenths of a degree, while those with mixing ratio about 10^{-9} and less, ΔT_R rises by several hundredths of a degree. F-11 ($CFCl_3$) and F-12 (CF_2Cl_2) increases cause the highest temperature rise among the other anthropogenic chlorofluorocarbons in the atmosphere (see below). Estimates for O_3, CH_4 and H_2O based on different radiative-convective models agree fairly well.

Almost all trace gases in the atmosphere interact photochemically with each other, and their concentration distributions are substantially determined by their photochemical sources and sinks. Their intensities depend on temperature and spectral fluxes of solar ultraviolet radiation, that are closely related to the content of ozone and other trace gases which absorb radiation, especially in the stratosphere. A complicated system of interactions is formed between photochemical, radiation and dynamic processes in the atmosphere with many feedbacks that cannot be described fully by the extant models, including three-dimensional general atmospheric circulation models.

An attempt was made to model this system by combining radiative convective and simple photochemical models into radiation-photochemical ones. Their photochemical parts include up to a hundred photochemical reactions between dozens of compounds involved in basic atmospheric cycles of oxygen, nitrogen, carbon, hydrogen and halogens (chlorine, fluorine, bromine) in the modern photochemistry of the troposphere and stratosphere (Karol', Rozanov, Timofeyev, 1983; Karol' et al., 1986b). Vertical transport of trace gases which is important in photochemistry is parametrized by turbulent diffusion with coefficient K_2 that changes with altitude and sometimes seasonally. Fluxes of ultraviolet radiation and temperature profiles computed by radiative convective models serve as input data for photochemical models, where the output is vertical profiles of trace gases concentrations, which are incorporated into radiative convective models. Less than ten successive iterations are practically necessary for establishing profiles of trace gas temperature and concentrations in a radiation-stable model with errors less than 0.01°C and 10%, respectively. These profiles take into account the interaction between radiation and photochemical processes in the atmosphere. This interaction somewhat compensates for the effects of external disturbances. Deviations in trace gas temperature and concentrations in the radiation-photochemical models are usually smaller than their separate estimates obtained in radiation and photochemical models for the same disturbances (Karol' et al., 1986b).

5.3. Atmospheric Cycles of Trace Gases

The atmospheric cycle of a trace gas is determined by the nature and intensity of its sources and sinks, their time and spatial variations as well as by the mean trace gas lifetime τ in the atmosphere. For the continuous sources and sinks under consideration, the mean lifetime is determined by the ratio of mean mass of the trace gas in the atmosphere M to the total intensity of its sink or source S

$$\tau = M/S.$$

Ground level air measurements of concentrations of previously known and of some new trace gases have significantly expanded in recent years (including the Southern Hemisphere); satellite data on their global concentration distributions in the stratosphere have become available. This made it possible to determine not only seasonal and latitudinal, but also annual variations in trace gas content in different parts of the atmosphere, and significantly extend our knowledge about their global atmospheric cycles.

Methane. Figure 5.1 illustrates the seasonal and latitudinal variation CH_4 mixing ratio in the surface air in 1983–1984, measured at 20 evenly distributed, primarily, island stations (Gammon, Steele, 1984), while Figure 5.2 presents the mean zonal CH_4 distribution in the stratosphere in January, April, July and October 1979–1981, measured by the Nimbus 7 satellite (Jones, 1984). These data represent not only the three-dimensional atmospheric CH_4 concentration field, but also characterize its considerable seasonal and annual variations, especially in the surface air, where winter and summer peaks occur in the high latitudes of both hemispheres. Distribution of the CH_4/CO_2 concentrations ratio in the surface air layer measured at the aforementioned monitoring network indicates that the basic CH_4 source is located within the continental zone northward of 40°N (Gammon, Steele, 1984). Estimates of components in the global atmospheric CH_4 balance (Seiler, 1984; Karol' et al., 1986) indicate predominant biological land sources of CH_4, two-thirds of them being anthropogenic by origin. According to these estimates, in the 1960's–1970's, the intensity of the biological components' annual increase was approximately 1.1–1.6%, while that of the nonbiological resources was 1.7–2.0%. Incorporating possible development scenarios for the world economy, Karol' et al. (1986b) maintain the forecasts that such an increase in CH_4 sources will be preserved up to the year 2000.

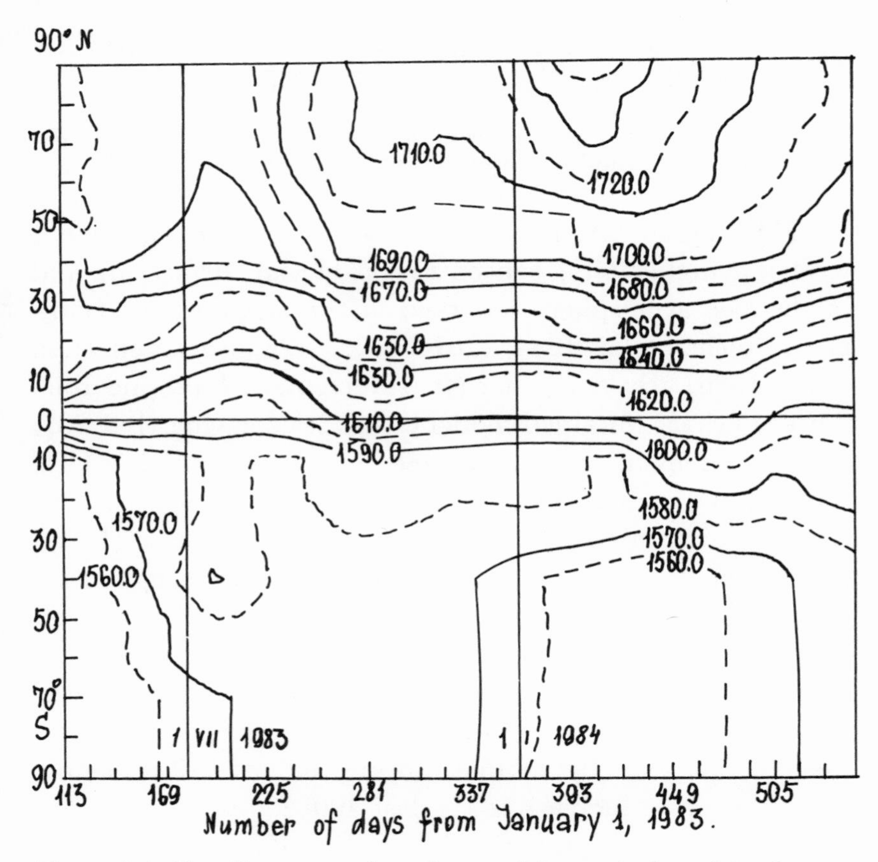

Figure 5.1. Zonally averaged methane mixing ratio (ppm) in the surface air layer according to measurements at GMCC station network (Gammon and Steele, 1984).

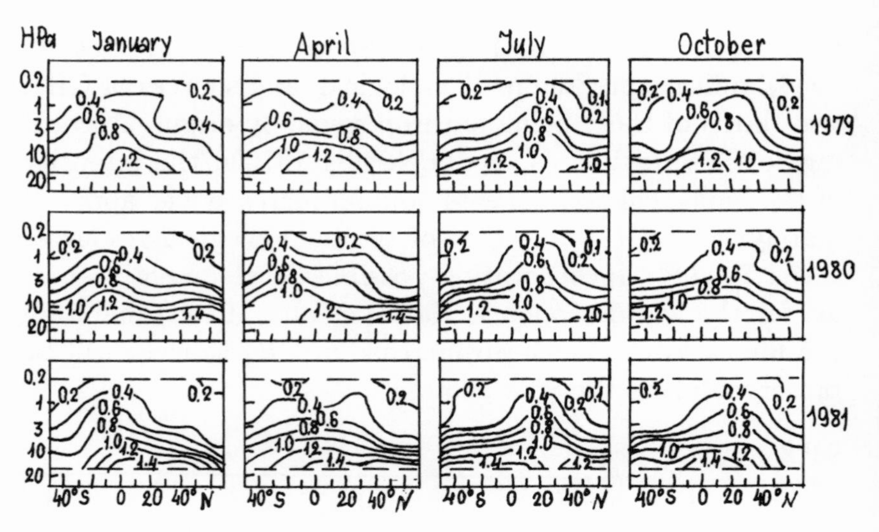

Figure 5.2. Zonally averaged mean monthly methane mixing ratio in the stratosphere (ppm) according to satellite data (Jones, 1984).

This higher intensity of CH_4 sources is in agreement with the higher CH_4 concentration in the surface air and in the free atmosphere observed in the last decade (Gammon, Steele, Seiler, 1984; Karol' et al., 1986b; Ramanathan et al., 1986). By analyzing measurements, Khalil and Rasmussen (1985) obtained model estimates showing that 70% of this increment, on the average, is caused by higher CH_4 source intensity while the remaining 30% is associated with its lower sink intensity because of possible decrease (by an average of 20%, since the early eighteenth century) in atmospheric content of hydroxyl OH, the basic CH_4 oxidizer. Precise and complicated measurements of CH_4 content in air bubbles contained in deep layers of ice sheets in Greenland and Antarctica revealed that the CH_4 mixing ratio was 0.6–0.8 ppm (by volume) during the last 2–3 millennia and it began to increase in the last 100–200 years when the Earth's population growth started (Stauffer et al., 1985).

Consequently, an increase in the rate of atmospheric CH_4 content rise can be expected in the future, especially considering the positive correlation between intensity of CH_4

sources and the Earth's surface temperature observed in subtropical swamps (Cess, Hameed, 1983; Seiler, 1984). Extensive permafrost melting in the taiga and tundra because of the expected global warming will probably also lead to a considerable increase in methane emission in the future. At the same time, the higher Cl and ClO content in the middle and upper stratosphere, caused by accumulation and photolysis of chlorofluorocarbons (details are presented below) makes for more intensive photochemical CH_4 discharge in the stratosphere by the following reaction:

$$CH_4 + Cl \rightarrow HCl + CH_3$$

and slower growth of CH_4 content in the stratosphere.

Non-methane hydrocarbon compounds. Many organic gases have been measured in the last 10 years: ethane, propane, butane, pentane, ethylene, acetylene and others of both natural (released by coniferous forests, isoprenes, terpenes) and anthropogenic sources (mainly cars) (Isidorov, 1985; Ramanathan et al., 1986). These gases have background concentration of several ppb and less, and are readily oxidized in the lower troposphere to yield CO and CO_2. Their mean lifetime in the atmosphere therefore, is no longer than several days or even hours. Their optical properties are still obscure. They are unlikely to affect significantly the radiation fluxes in the atmosphere, but noticeably contribute to atmospheric sources of CO (Report of the Meeting of Experts, 1982; Ramanathan et al., 1985, 1986). Carbon monoxide is not an active radiation trace gas, but has abundant anthropogenic sources and participates actively in photochemical reactions with other trace gases to alter significantly their atmospheric content.

Figure 5.3 illustrates the global latitudinal and seasonal distribution of total CO in an air column based on measurements under background conditions, mainly above oceans

(Carbon monoxide content..., 1985). The noticeable latitudinal and seasonal variation in CO content, which is 1.6–3.1 times greater in the Northern Hemisphere than in the Southern, and a third are from anthropogenic sources in the Northern Hemisphere. Combustion of fossil fuel and different biomass is a prime global CO source (Isidorov, 1985; Carbon monoxide content..., 1985; Karol' et al., 1986b). Almost half of the global CO source total intensity is produced by CH_4 oxidization by hydroxyl OH. This source is also subject to anthropogenic activity (through CH_4). However, the brief residence of CO in the atmosphere (about 2 months versus 10 years for CH_4) causes almost complete isolation of atmospheric CO cycles in the Northern and Southern Hemispheres. Each of these cycles has a seasonal variation in CO_2 troposphere background concentration, with maximum in late winter-spring and minimum in late summer-autumn. It is probably associated with seasonal variation in mean OH atmospheric content with summer peak and winter depression (Dvoryashina, Dianov-Klokov, Yurganov, 1984; Isidorov, 1985; Karol' et al., 1986b).

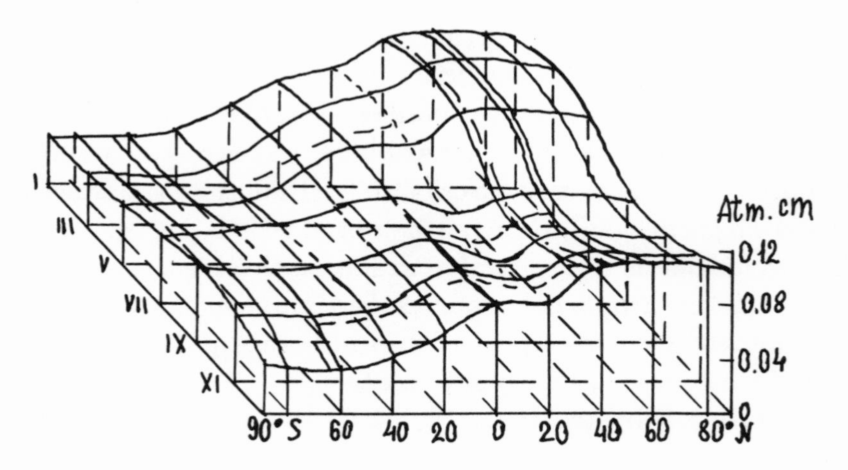

Figure 5.3. Seasonal and latitudinal distribution of total carbon monoxide content in the air column (Carbon monoxide content..., 1985).

The only known long series of measured total background CO content of air column (near Moscow) for the period 1970–1984 shows its steady rise by 1.5% per year approximately, by 1.7% during January–April, and by 1.4% during July–September (Carbon monoxide content..., 1985). A much less reliably based trend of 2–5% per year (Khalil, Rasmussen, 1984a) was found by measurements of CO surface air layer concentration on the U.S. west coast in 1979–1982. Almost the same trend is observed in short-term series of measured background CO concentration in the lower atmospheric layer at 2–3 sites in the Southern Hemisphere for the period 1980–1984 (Khalil, Rasmussen, 1984a; Seiler et al., 1984). However, spectroscopic measurements of total CO in an air column on the Antarctic coast in the summer of 1977 and in 1983–1984 revealed the same concentrations within the limits of measurement accuracy (Carbon monoxide content..., 1985).

Nitric oxides. Nitrous oxide N_2O is more evenly distributed in the troposphere of both hemispheres and has longer lifetime in the atmosphere than CH_4, although there is less data on the composition and total intensities of atmospheric N_2O sources and sinks. The anthropogenic contribution to N_2O sources is no more than 20% of their total intensity and comes from fuel combustion and denitrification of nitrogen fertilizers used in agriculture (Weiss, 1981; Karol', Rozanov, Timofeyev, 1983; Karol' et al., 1986b). More intensive output of anthropogenic N_2O sources probably led to the concentration increase of 0.15–0.4% per year observed since the late 1970's (Geophys. Monitoring, 1984–1986; Karol' et al., 1986). Satellite and balloon data on N_2O stratospheric concentration indicate that it is distributed similarly to CH_4 (see Figure 5.2). They also show peak N_2O concentration in the tropics caused by ascending fluxes in the equatorial zone, which intensify CH_4 and N_2O transfer from ground sources to stratospheric photochemical sinks (Karol', Rozanov, Timofeyev, 1983; Jones,

1984). These CH_4 sinks serve as major moisture sources in the middle and upper stratosphere, while the N_2O sinks are nitric oxide (NO_x) sources in the stratosphere and upper troposphere (the sum of NO and NO_2 is designated by NO_x).

The second significant NO_x source in the upper troposphere is emissions from worldwide aircraft engines that accounts for 0.6–0.9 MT of NO_2 annually, mainly, in the temperate latitudes of the Northern Hemisphere, while N_2O oxidation in the stratosphere produces 1.6–3.3 MT of NO_2 (Mahlman, Levy, Moxim, 1980; Karol', Rozanov, Timofeyev, 1983; Crutzen, Gidel, 1983). The available estimates indicate that the NO_x annual emissions by transport aircraft in the 1970's–1980's increased by up to 7% and will become a predominant NO_x source in the near future above the layer where precipitation eliminates the trace gases (Wuebbles, MacCracken, Luther, 1984; Karol' et al., 1986b).

The NO_x content in the lower troposphere "washed" by precipitation depends both on washout intensity and on strength of natural and anthropogenic ground sources: combustion of fuel and biomass, denitrification of nitrogen fertilizers, as well as NO_x formation during thunderstorms. According to Dignon and Hameed (1986), the intensity of anthropogenic NO_x sources (mainly from fuel combustion) in the Northern Hemisphere rose by approximately 4% per year from 1966 through 1973, by about 1% per year after 1973, and reached 22 MT of N per year in 1980; 95% of this emission occurred in the Northern Hemisphere with high peak in the 30–60°N zone. About 2 MT of N is produced each year by lightning, peaking in the tropics, and up to 10 MT of N is produced by the land biosphere with maximum in the middle and subtropical northern latitudes, zones of peak primary bioproductivity (Crutzen, Gidel, 1983; Karol' et al., 1986b).

Washout by precipitation reduces mean NO_x lifetime in the lower troposphere to several days (up to 1–2 weeks). Therefore, NO_x concentration is subject to high time and spatial variability, especially in cities, industrial areas and their environs (Dignon, Hameed, 1986). For this reason, it is difficult to reveal and to observe the evolution of NO_x background concentration in the surface air layer, and such data are not available yet. However, the observed rise in background ozone concentration in the surface air layer and in the free troposphere, where NO_2 is a main background source of ozone, indirectly indicates its growth in the 1970's (Crutzen, Gidel, 1983; Karol' et al., 1986b). Since the NO_x content determines the concentration of ozone, one of the most important trace gases, oxidant and toxicant, in the surface air layer, we need to study the atmospheric cycle and monitor the background NO_x levels in the surface air layer and in the free atmosphere.

Ozone. Sources and main ozone sinks are situated in the atmosphere and they are related to dozens of photochemical reactions between many minor constituents in the atmosphere where aerosols participate (Report of the Meeting of Experts, 1982; Karol' et al., 1986). Ozone is only formed by the reaction:

$$O_2 + O + M \rightarrow O_3 + M.$$

This describes ozone formation from atomic oxygen with molecules M of any atmospheric gas. Ozone is destroyed according to the following reactions:

$$\text{[1]} \qquad \left.\begin{array}{c} O + AO \rightarrow A + O_2 \\ O_3 + A \rightarrow AO + O_2 \\ \hline O + O_3 \rightarrow 2O_2 \end{array}\right),$$

where A designates radicals H, HO, NO, Cl, Br, that are the catalysts that destroy ozone faster than in the direct reaction (R.4). Atomic O is thus needed both to form and to disintegrate ozone. Its main source in the middle and upper stratosphere is photolysis of molecular oxygen O_2 by hard solar ultraviolet radiation with $\lambda < 0.24\mu m$. In the lower stratosphere and troposphere, where this radiation intensity decreases sharply, atomic oxygen is mostly formed through photolysis of $NO_2 + h\gamma \rightarrow NO + O$. This is the result of soft ultraviolet radiation with $\lambda < 0.42\mu m$ during oxidization of NO into NO_2 without participation of atomic oxygen in reactions with HO_2 and CH_3O_2. Photochemical ozone loss in the troposphere is the result of its catalytic breakdown in reaction [1] with A = HO, by the reaction:

$$O_3 + HO_2 \rightarrow HO + 2O_2.$$

In the end the total global intensities of ozone photochemical sources and sinks separately in the stratosphere and troposphere almost compensate each other. The ozone flux from the stratosphere via the tropopause is a small fraction of the total intensity of its stratospheric source. This flux (about 7×10^{14} g/year) is close to total intensities of global photochemical sources and sinks of ozone in the troposphere and its sink at the Earth's surface (Karol', Rozanov, Timofeyev, 1983).

Figures 5.4 and 5.5 illustrate deviations in total ozone X_3 and densities (O_3) at various stratospheric levels measured in Switzerland (Dutsch, 1985). While X_3 and O_3 have been decreasing since 1960 in the lower stratosphere, the layer of maximum O_3 density increased by 2–6% in 10 years, and an unusually sharp drop occurred at the end of 1982 and in 1983. According to satellite data, by January 1983 the mean monthly global X_3 values had dropped at most by 3.7%, and O_3 during the first months of the same year had decreased by tens of a percent near levels 100 and 50 gPa (16 and 20 km, see, also, Figure 5.5) (Heath and Schlesinger, 1985). This extreme X_3 drop (by 7%) in its longest measurement series in Aroza (Figure 5.4) was caused by the effect of products from the El Chichon eruption in April 1982 in Mexico at 17°N. Data from aerial and ground-based optical measurements made in the United States in 1983 revealed that the total content of HCl in the air column was 40% and of OH almost 30% above the normal seasonal level. This was caused by emissions of large amounts of water and chlorine compounds into the stratosphere when El Chichon erupted (Burnett and Burnett, 1984; Mankin and Coffey, 1984). These compounds, as well as volcanic aerosols evidently caused severe ozone destruction in the troposphere. It should be noted that the sharp X_3 depressions in 1963 and 1974 (see Figure 5.4) were also the result of volcano eruptions of Agung in Indonesia and of Fuego in Guatemala, respectively (Dütsch, 1985).

Several attempts to reveal long-term trends of ozone density variations in the middle and upper stratosphere, where, according to model estimates, ozone should be destroyed more intensively by Cl and ClO formed during pho-

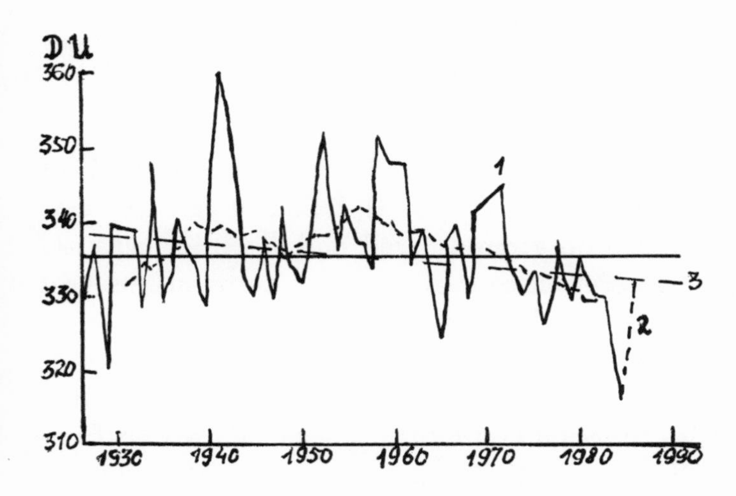

Figure 5.4. Mean annual total ozone content from measurements in Aroza (Switzerland) (1), 10–year running averages; (2), and the trend line; (3), (Dütsch, 1985).

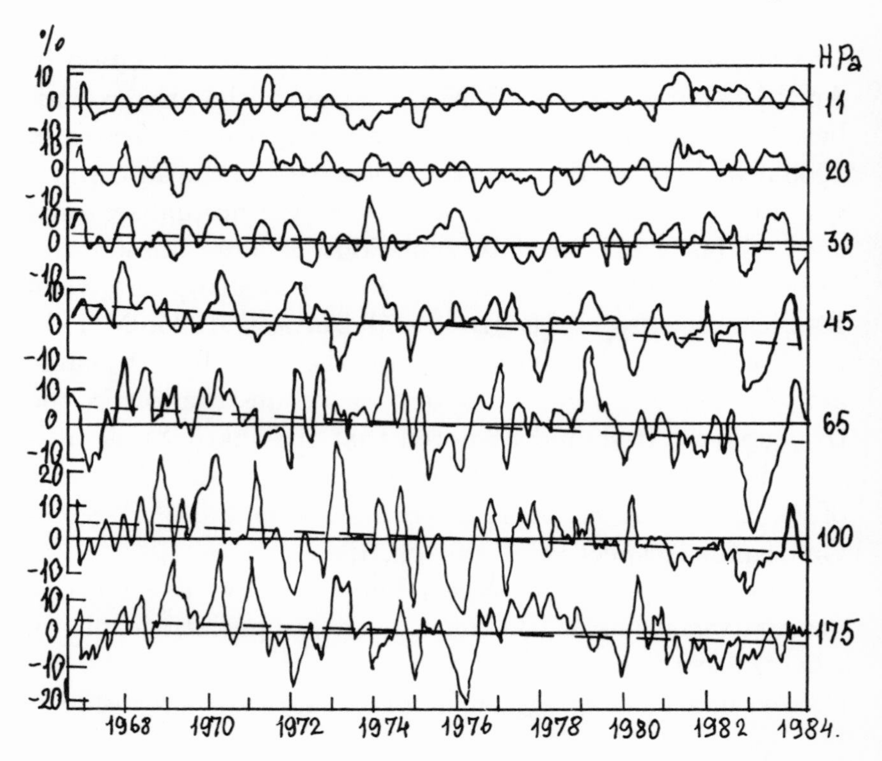

Figure 5.5. Variations in ozone density deviations from the average for the period 1967–1984 at various stratospheric levels from ozone sensing data in Payerne (Switzerland) with three-month running averaging (hatch line indicates trend lines) (Dütsch, 1985).

tolysis of freons and other chlorofluorocarbons (Karol' et al., 1986; Ramanathan et al., 1986), failed to give any definite results based on ground-based and satellite measurements. The observation time is too short to reliably reveal any trend (Angell and Korshover, 1983).

Recently published results from ground-based and satellite measurements of total ozone above the polar Antarctic regions (Shoeberl and Krueger, 1986) show that from the late 1970's to 1984–1985, the October X_3 minimum decreased by 30–40%, in the sub-Antarctic zone of peak X_3, it decreased by 20% while in the same latitudes but in other months (December-August) X_3 values do not have such a marked trend. There is still no definite explanation for the "ozone hole" phenomenon. It is possibly the result of dynamic processes, or the result of chemical reactions of chlorine compounds in the stratosphere from fluorocarbons (F-11 and F-12, see below) on the photochemical ozone sources and sinks in the Antarctic circumpolar regions, in the presence of reduced air exchange with other latitudes, low temperature and where the sun is low over the horizon in this region in October (Shoeberl and Krueger, 1986).

Systematization and analysis of ozone content measurements in the troposphere (Logan, 1985) revealed that intensive photochemical formation of ozone occurs above the populated and industrially developed areas in the middle latitudes of North America and Eurasia, resulting in the formation of a significant and lengthy summer peak ozone density in the lower and middle troposphere. A few measurements indicate a steady 1–2% annual increment in ozone density (more in summer) in the 1970's–80's. Ozone measurements in the surface air layer made by 5–6 stations in the GDR (some of them are situated far from cities) show a similar trend since the 1950's (Angell and Korshover, 1983; Logan, 1985; Feister and Warmbt, 1985).

The tropospheric ozone distribution above oceans and sparsely populated continental regions (including the Southern Hemisphere) depends to a large extent on its influx from the stratosphere, and there are too few observations to reveal any trend in its growth there (Logan, 1985; Geophys. Monitoring, 1984–1986). The tropospheric ozone content in the tropics shows strong spatial and seasonal variations. Ozone density during individual seasons (for example, in September–October in Brazil) is high and close to the levels in industrial regions in the middle and high latitudes of the Northern Hemisphere. These high values are probably associated with photochemical formation of ozone in the atmosphere polluted by smoke from the burning of tropical forests (Logan, 1985). Zonal annual averaging of tropospheric ozone distribution shows that its density in the boundary layer of both hemispheres is almost the same, but in the free troposphere of the Northern Hemispheric middle latitudes it is 30–40% higher than in the Southern Hemispheric middle latitudes (Karol' et al., 1983).

Chlorofluorocarbon compounds. Intense studies and measurements of atmospheric $CFCl_3$, CF_2Cl_2 (F-11, F-12) and CCl_4 content started in the 1970's. The current content along with methylchloroform (CH_3CCl_3) are continually measured at about ten ground-based recording stations belonging to the Geophysical Monitoring for Climatic Changes (GMCC) and Atmospheric Lifetime Experiment (ALE) systems (Geophys. Monitoring, 1984–1986; Prinn et al., 1983). The atmospheric content of newly detected chlorofluorocarbons, CHF_2Cl (F-22) and $CFCl_2CF_2Cl$ (F-113) was recently measured at less than 0.1 ppb by volume, while $CBrClF_2$ (F-12B1) was about 0.001 ppb by volume, but their annual growth exceeds 10% (Fabian et al., 1981). All these gases are of anthropogenic origin (although higher F-11 and F-12 concentrations have been found in gases emitted from volcanos on the Kuril Islands (Isidorov, 1985)), and their main sink is of ultraviolet radiation in the stratosphere. Hydrogen-containing chlorofluorocarbons are also destroyed by reacting with tropospheric hydroxyl and, therefore, have a shorter mean lifetime in the atmosphere (Report of the Meeting of Experts, 1982; Karol' et al., 1986b).

Table 5.2 provides the estimated rate of relative increase in basic chlorofluorocarbon concentration and their estimated mean lifetime in the atmosphere made using relatively long series of measurements from GMCC and ALE ground-based stations.

The higher rate of increment in F-11 and F-12 in the Southern versus the Northern Hemisphere probably reflects their lower rate of emission into the atmosphere after 1975 in the middle latitudes of the Northern Hemisphere, their basic global source zone, from which they are transported to the Southern Hemisphere. There is considerable disagreement

Table 5.2. Estimated Rate of Chlorofluorocarbon Concentration Increment in the Lower Troposphere (%/year) and Their Mean Lifetime (year) (Geophys. Monitoring ..., 1984–1986; Prinn et al., 1983)

Gas	Northern Hemi- -sphere*	Southern Hemi- -sphere*	Averaged over the globe**	Mean lifetime in years
$CFCl_3$	6.4	8.8	5.7	78
+				
CF_2Cl_2	4.6	7.4	6.0	81
CCl_4			1.8	52
CH_3CCl_3			8.7	10

*According to data from GMCC network for 1977–1983.
**According to data from ALE network for 1978–1981.

between the published estimates of F-11, F-12, CH_3CCl_3 emissions and their total measured content in the global atmosphere (Karol' et al., 1986b). This requires immediate global monitoring of atmospheric chlorofluorocarbons content, including CH_3Cl, CCl_4, CH_3Br that have natural sources (CCl_4 has both natural and anthropogenic sources) and introduce into the atmosphere Cl and Br atoms which are active participants in ozone photochemistry (Karol', Rozanov, Timofeyev, 1983). The restrictions imposed in some countries (e.g., in the United States) on the use of F-11 and F-12 connected with their prompt emission into the atmosphere lead to a greater increased usage and, consequently, to higher emissions of CH_3Cl, CH_3CCl_3, CHF_2Cl which have considerably shorter lifetime in the atmosphere (Fabian et al., 1981; Prinn et al., 1983). However, these chlorofluorocarbons enhance the greenhouse effect. Even though their current atmospheric concentration is lower, their concentration is rising faster than F-11 and F-12.

Therefore, if their present-day growth rates are maintained in the future, they may be comparable to F-11 and F-12 in their influence on the ozonosphere, on the thermal and radiation condition of the atmosphere.

Other radiative active trace gases. Radiation and climatic effects of stratospheric aerosols, especially products of the El Chichon volcanic eruption have recently been emphasized. Aircraft and ground-based measurements have noted a considerable solar radiation attenuation that reaches peak levels in the latitudes where the volcano is located (Mauna Loa Observatory, Hawaii, 20°N). Data from this observatory show a drop in the monthly direct solar radiation to 67% in May and June 1982, compared to the same time frame in 1981, while the monthly total radiation was 89 and 86% of the 1981 levels. The July 1982 levels were 71% for the total direct and 91% for the total radiation (Karol', 1984; Volcanos..., 1986). Almost all the Northern Hemispheric actinometric stations observed a significant attenuation in solar radiation starting in May 1982. Generalized data and models of global climatic fluctuations caused by this attenuation are discussed in the monograph Volcanos, Stratospheric Aerosol and Climate of the Earth (1986).

The influence of tropospheric aerosol on radiation and climatic elements was studied in geographic zones, where there is systematic loss of large masses of desert dust, including the tropical Atlantic which receives dust from the Sahara Desert, and the subtropical western Pacific which receives dust from the Gobi Desert (Kondrat'yev, Moskalenko, Pozdnyakov, 1982; Desert Dust, 1981).

The recently detected winter phenomenon of aerosol transport with a large amount of soot particles in the troposphere from the middle continental latitudes to the Arctic has been the subject of intensive investigation. These aerosol masses are accumulated in the lower Arctic troposphere (often in its inversion layers) and form a haze clearly seen in the high latitudes in spring when the Sun appears low above the horizon (Schnell, 1984; Shaw, 1985). The few available actinometric measurements in the Arctic (for the USSR territory only) indicate a lower atmospheric aerosol spring transparency as far back as the 1950s (Voskresenskiy, Marshunova, 1982), although it is probably increased to some degree in the last decade. The resulting measurements of aerosol concentrations and optical properties, and of the Earth's surface in individual Arctic regions (Alaska, Greenland, Spitsbergen) were the basis for model assessments of radiation heating in the lower troposphere of these regions as caused by the aforementioned spring aerosol haze under clear sky. It can be concluded from these results that aerosol absorption of solar radiation in spring causes warming of the lower troposphere, and its increased thermal radiation warms the Earth's surface and surface air layer, despite some decrease in solar radiation absorbed by Earth's surface. According to rather rough model estimates, the observed aerosol Arctic haze can increase ground layer air temperature T° by approximately 1°C in April-May and by 0.2–0.4°C during the polar night (McCracken, Cess, Potter, 1986).

Measurements of trace gas content in samples taken from aircraft in aerosol haze layers and in surface air over the regions of Alaska and Spitsbergen in spring 1983 demonstrated a considerable increase in content of chlorofluorocarbon and other organic gases compared to their content in the same regions in summer 1982 and in spring 1983, and beyond haze boundaries. In this case, the concentration of short-lived (less than one-year lifetime) trace gases was hundreds of times, and of long-lived tens of times higher than background levels (Khalil and Rasmussen, 1984b). Radiation and thermal effects of these layers with higher trace gas content have not yet been assessed.

The problem of higher background concentration of tropospheric aerosols of anthropogenic origin has been discussed extensively (Kondratiyev, Moskalenko, Pozdnyakov,

1982), however, there are still too few observational data. More or less lengthy series of measurements from several U. S. stations under background conditions do not indicate any trends (Bodhaine, 1983). It is possible that high spatial and time variability of sources and sinks of tropospheric aerosol and its short atmospheric lifetime of less than two or three months prevent formation of higher aerosol content for a lengthy period over extended regions, except for the aforementioned specific case of Arctic aerosol.

5.4. Estimated Expected Changes in Mean Gas Composition and Temperature Conditions of the Atmosphere

An assessment of the influence of anthropogenic changes in atmospheric gas composition on climate was urgently needed; therefore, several attempts were made to compute the evolution of active radiation trace gas content of the atmosphere through linear extrapolation of their current trends and subjective judgments of possible future variations. Examples of these extrapolations are found in a number of publications (Rasmussen and Khalil, 1981; Wuebbles, 1983, 1984; Ramanathan et al., 1985, 1986).

Since radiative, photochemical and dynamic processes and their interactions are essentially nonlinear, we do not have prior knowledge of the extent to which their linearization in these extrapolations is permissible. We also have to make allowances for possible significant variations in the nature and intensity of the future anthropogenic effect.

Models of radiation and photochemical processes in the troposphere and stratosphere involving trace gases, described in Section 5.2, can and are used to evaluate the expected changes in mean annual trace gas content and variations they cause in surface air temperature in the Northern Hemisphere. Various scenarios for the evolution of anthropogenic emissions of some trace gases into the atmosphere are used in this case.

Section 5.3 has already noted that the distributions served in CH_4 and N_2O in the troposphere of the Northern Hemisphere are approximately uniform; therefore, their evolution can be described by the transient one-dimensional model of atmospheric gas composition in the Northern Hemisphere with seasonally varying or mean annual parameters. The photochemical model with specified temperature profile T may also be used here, since the research results indicate slight changes in T even when the atmospheric CH_4 content quadruples (Karol' et al., 1986b). Concentration distributions of tropospheric chlorofluorocarbons of anthropogenic origin show a sharp peak in the middle latitudes of the Northern Hemisphere, where their primary source is located, and its intensity is increasing steadily. The expected evolution of chlorofluorocarbon distribution is therefore computed by a two-dimensional zonally averaged photochemical model with standard temperature distribution, that has almost no effect on the photochemical destruction of these chlorofluorocarbons in the stratosphere.

The expected increment in atmospheric CO_2 content can result in stratospheric cooling by several degrees, thus altering ozone concentration in the middle and upper stratosphere by tens of a percent, and, consequently, influencing the temperature of these layers. In order to estimate this effect on the expected changing atmospheric gas composition, these variations were also computed by a one-dimensional radiation-photochemical model with mean annual parameters and for a prescribed uniform increment in the CO_2 mixing ratio for the entire atmosphere (see Wuebbles, McCracken, Luther, 1984, or Table 3.3).

Analysis of the atmospheric trace gas levels in the next decades indicate their possible lower and upper limits that characterize the extent of their uncertainty, arising both from inaccurate evaluation methods and imprecise estimates of future anthropogenic change in source and sink intensities of trace gases. The most probable estimates of such changes should be used to obtain the most likely levels of trace gases content.

Table 5.3 presents estimates of the most probable increment rates for global CH_4 and N_2O emissions into the atmosphere and possible limits of their change, obtained by using a UN model of global economic development, prepared by V. Leont'yev (Karol' et al., 1986b).

Table 5.3. Expected Growth Rate of Global CH_4 and N_2O Emissions into the Atmosphere (%/year) up to the Year 2000 versus the 1980 Level

Period	CH_4	N_2O
1980–1990	0.5–1.8	0.2–0.6
1990–2000	0.6–2.0	0.2–0.8

Data in Table 5.3 were used to compute the extreme and most probable variants for evolution of atmospheric trace gas content using the aforementioned models. The estimated growth rates of emissions of other trace gases into the atmosphere are indicated in Table 5.4 and generalize the results published earlier (Rasmussen and Khalil, 1981; Wuebbles et al., 1983, 1984) as well as data from Table 5.3. The increment in CH_4 and N_2O emissions (Table 5.4) is also used in the calculations for the periods after the year 2000 that are not covered by data in Table 5.3.

Table 5.5 presents the mean annual mixture ratio of trace gases in the lower troposphere of the Northern Hemispheric extratropical latitudes and limits of their possible values measured in 1983, and expected every 5 years up to the year 2000, and every 10 years to the year 2030. Data in Table 5.5 need to be explained in the following way.

Carbon monoxide. Because the estimates of global components of atmospheric CO sources and sinks and their evolution are not sufficiently accurate, extrapolation of surface

Table 5.4. Rate of Increment in Trace Gas Emissions into the Atmosphere (%/year) Used in the Calculations

Variant	CH_4	N_2O	NO_x		F-11, F-12
			a	b	
Minimum	1.0	0	0	0	-3
Mean	1.4	0.2	2	7	0
Maximum	2.0	0.4	5	7	5

Notes: a, ground source; b, emissions of transport aircraft engines.

background CO concentration with zero growth was used for the lower boundary estimate, and with 2% and 5% growth per year for the most probable estimate and for the upper boundary. As shown above, these growth rates approximately equal the rate of increment in background CO concentration evaluated by measurements from localities near Moscow (1970–1984) and individual sites in the middle latitudes of the Southern Hemisphere (1979–1982) (Seiler et al., 1984; Carbon Monoxide Content..., 1985).

Methane. The most probable annual 1% increase in concentration up to the year 2000 listed in Table 5.5 agrees with the global source growth rate indicated in Table 5.3, and with the lower limit of the currently observed growth rate in globally averaged CH_4 concentration (Gammon and Steele, 1984; Seiler, 1984; Khalil and Rasmussen, 1985). The upper limit of CH_4 growth by 2.4% per year has already exceeded the indicated upper limit of CH_4 source intensity growth rate (2.0% per year) during the period 1980–2000 (see Table 5.3). About 20% of the CH_4 content increase in this case is caused by reduced intensity of its sink: the model mean concentration of hydroxyl (OH) in the tropo-

Table 5.5. Mean Annual Mixture Ratios of Trace Gases in the Lower Troposphere for the Middle Latitudes of the Northern Hemisphere Measured in 1983 and Expected in 1985-2030 (in parentheses: possible range)

Gas	Unit	1983	1985	1990	1995	2000	2010	2030
CH_4	ppm	1.70	1.72	1.77	1.83	1.90	2.2	2.9
		1.62-1.73	(1.65-1.75)	(1.73-1.82)	(1.73-1.90)	(1.77-2.0)	(2.0-2.4)	(2.4-3.6)
CO	ppb	130	139	154	170	188	230	340
		130-144	(130-150)	(130-197)	(130-250)	(130-300)	(130-490)	(130-1300)
N_2O	ppb	302	306	315	327	335	350	370
		298-306	(304-310)	(307-325)	(311-345)	(315-360)	(330-380)	(340-420)
NO_2	ppb	0.8-1.0	(0.86-1.1)	(0.9-1.4)	(0.9-1.8)	(0.9-2.3)	(0.9-3.2)	(0.9-6.4)
$NO_2 \times 10^{14}$	mol/cm^2*	1.3	(1.3-1.75)	(1.3-2.0)	(1.3-2.4)	(1.3-2.9)	(1.3-4.1)	(1.3-6.8)
F-11	ppt	200-240	280	380	480	570	840	1850
			(270-320)	(360-430)	(440-590)	(550-780)	(740-1400)	(1300-4500)
F-12	pp^+	340-380	360	540	690	850	1260	2800
			(340-400)	(420-600)	(510-820)	(640-1020)	(860-1800)	(1500-5800)
CCl_4	pp^+	140-180	(160-200)	(200-240)	(230-270)	(280-320)	(350-400)	(480-570)
CH_3CCl_3	pp^+	100	110	130	160	200	360	650
		90-120	(100-130)	(110-150)	(140-210)	(170-270)	(230-480)	(420-1600)

*Note: Content of the 5 to 15 km layer.

sphere decreases by 10–20% in the year 2000 and by 30–40% in the year 2030 as compared to 1980, which is in line with our estimates and those made by other authors (Khalil and Rasmussen, 1985). The estimated rise in CH_4 content in the twenty-first century is probably underestimated on the whole, since the possible increment in CH_4 emissions into the atmosphere from swamps and tundra due to expected global climate warming is ignored (Gammon and Steele, 1984; Karol' et al., 1986b).

Nitric Oxides. Oxidization of N_2O into NO_x is a basic NO_x source not only in the stratosphere, but also in the upper troposphere with an intensity of $(1–3) \cdot 10^{11}$ gN per year in the Northern Hemisphere by model estimates. NO_x emissions from transport aviation engines throughout the world in 1980 were $(2–3) \cdot 10^{11}$ gN and increased approximately by 7% per year (Wuebbles, McCracken, Luther, 1984). The higher intensity of this source resulted in the emission of $3 \cdot 10^{11}$ gN per year in 1985 and a 2.5 times increase by the year 2000 is expected. The intensity of land, natural and anthropogenic NO_x sources is indicated in Table 5.3 and is $(3–4) \cdot 10^3$ gN per year.

Chlorofluorocarbons. The mean annual mixing ratios of $CFCl_3$, CF_2Cl_2, CCl_4, CH_3CCl_3 in the 0–1 km layer, averaged over the 40–60°N latitude zone, were computed by the two-dimensional photochemical model. The most probable values were obtained on the assumption that before the year 2000, the intensity of emissions of these gases into the atmosphere will remain at the mean 1978–1982 level (Prinn et al., 1983):

Gas	F–11	F–12	CCl_4	CH_3CCl_3
Emission, 10^3 T/yr	260	400	150	550

The average annual 5% increment in chlorofluorocarbon emissions observed during the 1970's is adopted as the upper limit in Table 5.4. The lower limit is the annual 3%

decrease in F-11 and F-12 emission. Some available data indicate that such a decrease occurred in 1975–1979 (Prinn et al., 1983). According to Prinn's estimates an annual 1.5% decrease for CCl_4 (the maximum smoothed annual decrease in CCl_4 emission for the period 1958–1980) was taken as the lower limit. Calculation of CH_3CCl_3 concentrations defined the intensity of its photochemical destruction for the mean annual hydroxyl concentration (OH) field, obtained using a two-dimensional photochemical model for standard present-day conditions.

Ozone. Estimates of expected, but inadvertent anthropogenic changes in ozone content (mainly, from chlorofluorocarbon emissions) have been made in the United States since the late 1970's. The assessed drop in the total ozone content with stable conditions and preservation of the current rate of F-11 and F-12 emissions into the atmosphere changed from 20% by the 1977 estimate, to 2–4% according to the last 1984 report (Strat. ozone, 1984). The primary reason for this lower value is that the model photochemical blocks and the rates of a number of photochemical reactions, including photodissociation, were refined. The latter estimate of lower total ozone is 2–4% below the actual error of model computation, and actually means that the current total ozone level will be maintained in the future.

Whereas total ozone content variations mainly influence only the fluxes in biologically active ultraviolet (UV) solar radiation on the Earth's surface, the vertical ozone distribution changes are important for estimating climatic conditions, as well as the toxic effect of ozone in the surface air layer on biological systems, and therefore should be studied in detail.

The present-day ozonospheric models take account of the combined influence of the aforementioned altered concentrations of all gases participating in photochemical formation and destruction of ozone. In the United States, such estimates have been made by one-dimensional radiative and photochemical models. Wuebbles et al. (Wuebbles, Luther, Penner, 1983) used such a model to compute the relative variations in mean annual total ozone ΔX_3 and its density $\Delta(O_3)$ at 20 and 40 km levels for individual periods of 1985–2030 (Table 5.6).

Table 5.6 and Figure 5.6 also present relative changes in total ozone and its density that were obtained by one-dimensional radiative and photochemical model. It took into account variations in the mean annual trace gas mixing ratios for the Northern Hemisphere (see Table 5.5) and the influence of higher atmospheric CO_2 level on air temperature (Karol' et al., 1986b).

In this case, the chlorofluorocarbon concentrations presented in Table 5.5 for the 40–60°N zone have been converted into globally averaged values by integrating them over the entire atmospheric layer of the Northern Hemisphere. This does not cause any noticeable errors within the field of O_3 changes, since latitudinal variations in chlorofluorocarbon concentrations in the stratosphere, where they influence ozone, are insignificant, as seen from Chapter 5 (Karol' et al., 1986b). Future adjusted two-dimensional distributions of ozone and temperature changes can be made using combined two-dimensional photochemical and energy-balance radiation and convective models described in the study of Karol'.

Table 5.6 and Figure 5.6 show that our estimates and those made by Wuebbles et al. (1984) indicate practically the same total ozone content up to the year 2000 within errors of its model computations. By the year 2000, the O_3 drop at the 40 km level will not exceed 25%, and it will reach 40%, according to the photochemical model estimates that ignore altered air temperature because of the greenhouse effect (these estimates are indicated in Table 5.6 and Figure 5.6 in parentheses). According to the radiation and photochemical model of Wuebbles et al. (1983), strato-

Table 5.6. Computed Model Estimates (%) of Changes in the Mean Annual Ozone Content in the Northern Hemisphere versus the 1980 Total Ozone X_3 and to Ozone Density Δ (O_3) at Various Atmospheric Levels in 1985–2030 (in parentheses: values obtained without consideration for temperature variations)

Value	Level km	1985	1990	1995	2000	2010	2030
$\Delta X_3/X_3$		0.5(1.1) 0.0–1.0	0.7(1.8) 0.2–1.2	1.0(3.2) 0.0–2.0	1.4(4.8) −0.1–2.6	2.3(9.2) −0.2–5.8	5.8(14.1) −1.0–12.6
$\Delta(O_3)_0$	0	5.0(6.7) 0.5–9.0	6.1(9.6) 0.9–8.2	8.8(15.4) 1.1–16	11.6(22.3) 1.8–24	26.5(38.6) 3.3–51	53(92) 5–243
$\Delta(O_3)_{20}$	20	−1.5(−1.7) −1.9–−1.4	−0.9(−1.3) −1.2–−0.1	−0.2(−0.7) −1.0–0.9	0.4(0.0) −1.2–1.8	2.5(0.9) −1.8–6.2	5.8(2.3) −3.1–10.3
$\Delta(O_3)_{40}$	40	3.3(−1.2) 1.3–2.1	−1.0(−4.9) 0.1–1.2	−6.9(−13) −0.4–−8	−8(−25) −0.9–−12	−20(−61) −4.0–−32	−30(−82) −12–−74
	Estimates made by Wuebbles, Luther, Penner (1983) compared to the 1911 level						
$\Delta X_3/X_3$		0.66(−0.9)	0.76(−1.1)	0.65(−1.50)	0.55(−1.8)	0.41(−2.4)	0.23(−3.5)
$\Delta(O_3)_{20}$	20	1.1	1.3	2.0	2.5	3.3	6.8
$\Delta(O_3)_{40}$	40	−7.4(−9.9)	−9.5(−12.7)	−11.0(−13.4)	−12.3(−14.1)	−14.6(−16.1)	−17.9(−20.0)

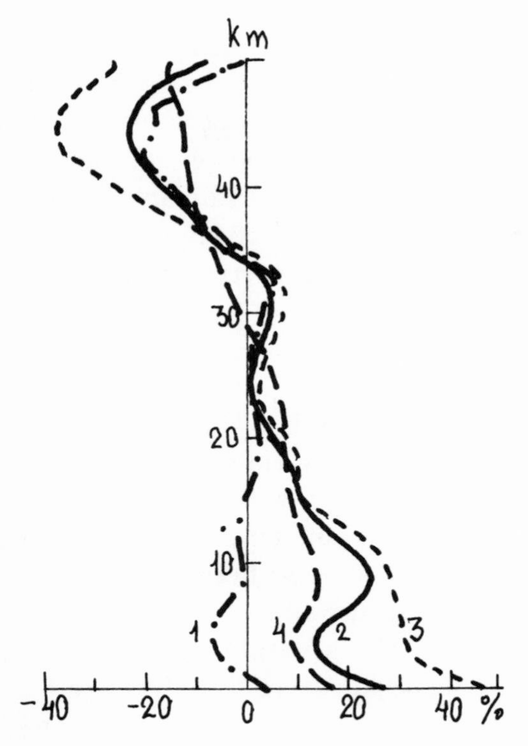

Figure 5.6. Vertical profiles of altered mean annual ozone density in 2000 versus 1980 from photochemical model calculations. 1, Lower boundary; 2, mean; 3, upper boundary; 4, mean from radiation and photochemical model calculations.

spheric cooling at the 40 km level in the years 2000 and 2030, compared to 1980, is 2.6 and 5.9°C (1.2 and 2.3°C without the effect of higher CO_2 content). The radiation and photochemical model of Karol' et al. (1986b) yields 3.2°C in 2000 and 8.4°C in 2030. Stratospheric cooling reduces the rate of ozone chemical destruction by nitric and chlorine oxides, and partially lowers the rate of O_3 decrease there. The greenhouse effect in the troposphere and lower stratosphere is accompanied with smaller temperature variations (according to our estimates, in the troposphere by the year 2000 $\Delta T = 0.7$°C (0.4–1.0°C).) This leads to a smaller difference between the estimates of $\Delta O_3/O_3$ at the Earth's surface and at the 20 km level made by using radiative-photochemical and photochemical models, i.e., with and without temperature variations taken into account. Since ozone density is higher in the lower atmosphere, the increased ozone content at the 20 km level and in the troposphere almost compensates for the relatively greater decline in ozone mixing ratio in the upper stratosphere and preserves the total ozone content.

The mixing ratio of total odd chlorine (Cl_y) [(Cl) + (ClO) + (HCl) + (HOCl) + ($ClONO_2$)] in the stratosphere by the years 2000 and 2030 will reach 2.7 and 7.6 ppm, respectively, for the main scenario variant, and for the maximum scenario variant in 2030 it will reach 12 ppm, because of greater chlorofluorocarbon emissions into the atmosphere. The higher (Cl_y) stratospheric content by almost an order compared to the current level causes a drop in ozone concentration several times within its peak mixing ratio layer (above the 36–38 km level) (see Table 5.6). However, this decrease has a small effect on the total ozone content X_3, consequently, there is no sharp drop nor significant decrease in ozone density in its peak layer (between 20–30 km levels), that is, the nonlinear effect obtained by Cicerone et al. (1983). As shown by Herman and McQuillan (1985), this nonlinear effect, simulated by a photochemical model which neglects the temperature change influence on the photochemical reaction rates is highly exaggerated because of inaccuracies in the photochemical scheme. Cooling of the stratosphere also decelerates ozone destruction by nitrogen and chlorine catalysts, since the rates of corresponding reactions are very dependent on temperature. Comparison of results calculated by the photochemical and radiation models reveals thermal moderation of the decline in stratospheric ozone concentration and its tropospheric increase, caused by the greenhouse effect, a negative feedback revealed by the radiative and photochemical model.

The $(O_3)_o$ level in the surface air layer rises 1.2% annually, peaking at 2%, depending more on the growth rate of NO_x and less on CH_4 and CO tropospheric growth rates. A higher $(O_3)_o$ level can noticeably intensify the toxic effect of ozone on the biosphere, especially in the tropical and subtropical regions, where the background ozone content is high in the surface air layer, and can aggravate the smog in major cities.

Assessing the impact of thermal inertia of the ocean. In order to assess the possible changes in climatic characteristics caused by magnified global anthropogenic impact, we primarily have to take into account thermal inertia of the ocean, that absorbs a portion of the heat formed in the lower atmosphere as a result of the greenhouse effect, and thus reduces and moderates temperature increment ΔT_S. Methods for estimating this deceleration Δt or the extent of lower temperature increment $\Delta(\Delta T_S) = \Delta^2 T_S$ are given in Section 6.2 of this monograph.

Chapter 6 demonstrates that the ocean is thermally divided into upper quasihomogeneous 100–200 m layer, thermocline layer down to depth 1–1.5 km, and deep layers. The upper layer temperature is close to that of the oceanic surface, while the deep oceanic temperature is several degrees centigrade at all latitudes. Heat therefore flows, albeit slowly, into the deep ocean via the thermocline, since oceanic stratification is stable by and large: cold water is at the bottom and warm is above. It is reasonable to use for the evaluation a simplified model of oceanic thermal inertia input to the greenhouse effect in the ocean-atmosphere system, proposed by Byutner (1983). In this model the deviation $\Delta T = T - T_0$ in temperature of the upper quasihomogeneous oceanic layer from its initial value T_0 because of the greenhouse effect is determined by the thermal balance equation of the upper quasihomogeneous oceanic layer:

$$hcd\Delta T/dt = \Delta P - B_1 \Delta T; \quad B_1 = const. \tag{5.1}$$

where t is time, c is specific heat of water, h is depth of the upper quasihomogeneous layer, ΔP is heat influx from the atmosphere, B_1 is the parameter of mean annual rate of heat efflux from the upper quasihomogeneous layer into the deep ocean. The following expression is proposed to determine ΔP:

$$\Delta P = \Delta F(t) - B_0 \Delta T, \tag{5.2}$$

where $\Delta F(t)$ is the change in radiation heat influx to the ocean surface, B_0 is the parameter of sensitivity of the model radiation temperature to the change in radiation mode, associated with this effect; B_0 is defined, for instance, by the formula:

$$B_0 = \Delta F / \Delta T_R \approx \partial F / \partial T_R + S_0 \partial \alpha / 4 \partial T_R, \tag{5.3}$$

derived by linearization of the known radiation equilibrium equation of the Earth-atmosphere system $F = S_0(1 - \alpha)/4$. Here, F is longwave radiation flux leaving the system, S_0 is the solar constant and α is planetary albedo. It is also assumed that ΔT in the upper quasihomogeneous layer and in the sea surface air layer ΔT_S are the same.. Actually, $\Delta T_S < \Delta T_R$ and $\Delta P > 0$. After substituting all these relationships into equation (5.1), its solution with zero initial condition: $\Delta T_S = \Delta F = 0$ and t $= 0$ is

$$\Delta T_S(t) = e^{-\gamma t} \int_0^t \Delta F(t') e^{\gamma t'} dt'; \gamma = (B_0 + B_1)/c. \tag{5.4}$$

The rise in ΔF is directly proportional to the content of optically thin absorbing layers in the air column discussed in Section 5.1, and using the exponential approximation of ΔF growth with time,

$$\Delta F = a(e^{bt} - 1); \quad a, \quad b = \text{const}, \tag{5.5}$$

we arrive at the following expression for the difference in the temperature increment $\Delta^2 T_S$ taking formula (5.3) into account

$$\Delta^2 T_S = \Delta T_S - \Delta T_R = -\delta \Delta T_R - \epsilon \Delta T_{R2}. \tag{5.6}$$

Here

$$\delta = 1 - \frac{B_0}{hc(b+\gamma)} = \frac{b + B_1/hc}{b+\gamma};$$
$$\epsilon = \frac{b(1-e^{-\gamma t})}{(b+\gamma)(1+B_1/B_0)}. \tag{5.7}$$

$\Delta T_{R2} = a/B_0$ is the variation in radiation surface temperature, resulting from doubling of atmospheric trace gas content ($e^{bt} = 2$). In addition to $\Delta^2 T_S$, it is also important to assess the lag time Δt in temperature increment ΔT_S from the radiation temperature increment ΔT_R. Assuming, as in (5.5)

$$\Delta T_S = \Delta T_{R2}(e^{b(t-\Delta t)} - 1) \tag{5.8}$$

we find for $T >> \gamma^{-1}$:

$$e^{b(t-\Delta t)} = (1-\delta)e^{bt} + \delta - \epsilon, \tag{5.9}$$

from which, in the case of $B_1 = 0$, $\delta = \epsilon$ it follows:

$$\Delta t = b^{-1} \ln 1/(1-\delta). \tag{5.10}$$

The same formula is approximately correct for t in which

$$e^{bt} >> \delta - \epsilon = 0.2.$$

Numerical values of parameters for the mean annual mode of the upper oceanic layer are: $ch = 2.1 \times 10^8$ J/(m$^2 \cdot$°C) for the Northern Hemisphere and $3.14 \cdot 10^8$ J/(m$^2 \cdot$°C) for the Southern Hemisphere, which corresponds to mean thicknesses of the upper layer of 51 and 77 m. $B_1 =$

0.2–0.6 W/($m^2 \cdot °C$), but it is possible as a limiting case, that $B_1 = 0$, i.e., heat exchange between the upper quasihomogeneous layer and the deep ocean stops if static stability of the upper oceanic layers increases as they are warmed (Byutner, 1983). This value is used below for the radiation and convective model in this work $B_0 = 1.6$ W/($m^2 \times °C$).

Values of δ and Δt for b = 0.004, 0.02 and 0.06 correspond approximately to mean increment in CO_2 , N_2O , CH_4 , O_3 and chlorofluorocarbon contents in the Northern Hemispheric troposphere.

The magnitude of heat transfer rate B_1 from the upper quasihomogeneous layer into the deep oceanic layers has a significant effect on $\Delta^2 T_S$ and Δt estimates. In the upper layer, thermally insulated from the deep ocean, Δt is approximately 4 years, regardless of the b values and equals the relaxation time γ^{-1} of the initial condition in (5.4) for the Northern Hemisphere. When $B_1 = 0.4$ W/($m^2 \cdot °C$), Δt ranges from 6.8 years for $b = 0.6$ for chlorofluorocarbons to 10 years for $b = 0.004$ yr^{-1}. In the Southern Hemisphere, for $h = 80$ m and for $b = 0.004$ year^{-1}, Δt is almost double that in the Northern Hemisphere. The magnitude of ΔT_S, relative to radiative ΔT_R determined only by the atmosphere, decreases more rapidly when b and B_1 are larger, i.e., to a larger extent for more rapidly growing greenhouse effect and for a higher rate of heat exchange between the upper layer and the deep ocean.

Assessing expected temperature variations. Model estimates of variations in thermal atmospheric conditions, that are caused by the aforementioned expected rise in trace gas content in the atmosphere up to the year 2030, are the basis for climatic change forecasting. The estimated limits for expected increment in radiation active trace gas content in the atmosphere (see Tables 5.5 and 5.6) are used to obtain the lower and upper limits of expected 10–year changes in mean annual temperature ΔT_S of the surface air layer in the Northern Hemisphere from 1970 to 2000 and for the year 2030.

The estimated rise in "radiation" temperature ΔT_R for the decade 1970–1980 was obtained earlier from the actual measured increase in gas content (except O_3) in the middle latitudes of the Northern Hemisphere (Lacis et al., 1981). These values are presented in Table 5.7 along with corresponding ΔT_S limits, taking into account thermal inertia of the surface and deep layers of the ocean. For the decades from 1980 to 2000 and for the year 2030, Tables 5.8–5.9 indicate the limits of expected altered gas content in the troposphere and the corresponding ΔT_R and ΔT_S for each gas and for the sum, including O_3. Estimates for this were taken from Table 5.6.

Table 5.7. Variations in Trace Gas Mixing Ratio in 1970–1980 in the Lower Troposphere of the Northern Hemisphere and Corresponding Increases in the Surface Air Temperature

	Mixing Ratio, ppb		Increment		
Gas	1970	1970–1980	%/yr	ΔT_R	ΔT_S
O_3	30	0–5	0–1.6	0–0.10	0–0.05
CH_4	1500	150	1, 0	0.032	0.029–0.030
N_2O	295	6	0.2	0.016	0.004–0.005
F–11, F–12	0.170	0.325	11	0.054	0.039–0.041
TOTAL:				0.10–0.20	0.07–0.13

The last columns of Tables 5.7–5.9 contain upper and lower limits of the surface air temperature increment, calculated by formulas (5.6)–(5.7) with regard for thermal inertia of the ocean and the magnitude of b, the parameter of

Table 5.8. Expected Variations in Trace Gas Mixing Ratios in 1980–1990 in the Lower Troposphere of the Northern Hemisphere and Corresponding Increments in Surface Air Temperature

	Mixing Ratio, ppb		Increment		
Gas	1980	1980–1990	%/yr	ΔT_R	ΔT_S
O_3	35	0–7	0–2.0	0–0.20	0–0.13
CH_4	1650	80–200	0.5–1.2	0.02–0.05	0.003–0.03
NO_2	300	6–25	0.2–0.8	0.016–0.067	0.001–0.05
Chlorofluoro–carbons	0.76	0.35–0.50	5–7	0.05–0.11	0.04–0.07
TOTAL:				0.09–0.43	0.04–0.28

gas content growth. The lower limits were evaluated with $B_1 = 0.4$ W/(m^2·°C) for the rate of heat transfer from the upper quasihomogeneous layer into the deep ocean, when $\Delta^2 T_S$ is the maximum, and the upper limits with $B_1 = 0$, i.e., the upper layer is thermally insulated, and $\Delta^2 T_S$ is the minimum. Two scenarios are considered in Tables 5.7 and 5.8 for tropospheric ozone: constant ozone concentration during the 1970's, recorded by several measurement systems, for example, by the US system of GMCC, and higher ozone content during the 1970's–1980's, rising at a mean annual rate of about 2%, noted in Geophys. Monitoring (1984–1986); Logan (1985); Feister and Warmbt (1985).

As seen from Table 5.9, the 1970–2030 mean annual temperature increment due to trace gases, averaged over the extratropical zone of the Northern Hemisphere ΔT_S is estimated to be 1.3°C (0.7–2°C); approximately 30% of this increase occurred in the twentieth century. The higher temperature is induced by the greenhouse effect, mainly because of two gases, chlorofluorocarbons and tropospheric ozone. The N_2O tropospheric content will increase during this period by 30–40%, chlorofluorocarbon content 15–50

Table 5.9. Variations in Mean Annual Mixing Ratio of Trace Gases in 1970–2000 and in 2000–2030 in the Lower Troposphere of the Northern Hemisphere and Corresponding Increases in Surface Air Temperature

	Mixing Ratio, ppb		ΔT_R	
Gas	1970–2000	2000–2030	1970–2000	2000–2030
O_3	5–16	5–35	0.15(0.10–0.31)	0.34(0.10–0.67)
CH_4	270–500	600–1600	0.08(0.06–0.11)	0.25(0.12–0.37)
N_2O	20–75	30–60	0.12(0.06–0.19)	0.12(0.07–0.18)
Chlorofluoro-carbons	1.3–2.0	2.1–10.0	0.20(0.16–0.26)	0.47(0.28–1.23)
TOTAL:			0.55(0.38–0.87)	1.18(0.57–2.45)

	Mixing Ratio, ppb		ΔT_S	
Gas	1970–2000	2000–2030	1970–2000	2000–2030
O_3	5–16	5–35	0.11(0.07–0.25)	0.27(0.07–0.57)
CH_4	270–500	600–1600	0.06(0.04–0.08)	0.20(0.10–0.31)
N_2O	20–75	30–60	0.11(0.04–0.18)	0.11(0.06–0.17)
Chlorofluoro-carbons	1.3–2.0	2.1–10.0	0.12(0.10–0.17)	0.33(0.21–1.00)
TOTAL:			0.40(0.25–0.68)	0.91(0.44–2.05)

Note: Value limits are indicated in parentheses.

times, and the content of all other considered gases (except for N_2O) will approximately double compared to the 1970 levels. The wide range in expected tropospheric ozone, a result of insufficient knowledge of its cycle should be emphasized. Comparison of estimated ΔT_S from Tables 5.7–5.9 with similar evaluations of higher trace gas content made by Ramanathan et al. (1986) shows that the latter are 1.6 times lower than the mean for 1970–2000, and 1.9 times lower for the period 2000–2030, but are close to the lower limits of the tabular estimates. The reason for this discrepancy is that the $(\Delta T_S)_0$ values estimated by Ramanathan et al. were determined by a one-dimensional model without any positive feedbacks, which increase ΔT_S by no less than 1.5 times, according to estimates made by the authors. In Ramanathan's scenario, the growth rate of chlorofluorocarbon atmospheric content is also 3% per year, which is approximately half of the current level and is adopted as the mean estimates in Tables 5.7–5.9. Estimating model computation precision of the trace gas impact on climate is rather complicated and has been poorly investigated.

These forecasting estimates show that it is possible to apply radiation energy balance and radiative photochemical models in analyzing trace gas impact on climate. However, the tabular values are not conclusive, because the external perturbations incorporated into the model are inaccurate, and the actual models are erroneous. The accuracy required for computing changes in thermal conditions and trace gas content (0.01°C and some percentage of the mixing ratio), that was indicated in Section 5.2, is only a part of the total error in estimated changes in climatic characteristics. It is rather difficult to determine the total error of these computed estimates.

A reliable forecast of growth trends in atmospheric trace gas content, in addition to possible errors in computing the dependence of thermal conditions on the trace gas content is very important for accurate estimation of future climatic changes. This problem is considered in Chapter 11.

CHAPTER 6

SIMULATION OF CLIMATIC SENSITIVITY TO ATMOSPHERIC GAS COMPOSITION CHANGES

6.1. Modeling Climatic Changes

Introduction. The climatic system, which includes the atmosphere, the ocean, the land surface with biota, and the cryosphere, is so complex that any description will be merely an approximation. We therefore speak only of the climatic theory models. They have been developed in the last 20 years and are divided into classes, depending on the detailed spatial and time description of the system and the number of the processes used in this description.

In order to estimate climatic change using these models, we must first assess the altered radiation fluxes within the system. Variations are found in the fraction of solar energy absorbed by the system (for example, volcanic aerosol can increase albedo of the system, and thus decrease this fraction) and in the ratio between the portion of solar energy absorbed by the atmosphere and reaching the Earth's surface. The climate changes when an asteroid falls to Earth and raises large amounts of dust that absorb solar radiation; when there are dust storms on Mars, or if a large-scale nuclear conflict should contaminate the atmosphere by fires (Budyko, Golitsyn and Izrael', 1986). In all these cases, if solar radiation is absorbed high enough above the surface, the upper atmospheric layers are heated, while the Earth's surface is cooled.

The conditions of thermal radiation absorption at the Earth's surface and in the atmosphere itself can also vary, i.e., this radiation will escape into space from the higher atmospheric layers because of higher concentration of atmospheric gas components that absorb thermal radiation. The so-called greenhouse effect (Kondratiyev and Moskalenko, 1984) would then increase and the Earth's surface would be heated. When the concentration of such absorbing substances decreases, the temperature trend is reversed, and the Earth's surface becomes cooler.

Solar luminosity can vary as well. However, for the 100-year time scale, this value may vary at about 0.1% or less (Bolin et al., 1986). Such variations will alter mean global surface air temperature $\overline{T}_S$ approximately 0.1–0.15°C (or less), since a 1% change in solar constant results in 1–1.5°C change in $\overline{T}_S$ (Budyko, 1980; and this chapter).

Characteristics of the Earth's orbit around the Sun vary under the influence of other planets over thousands of years. This leads to seasonal and latitudinal redistribution of the solar radiation that enters the upper atmosphere. A decrease in the total solar radiation in summer in high latitudes of the Northern Hemisphere (it increases in winter), given the current position of the continents and oceans, promotes glaciation (North et al., 1983; Berger et al. (eds.), 1984). This is now considered the major trigger for the onset of the Quaternary ice ages. They are also usually accompanied by lower atmospheric CO_2 concentration (Lorius et al., 1985). The mechanisms that reduce CO_2 concentration to 180–200 ppm during peak glaciation are still obscure, but they are probably related to oceanic processes.

For the hundred-year time interval of possible climate scenarios, all the indicated climate-changing factors can be considered external to the climatic system. Thus, particular estimates of solar and thermal radiation fluxes in the atmosphere and their changes are the initial basis for all climatic theory models. Current calculations of radiation fluxes based on the main absorption bands for atmospheric gases, their possible overlapping, and other effects are fairly accurate (Radiation Algorithms ..., 1983; Kondrat'yev, 1980; Ramanathan et al., 1985). However, errors in these calculations introduce some uncertainties into the estimates of future climate.

All the models require the inclusion of changes in solar thermal radiation, even when a comprehensive spatial and temporal description of the system and a detailed consideration of physical processes allow the hierarchy of climatic models to be used. A detailed spatial and temporal description of climate and its changes due to variations in certain external factors requires the use of complete three-dimensional models of general atmosphere-ocean circulation, which take into account both interface and land processes. Integration of the equations of such models for a lengthy period is a time-consuming process even for powerful computers. Simpler climatic models have hitherto been used, therefore, to study the role of physical processes in the climatic system, to estimate certain parametrizations of physical processes, and to analyze the results of more complicated models or observational data. We will now briefly describe these models.

Globally averaged climatic model. In its simplest version, the relationship of balance between incoming solar and outgoing thermal radiation is the main equation of this model:

$$\pi r^2 S_0(1-\alpha) = F\uparrow = 4\pi\sigma T_e^4 r^2, \tag{6.1}$$

where r is the planet's radius (here, the Earth), $S_0 = 1372 \pm 4\ \mathrm{W/m^2}$ is the solar constant value (the indefiniteness reflects differences in this measured value over the last several years), α is the reflectivity coefficient (albedo) equal to 0.30 ± 0.01, and $\sigma = 5.67 \cdot 10^{-8}\ \mathrm{W/(m^2 \times K^4)}$ is the constant in the Stefan-Boltzmann radiation law. We note that in (6.1) solar radiation arrives at the area of a circle with r radius, while thermal radiation $F\uparrow$ escapes from the entire area of the sphere $4\pi r^2$.

The known temperature scale is derived from (6.1):

$$T_e = [S_0(1-\alpha)/4\sigma]^{1/4}, \tag{6.2}$$

where T_e is the effective temperature of outgoing thermal radiation equal to 255 K (or -18°C) for the Earth. The mean Earth's surface temperature T_S is close to 14°C. The 32°C difference between these temperatures is the result of the so-called greenhouse effect, meaning that the atmosphere is comparatively transparent for most of the solar radiation spectrum, but is a good absorber of thermal emission of the surface and lower tropospheric layers. The main absorbers in order of their importance are water vapor, CO_2 , and, to a far lesser degree, methane, CFCs, ozone, and nitrous oxide (N_2O). Higher concentration of any of these gases raises the altitude from which the radiation escapes the atmosphere into space. The vertical temperature gradient varies slightly when the temperature changes (Mokhov, 1983), which leads to higher temperature of the Earth's surface and the troposphere.

Budyko (1968) was the first to notice that a linear equation is a good approximation of the dependence of the intensity of outgoing thermal radiation on surface air temperature (which is of prime importance for us):

$$F \uparrow = A + b_0 T_S, \tag{6.3}$$

where A and b_0 are dimensional constants. Equation (6.3) describes the observational data well, because the water vapor distribution over latitude is inhomogeneous as the temperature drops while latitude rises. Since its nature is related to water vapor (Mokhov and Petukhov, 1978), relationship (6.3) implicitly takes into account one of the major mechanisms of positive feedback in the Earth's climate: increases in atmospheric water content when temperature rises. The coefficients in (6.3) have been repeatedly derived by using satellite measurements of radiation balance components and surface temperature (e.g., North et al., 1981). Agayan, Golitsyn and Mokhov (1985) used the most complete and diverse data to find A = 206 W/m^2, b_0 = 1.8 W/(m$^2\times$°C).

By changing the radiation balance at the upper troposphere, we find from (6.3)

$$\Delta F \uparrow = b_0 \Delta T_S. \tag{6.4}$$

Numerous calculations show that doubling the CO_2 atmospheric concentration alters the outgoing thermal radiation at the upper troposphere $\Delta F \uparrow$ by 4.2 W/m^2 (Dickinson, 1986; see Bolin et al., 1986). We can easily determine surface temperature changes in this case $\Delta T_S = \Delta F \uparrow / b_0 =$ 2.3°C. It is further seen that this estimate is in the range of the current spread of estimated climatic sensitivity with double atmospheric CO_2 concentration, namely, in the lower half of this dispersion interval. This is easily understood, since the estimate did not take into account any feedbacks in our climatic system (except for water vapor). It is sometimes stated that a 2% solar constant increment is approximately equivalent to a double CO_2 atmospheric concentration. This can be quantitatively confirmed by the simplest radiation balance model used here. If solar energy influx actually changes, the system's albedo does not vary, and the mean energy influx increment is $1372 \cdot 0.7 \cdot 2 \cdot 10^{-2}/4 = 4.8$ W/m^2. Since $\Delta S = \Delta F \uparrow$, then we obtain from (6.4) that in this case the surface air temperature will rise by 4.8 W/m^2: 1.8 W/(m^2×°C) = 2.7°C. The CO_2 doubling thus yields a slightly lower effect (by 12.5%), than a 20% solar constant increase.

One of the important feedbacks is the dependence of the Earth's surface albedo on temperature T_S (Budyko, 1968). Dickinson used different model calculations and observational data (Lian and Cess, 1977; Wetherald and Manabe, 1981; Washington and Meehl, 1983, 1984; Robock, 1983; Hansen et al., 1984; Spelman and Manabe, 1984) to estimate the intensity of this relationship at -0.4 ± 0.15 (W/(m^2×°C)). The addition of this term to b_0 increases ΔT_C up to 3 ± 0.3°C (in contrast to Dickinson, see Bolin et al., 1986). We took only one standard deviation in the spread of estimated albedo-temperature feedback, since two standard deviations formally accepted by Dickinson, bring ΔT_c beyond the ordinary range with a relatively small number of estimates.

Consideration of only one positive feedback thus places us in the middle of the range $\Delta T_c = 1.5-4.5°C$ adopted in a number of surveys. Calculation of other positive feedbacks contributes to further increases in ΔT_c, however, since there are also negative feedbacks limiting this increment, we conclude this discussion by mentioning the studies by Mokhov (1981), Hansen et al. (1984), and Schlesinger et al. (1985). They assess climatic sensitivity to double CO_2 atmospheric content, including the case where several feedback mechanisms exert their effects simultaneously.

The globally averaged climate model expressed by equation (6.3) can be re-written so that it takes into account the transient effects. This summarization can be simply expressed as

$$C\frac{\partial \Delta T_S}{\partial t} + b_0 \Delta T_S = \Delta F \uparrow, \qquad (6.5)$$

where C is the thermal inertia parameter, for example the specific heat of the upper mixed oceanic layer. Equation (6.5) was used by Golitsyn and Demchenko (1980) to evaluate fluctuations in temperature and snow-ice cover area with the given radiation balance variations measured from satellites. It was found that radiation balance fluctuations correctly explain the root-mean-square values of mean hemispheric temperature fluctuations and interannual variability of snow-ice cover area. Later, Demchenko (1981, 1982, 1984) successfully used equation (6.5) to describe latitudinal variations in temperature dispersion.

Equation (6.5) can also take into account heat transfer from the quasihomogeneous layer into the deep ocean, if the diffusion term $K\partial^2 \Delta T/\partial Z^2$ is included in this equation.

According to different estimates, K is close to 2 $cm^2/sec.$ (Schlesinger et al., 1985). In this statement of the problem, equation (6.5) is solved for the upper oceanic layer, and equation

$$\frac{\partial \Delta T}{\partial t} = K \frac{\partial^2 \Delta T}{\partial Z^2} \tag{6.6}$$

for the deep ocean if temperature at the upper boundary determined from (6.5) is specified. Analytical solution to this problem is rather cumbersome and is omitted (see Wigley and Schlesinger, 1985).

The authors used this solution to obtain ΔT_S for the 1850–1980 period depending on CO_2 atmospheric content in 1850 (P_a^0) for a wide range of ΔT_C values with two values of K: 1 $cm^2/sec.$ and 3 $cm^2/sec.$

If P_a^0 is fixed at the level of 280 ppm (Chapter 3), then for ΔT_c ranging from 1.5°C to 4.5°C, the value of ΔT_S, according to Wigley and Schlesinger, lies within 0.32–0.62°C, when $K = 1$ cm^2/sec, and within 0.26–0.46°C, when $K = 3$ $cm^2/sec.$

It will be seen that the value $K = 1$ cm^2/sec agrees better with empirical data on heat exchange and temperature distribution within the ocean. The role of oceanic thermal inertia is treated more thoroughly in Section 6.2.

Energy-balance climatic models. The balance between incoming solar radiation and outgoing longwave radiation can be obtained for each latitudinal zone. However, this balance is not fulfilled for an individual latitude, even if vertical convection is taken into account. The Earth is heated much more in low than in high latitudes because of its spherical shape. This variance in heating different latitudes leads to fluxes that transfer sensible and latent (water vapor) heat, from the tropics to the poles. Oceanic currents transfer nearly as much heat as the atmosphere. These processes of energy transfer are described by energy-balance models in a simple manner by different means. The simplest way to describe them was first suggested by Budyko (1968).

According to Budyko's technique, the term $\gamma[T(\phi) - \overline{T}]$, where ϕ is latitude and $\overline{T}$ is the mean global (or hemispheric) temperature, is introduced into an equation similar to (6.1), but derived for each latitudinal belt. The proportionality coefficient γ is of the same dimension as b_0, i.e., $W/(m^2 \times {}^\circ C)$, and is selected empirically based on the best fulfillment of energy balance for each latitudinal belt. Mokhov (1981) applied this technique and used satellite data on energy balance at the upper atmospheric boundary to find $\gamma = 3.8\ W/(m^2 \times {}^\circ C)$ for the Northern Hemisphere. Another way to represent this transfer (North, 1975) is to assume that heat flux is proportional to $D\partial T/\partial y$, where D is the diffusion coefficient. Heat flux divergence will then be $(\partial/\partial y)(D\partial T/\partial y)$ and this term should be introduced into the energy-balance equation. North (1975) suggested solving the resulting equation by the Legendre polynomial expansion (eigenfunction of the Laplace operator on a sphere). He showed that if the expansion is limited to the first two polynomials, then both descriptions become equivalent. In this case, $D = \gamma/6b_0$ and taking into account γ obtained by Mokhov, $D = 0.34$.

The most important feedback considered in climatic energy-balance models is the dependence of the system's albedo on the snow and ice cover area. Approximately half of the solar radiation reaching the top of the atmosphere is absorbed by the Earth's surface, and about 20% by the atmosphere (Liou, 1984). Therefore, the Earth's surface albedo is an important parameter in determining the actual quantity of absorbed solar radiation. The surface albedo α_s can vary from 0.01 for calm water surface, when the Sun is at the zenith, to 0.95 for fresh snow when Sun's rays are inclined (the solar zenith angle is an important parameter in determining the value of α_S). The snow- and ice-covered areas on the whole have far greater values of α_S than those without snow and ice. The larger the area covered by snow and ice, the less solar radiation is absorbed by the surface and the lower their temperature, and vice versa. Simple energy-balance models incorporate this by introducing α_S dependent on temperature. Budyko (1968), and then Held and Suarez (1974) believe that albedo α_S is a discontinuous function of temperature, assuming $\alpha_S = 0.6$ at $T_S \leq -10°C$ and $\alpha_S = 0.2$ at larger values of T_S. Sellers (1969) considers albedo α_S to be a diminishing function of temperature with some upper and lower limits for the areas completely covered with snow and ice, and those free of them. This positive feedback makes the climatic system unstable relative, for instance, to a significantly lower solar constant. A drop by several percentages can result in total glaciation of the Earth (Budyko, 1968; Held and Suarez, 1974; North, 1975).

We have already estimated the intensity of this feedback. For seasonal temperature variation models (including some general circulation models, discussed below), it is appropriate to determine this feedback only for the mean annual conditions. Mokhov (1981) noted that in this case the seasonal albedo fluctuations can lag behind the temperature variations. This decreases the feedback intensity and lowers the sensitivity of climatic models that incorporate seasonal variations compared to the models for mean annual conditions, as mentioned in a number of calculations (Manabe and Stouffer, 1980).

Radiation-convective models (RCM). These models calculate the temperature change with height Z, whereas in the horizontal direction, the temperature is considered to be globally-averaged. In this case, we are able to thoroughly investigate the role of clouds, water vapor, the stratosphere, and corresponding feedbacks. Stratospheric temperature is determined to a considerable extent by the local radiation balance condition between heating due to the absorption of solar radiation by ozone and cooling because of longwave emission (Fels et al., 1980). Absorption of thermal radiation from the Earth's surface by water vapor, carbon dioxide and ozone is important in the lower troposphere.

The tropospheric profile $T(Z)$ depends on the transfer of heat by dry and wet convection, rather than on radiation balance. Beginning with the first radiation-convective model by Manabe and Wetherald (1967), convective transfer has been parametrized by restricting the smallest vertical temperature gradient to the value $dT/dZ = -6.5°C/km$, i.e., if the model predicts a smaller gradient, then it is considered to be equal to $-6.5°C/km$. This value of the critical gradient is used in many studies (e.g., Karol', Rozanov, 1982; Turco et al., 1983) to calculate climatic changes caused by smoke from nuclear war fires. It should be mentioned that a more detailed analysis shows that the mean vertical temperature gradient in the Earth's atmosphere is a decreasing function of surface air temperature T_S and varies from -6.5 to -5.0°C/km (Mokhov, 1983). Attempts have been made to use the wet adiabatic gradient (Rowntree and Walker, 1978). Radiation-convective models are reviewed by Karol' and Frol'kis (1984) (see also Ramanathan and Coackley, 1978).

When these models are used, positive feedback on atmospheric water vapor content has to be considered in explicit form. This was already done in the first radiation-convective model by Manabe and Wetherald. These authors initially examined changes in T_S at a fixed water vapor concentration and found that ΔT_S corresponds to the value of $b_0 = 3.7 \text{ W}/(\text{m}^2 \times °\text{C})$ in equation (6.4). However, according to the Clapeyron-Clausius equation, as temperature rises, the pressure of saturated vapor increases, and there will be more water vapor in a warmer atmosphere. It will therefore be more realistic to assume that relative humidity is fixed while temperature changes. In this case, Manabe and Wetherald found that $b_0 = 2.2 \text{ W}/(\text{m}^2 \times °\text{C})$. The difference of $1.5 \text{ W}/(\text{m}^2 \times °\text{C})$ characterizes the intensity of positive feedback, since a greater amount of water vapor absorbs outgoing heat radiation more strongly, and thus enhances the greenhouse effect.

However, the relative humidity field also varies during the actual climatic changes due to the three-dimensional atmospheric dynamics. This is important for estimating variability of the cloud cover that influences the atmospheric radiation balance. The following characteristics of clouds are very variable: their spatial distribution, optical thickness, height of their upper boundary, water content and sizes of droplets. All these properties of clouds affect both solar radiation and outgoing thermal radiation. Studies using radiation-convective models have revealed both positive and negative feedbacks that appear due to radiation and cloud cover interactions (Schneider and Dickinson, 1974; Ramanathan and Coakley, 1978). Golitsyn and Mokhov (1978) used the energy-balance model to study how variations in ceratin cloud properties can influence climate.

There are data showing that on the average for the Earth as a whole, increasing cloud cover causes not only an increment in the climatic system albedo, but also in longwave radiation absorption. As a result, the influence of these opposite factors on the T_S value is almost entirely compensated for (Budyko, 1975; Cess, 1976; Wetherald and Manabe, 1980; Cess et al., 1982). This compensation does not occur on a regional scale, however (Manabe and Wetherald, 1980; Meleshko, 1980). This question should therefore not be considered definitely resolved. One program in the World Climate Research Program (Scientific Plan..., 1984), The International Project on Satellite Cloud Climatology (ISCCP), was organized in 1983 to address this problem.

There are climatic models that describe zonal temperature distribution by using an energy-balance model and vertical temperature distribution in a radiation-convective approximation (Karol' and Frol'kis, 1984).

Three-dimensional climatic theory models. Three-dimensional climatic theory models usually imply general atmospheric models with corresponding boundary conditions at the Earth's surface. Since the ocean occupies 71% of the Earth's surface area and has enormous thermal inertia, a description of the atmosphere-ocean interaction, in particular, in time scales of ten years and more, is a necessary element of climatic theory. Climatic models that explicitly describe heat transfer by ocean currents have appeared just recently (Bryan et al., 1982; Spelman and Manabe, 1984).

The models of atmospheric general circulation describe spatial and temporal changes in the atmospheric wind field, incorporate the transport equation for thermal energy, water vapor and water droplet amount with regard for phase conversions and treat precipitation. The thermal transport equation considers the vertical transport of solar and longwave radiation (as well as their absorption and scattering), changes in heat during phase water conversions, thermal redistribution by air motion, parametrization of dry and wet convection or turbulent redistribution of thermal energy, and moisture as well. The equation for water vapor takes into account the source of evaporation and evapotranspiration from the surface (and sometimes evaporation of water cloud particles), while the equation for water droplet amount makes parameters for the process of its formation from water vapor and precipitation as rain or snow.

At the lower boundary, parameters for fluxes in momentum, heat and moisture are found with the help of turbulent boundary layer models taking into account stratification (the most advanced models, including the model of the European Center for Medium Range Weather Forecast, use the parametrization based on the Monin-Obukhov theory (1954; see also Yaglom, 1977)). Atmospheric models still have much more detail than the formation of sea ice as well as various processes on the land surface, including evolution of the snow cover (its growth and decrease over time, and corresponding albedo variations) because they have been insufficiently studied.

In particular, most of the models describe the ocean very schematically. The first models (Manabe and Wetherald, 1975; Wetherald and Manabe, 1975) simulated the ocean as a "swamp", i.e., it served as the source of moisture, although its specific heat was assumed to be as low as for the land (e.g., in the modern model of general atmospheric circulation of the British Meteorological Service, the specific heats of the land and a 0.7-meter thick water layer are equivalent. The models with mixed upper oceanic layer (Manabe and Wetherald, 1980; Wetherald and Manabe, 1981; Hansen et al., 1983; Hansen et al., 1984; and others) ascribe to the ocean a 67–70-meter water layer of thermal inertia. This is obtained by averaging the upper quasihomogeneous layer for all seasons and oceans. The drawback of these models is that they ignore thermal transfer by the ocean from the low to high latitudes, which can reach almost half of the actual total thermal flux from the tropics to the polar latitudes (Miller et al., 1983). As a result, either the total atmospheric thermal transfer is reduced in the general atmospheric circulation model, and then the contrast is higher between the tropical and high latitudinal temperatures, or interlatitudinal heat transfer is realistic, however, this is only possible if large-scale eddy transport is more intensive than the extant. Fixed thermal transport in the upper oceanic layer is sometimes specified in order to overcome this problem (Hansen et al., 1984).

There are still no general circulation models for the ocean that are comparable in detail with those for the atmosphere. This is because the major oceanic currents like the Gulf Stream and Kuroshio are about 100 km wide, and the synoptic eddies in the ocean are the same size. Models with grid spacing an order smaller than for the atmosphere are thus needed to describe them. We must wait for more detailed description of the ocean and for the next computer generation, that is an order or two more powerful than the current best computers. The extant oceanic models have approximately the same spatial resolution as the general atmospheric circulation models, and larger turbulent viscosity coefficients that severely smooth out the velocity fields (Bryan et al., 1982; Spelman and Manabe, 1984). These studies have integrated the general circulation models for both media together, however, for the period of several decades. The sensitivity of high latitudes to atmospheric CO_2 concentration variations in this case decreases compared to the models that ignore thermal transport in the ocean.

Description of the hydrological cycle and its changes because of general climatic variations is one of the urgent problems in climatic theory models. If a climatic model takes seasonal variations into account, it describes snowfall and its accumulation in winter, melting in spring, and rains in summer and autumn. Water reaching the surface partly evaporates and partly penetrates into the soil. Shukla and Mintz (1982) used general atmospheric circulation models for two numerical experiments: one for land surface saturated with moisture, and another for a dry surface with much lower precipitation in the interior regions of continents. A conceptually similar numerical experiment was conducted by Yeh et al. (1983), who showed that consideration for early snow melting in climatic models results in higher aridity in the interior regions. A review of climatic model sensitivity to different parametrizations of hydrologic and other processes on the land surface can be found in Mintz (1984) and in some special surveys (WCP–46, WCP–76, etc.).

The models still describe surface runoff very schematically. Soil is assumed to absorb a finite amount of water (usually equivalent to a 15–cm water layer), and excessive water is directed to the surface runoff. Evaporation from the soil is calculated on the assumption that the soil is completely wet, until the water reserve in the soil declines to a fraction of the critical. When water reserves in the soil are lower, evaporation is assumed to be linearly proportional to the remaining reserve (Budyko, 1971). Calculations with a similar description of hydrologic processes on the land surface can serve only as the first approximation in estimates of future climatic changes. A specific project in the World Climate Research Program, the so-called "Hydrology and Atmospheric Processes Experiment" (HAPEX) aims to study the relationship between the hydrology of surfaces of different types and meteorological parameters.

General atmospheric circulation models should primarily reproduce the seasonal and geographical variations of the modern climate. This is the main goal of modeling. This is practically accomplished by "adjusting" some numerical model parameters (e.g., by selecting the degree of "blackness" for the thermal spectral range of cirrus clouds. This introduces an element of uncertainty when these "adjusted" parameters are used under varied conditions, e.g., with higher CO_2 concentration or higher solar constant. The best test of the model would be to use it to describe climate under modified conditions.

One of the few experiments of this kind was reproduction of the climate 9,000 years ago (Kutzbach and Otto-Bliesner, 1982), when the intensity of the Indian monsoon was much stronger than at present, there was a savannah instead of the Sahara, etc. At that time the Earth's orbital elements were such, that our planet was in the perihelion during the Northern Hemisphere summer (now, during the winter). This hemisphere thus received 7% more solar energy in the warm season than now. The climate model, with the seasonal insolation adjusted to 9000 years ago, simulates the climate of Africa and Eurasia which is in agreement in main features with paleoclimatic data for that period.

A series of experiments have been carried out (Gates, 1976; etc.) that reconstructed the Earth's general circulation at the peak of the last glaciation 18,000 years ago.

However, in these experiments the glacial shields of North America and Scandinavia were fixed together with the extent of sea ice. The computed patterns of the atmospheric circulation differ from the present ones. These are not the climatic theory experiment in a rigorous sense because they do not address the causes and evolution of glaciation but are mere snapshots of circulation patterns at different lower boundary conditions which may be compared with some paleoclimatic data.

The Earth's climate undoubtedly depends on the position of continents and oceans. The absence of continents in high latitudes, the more so at the poles, allows the penetration of warm tropical water there contributing to general climatic warming as now the western part of the Arctic Ocean is heated by the Gulf Stream. Barron and Washington (1984) modelled the Cretaceous climate, taking into account the position of continents 10^8 years ago. Climate was warmer than the current, but noticeably colder than in the Cretaceous period. The authors thus drew the conclusion that the higher atmospheric CO_2 content at that time was possibly an additional source of heating the Earth's surface (see Budyko, Ronov and Yanshin, 1985).

Some authors (Manabe and Wetherald, 1975; Wetherald and Manabe, 1975; Hansen et al., 1983, 1984) have experimented with modeling climatic variations both with double CO_2 content in the atmosphere and 2% higher solar constant. These studies detected that zonal changes in the Earth's surface temperature in both cases were similar to each other. A good geographical distribution of these changes is similar to that observed in current, much shorter climatic variations. It is precisely these changes which are magnified as latitude increases, and they are greater in winter than in summer. The nature of latitudinal and seasonal climatic changes thus seem not to depend strongly on the causes for these variations.

Changes in the Earth's surface temperature with varying atmospheric CO_2 *concentration.* Schlesinger (1983) made the most detailed survey of studies analyzing climatic sensitivity to altered atmospheric CO_2 concentration. He not only discussed general circulation models, but also simplified climatic models. The climatic sensitivity to doubling atmospheric CO_2 concentration, i.e., ΔT_C, generally ranged from 1.5 to 4.5°C, although some earlier results obtained from radiation convective models exceeded these limits.

The spread of ΔT_C values remained the same even by the end of 1985 (Bolin et al., 1986). Dickinson (ref. Bolin et al., 1986) has interpreted these limits as 1.5–5.5°C depending on the intensity of the feedbacks and uncertainty of our knowledge about them based on a zero-dimensional model for globally averaged temperature (see above). However, the Villach Conference (Austria, October 1985), that discussed the climatic change problem in relation to variations in atmospheric chemical composition, after prolonged discussions retained the former limits 1.5–4.5°C for the ΔT_C value. This uncertainty is associated with lack of knowledge or unrealistic description of heat transfer in the ocean both in horizontal and vertical directions, atmosphere-ocean interaction, with uncertainties in cloud modeling and cloud-radiation flux interactions, of land surface processes, including hydrological, of temperature-snow-ice cover-albedo feedbacks, as well as temperature-water vapor and cloud cover feedbacks in the atmosphere. Consideration, or lack of it, for some of these enumerated processes, and differences in their parametrizations produce changes in the values of ΔT_C that describe sensitivity of the mean global temperature to a doubling of the atmospheric CO_2 content.

Table 6.1 presents values of ΔT_C obtained by the most representative models and a brief description of the models. We will explain. Topography, i.e., a prescribed location of oceans and continents (the latter with orography), can

Table 6.1. Characteristics of Climatic Models and Their Sensitivity to Higher Atmospheric CO_2 Content

Model	Topography	Specific heat of upper quasihomogeneous layer, depth equivalent, m	Insolation	Cloud cover	Sea Ice	ΔT_C^0, C	
						$2CO_2$	$4CO_2$
3-dimensional models							
Manabe and Wetherald (1975)	sector	0	annual	fixed	computed	2.9	—
Manabe and Stouffer (1980)	realistic	68	seasonal	fixed	computed	—	4.0
Manabe and Wetherald (1980)	sector	0	annual	computed	computed	3.0	5.9
Bryan et al. (1982)	sector	computed	annual	fixed	computed	—	5.0
Hansen et al. (1984)	realistic	65	seasonal	computed	computed	4.2	—
Washington and Meehl (1983)	realistic	68	annual	computed	computed	1.3	2.7
Washington and Meehl (1984)	realistic	68	seasonal	computed	computed	4.2	—
Schlesinger et al. (1985)	realistic	computed	seasonal	computed	computed	2.8	—
Spelman and Manabe (1984)	sector	computed	annual	fixed	computed	—	5.6
Mitchell and Lupton (1984)	realistic	70	annual	fixed	computed	—	4.0
2-dimensional models							
Dymnikov et al. (1980)	idealized	0	annual	fixed	fixed	2.0	—
MacCracken (1981)	idealized	0	seasonal	computed	fixed	1.7	—
Petukhov, Manuylova (1984)	idealized	computed	annual	computed	computed	3.0	5.8

be realistic to within the computational grid; idealized as in two-dimensional models, where the percentage of ocean and land area is prescribed for each latitudinal belt; and sector as in some models by Manabe and his colleagues. In the latter case, the sphere is divided into 6 parts by meridians spaced at 60°. The calculation is made for two adjacent sectors, one of which is considered to be land (soil moisture content is calculated), and the other to be the ocean (moisture content is unlimited). Conditions of periodicity are set for both boundaries of the whole 120° sector. Specific heat of the upper quasihomogeneous oceanic layer is specified by its depth in meters (remember that specific heat of water volume 1 m^2 in area and 1 m high is 4.2×10^6 J). When the column in Table 6.1 indicates that specific heat is calculated, this means that the model incorporates heat transfer by currents as in general circulation models, or heat transfer into the deep ocean (Petukhov, Manuylova, 1984).

The majority of the models take mean annual insolation. This can alter the sensitivity of some feedbacks, first of all of the temperature and albedo of the snow and ice cover. Seasonal variations of these values always lag behind in phase, thus weakening this feedback (Mokhov, 1981), and diminishing the value of ΔT_c in models with seasonal variations as compared to mean annual conditions (see Manabe and Stouffer, 1980; Manabe and Wetherald, 1980). The results obtained by Washington and Meehl (1983, 1984) are less understandable. Their seasonal model, on the contrary, was three times more sensitive than the mean annual version (see also the discussion of these results in Bolin et al., 1986). It is obvious that great importance should not yet be attached to the smaller value.

A characteristic feature of the model sensitivity is that the logarithmic dependence of mean global temperature change on the CO_2 concentration is preserved well (Manabe and Wetherald, 1980; Washington and Meehl, 1983;

Petukhov, Manuylova, 1984). This dependence (Augustsson and Ramanathan, 1977) is associated with saturation of the main 15 μm-absorption band for CO_2 . Therefore, a higher CO_2 content identifies thermal radiation absorption only in the wings of this band.

At the same time, the dependence of mean global temperature change on the concentrations of other greenhouse gases in the atmosphere can be different. This depends on whether their absorption bands overlap in the infrared spectral range, how far they are from saturation, and so on. For example, CFCs have absorption bands still far from saturation in the 8–13 μm window of transparency. Therefore, temperature variations resulting from elevated CFC content in the atmosphere should be directly proportional to these concentrations.

According to Table 6.1, ΔT_c values lie within 2–4.2°C for three-dimensional models, if the first study by Washington and Meehl (1983) is not considered. It has already been mentioned that different estimates of sensitivity stem from inadequate knowledge of many processes in the climatic system.

For practical application of information on climatic changes, we have to know not only variations in temperature, but first of all in precipitation, soil moisture content and surface runoff. We will dwell on the conclusions for this problem drawn from climatic models. Four special studies (Manabe et al., 1981; Mitchell and Lupton, 1984; Hansen et al., 1984; Washington and Meehl, 1984) found that the hydrologic cycle for the world as a whole intensifies when CO_2 concentration rises, especially in higher latitudes, thus causing more precipitation and surface runoff. This pattern is revealed through zonal averaging. At the same time, these effects can be manifested quite irregularly in geographical and seasonal distributions, and while there is a mean zonal increase in precipitation, there could be areas and seasons with lower precipitation and soil moisture content. The first two of the four studies show that a 4°C increase in the mean global temperature is accompanied by 1–2 cm increase in soil moisture content northward of 30–35°N, except for summer, when it drops by 1–1.5 cm. The increase in surface runoff in this case is 1 mm/day in the same area, except for the 40–60°N belt, where the decrease in surface runoff reaches 0.5–1.5 mm/day during April–July. According to Mitchell and Lupton's model, for the same change of mean global temperature (4°C), the increase in moisture content is 1–2 cm in summer over the European USSR, north of 55°N, diminishing somewhat south of this latitude. Washington and Meehl's model, instead of higher summer aridity, found a 1–2 cm rise in soil moisture content for the whole warm season over the entire USSR territory, except for the Far East, a 5–6 cm rise in winter (a water equivalent of snow) over the European USSR. Hansen et al. obtain similar results, although for summer they can be interpreted as a lack of statistically significant changes in moisture content versus the current levels.

All the models give a lower soil moisture content in the tropics. Figure 6.1 illustrates the annual variation in zonally averaged soil moisture content. Contradictory results for summer call for further investigations. For the present, all these models describe hydrologic processes very schematically with averaging over 4 or 5 years of integration, which is clearly insufficient to obtain statistically significant climatic changes. Modeling of land surface hydrology should undoubtedly be improved. Nevertheless, general conclusions from these models on the whole area with extant hypotheses regarding varying precipitation induced by comparatively small temperature changes (Drozdov, 1985), as well as with paleoclimatic data and some up-to-date data. Barnett (1986) thus points to a long-term trend in our century of higher total annual precipitation over Europe and Western Asia (remember that the mean global temperature

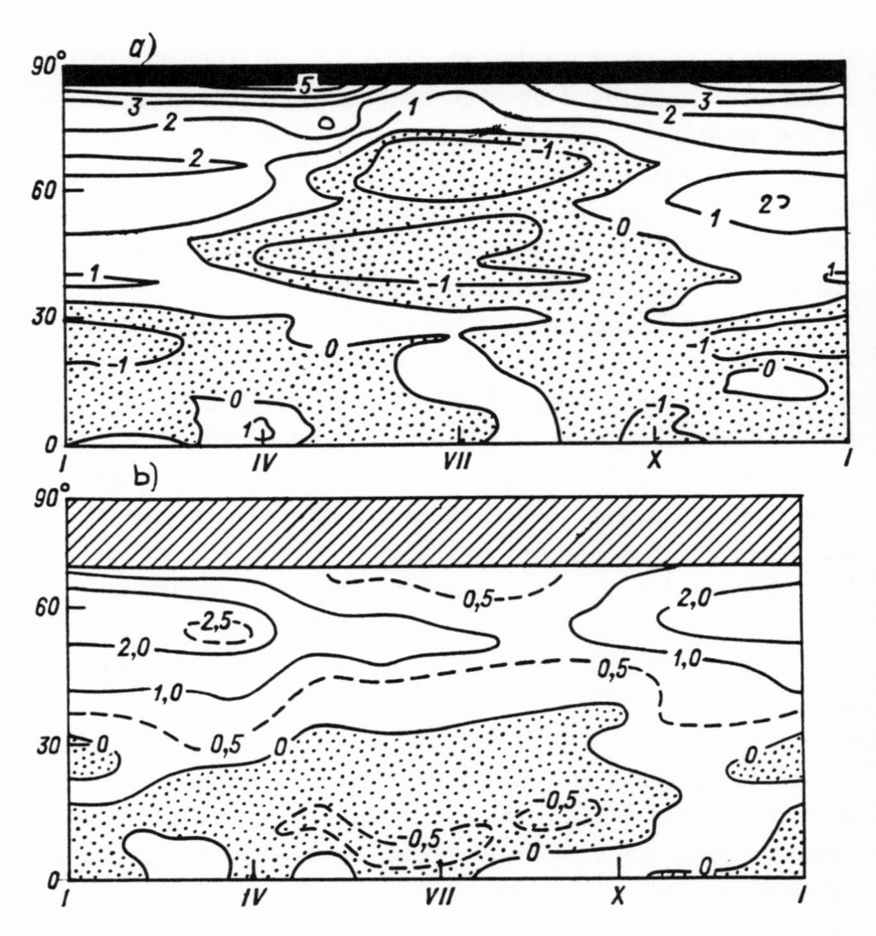

Figure 6.1. Annual variations in soil moisture at different latitudes during global warming obtained in calculations by models [Manabe et al., 1981 (a), and Washington and Meehl, 1984 (b)]. Figures near isolines are the change in soil water content (cm).

has increased by 0.5°C in the last 100 years) and lower precipitation over Africa. Apasova and Gruza (1982) point out that since 1890 there has been more precipitation over the Northern Hemispheric continents in winter and lower in summer. The current statistical significance of precipitation trends is still low however, because of strong interannual variations, data that are inhomogeneous, unrepresentative and incomplete, etc. These circumstances force us to make palaeoclimatic reconstructions for warmer epochs.

Such data (see Chapter 8) are available for the period of the Holocene optimum (5–6 Ma ago), Mikulino (Eemian) interglacial (120–130 Ma ago), when the mean global temperature T_S was 1–2°C higher than the current level, and for the Pliocene (3–4 Ma ago), when T_S was 3–4°C above the present. Palaeoclimatic reconstructions show that for all warm periods in the middle and high latitudes, precipitation was higher as compared to its current values.

The problem of changes in precipitation during warming is treated in more detail in subsequent chapters.

6.2. Influence of Climatic System Thermal Inertia on Climatic Sensitivity

Steady-state climatic models which study the effects of changing atmospheric chemical composition on T_S provide estimates of climatic sensitivity, i.e., surface air temperature variation when the composition of the atmosphere or solar radiation absorbed by the planet differs from the current.

If specific heat and heat conductivity of the upper land and ocean layers were close to zero then the climatic system should be considered essentially inertialess. In this case, if there is a time-dependent influx of energy $\Delta F(t)$ at the upper boundary of the atmosphere, temporal variations in temperature $\Delta T_{SR}(t)$ would be described by equation (6.4) or by an analogous relationship:

$$\Delta T_{SR}(t) = T_{SR} - T_S^0 = \Delta F / B_0, \qquad (6.4a)$$

where B_0 differs from b_0 because it considers the relationship between planetary albedo α and surface air temperature T_S (see formula (5.3)).

Under real conditions, the change in surface temperature $\Delta T_S(t)$ should depend on the complete spectrum of relaxation times of the climatic system, therefore, temperature shifts in the layers closest to the Earth's surface depend on the characteristic temporal scale of changing heat influx to the surface. During daily variations in short- and longwave radiation fluxes, surface temperature T_S and surface air temperature T_a change in a different manner, since there is a large deviation from equilibrium in the surface layers when the oscillation frequency is $\omega = (2\pi/86,400)$ $\sec^{-1}$. During seasonal variations in incoming solar radiation fluxes, this difference is smaller; it is especially small above the ocean, therefore, when evaluating thermal inertia of the climatic system, as a first approximation, we may consider T_S equals T_a, and evaluate the mean deviation of temperature T_S from temperature T_{SR} of the inertialess climatic system. Budyko (1971) made this estimate on the basis of an empirical relationship between the intensity of the outgoing longwave radiation and average monthly values T_S (see formula (6.3)).

The mean seasonal variation in absorbed solar radiation for the hemisphere can be represented as follows

$$I(t) = (S_0/4)(1 - \alpha)(1 + 0.32 \sin \omega t). \qquad (6.7)$$

Here S_0 is the solar constant (1372 W/M^2), α is planetary albedo, whose annual mean is assumed to be 0.30. It follows from formulas (6.3) and (6.7) that temperature T_{SR} at which the equilibrium condition $F(t) = I(t)$ is met for radiation at the top of the atmosphere should have seasonal variations of approximately 50 K. The actual amplitude values for seasonal variations of surface air temperatures of the planet are much smaller. Figure 6.2 illustrates their values for the surface air layer above land at various latitudes, for surface layer of the ocean, and surface air layer above the ocean according to Strokina (1982). The same graph presents amplitudes for seasonal variations in tropospheric air temperature at an altitude of about 6 km by Goody (1958). It is about 7°C, which is close to the mean amplitude of surface air temperature variations for the Northern Hemisphere, where 39% of the area is continents with

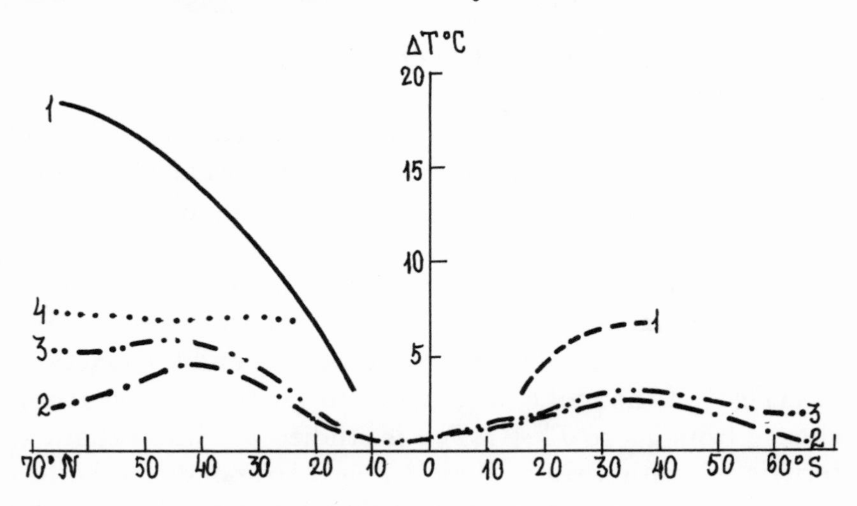

Figure 6.2. Amplitudes of seasonal temperature variations in temperature depending on latitude. 1, surface air layer over land; 2, surface oceanic water; 3, surface air layer over ocean; 4, troposphere at a height of 6 km.

temperature oscillations of 13.7°C, and 61% is oceans with much smaller amplitudes of 1.9°C. The effective specific heat of the entire Northern Hemisphere was estimated by Vinnikov and Groysman (1981, 1982, see also Chapter 7) from the indicated empirical data. Byutner and Shabalova (1985) calculated the characteristic relaxation times for land and ocean surface layers when they respond to seasonal isolation variations. These relaxation times for land were dependent on thermophysical properties of soil but, on the average, were short (about a tenth of a year). For the ocean, the characteristic relaxation time τ_1 for the temperature deviation in the upper quasihomogeneous layer from the mean annual value equaled about 3 years for the Northern and 5 years for the Southern Hemispheres. A similar estimate of the corresponding relaxation time $\overline{\tau}$ for the entire Northern Hemisphere, obtained by Budyko (1971) was $\overline{\tau} = 2.5$ yrs., i.e., close to τ_1. A conclusion can thus be drawn that even for the Northern Hemisphere, inertia of the climatic system, when it responds to seasonal insolation variations, depends mainly on properties of the ocean, especially in the Southern Hemisphere.

When the climatic system responds to external perturbation $\Delta F(t)$ with a characteristic time scale of several decades, its inertia, i.e., its relaxation time, should be longer (Budyko, 1979b), since the processes of heating or cooling should affect the deeper rather than upper oceanic layer.

Climatic models most often use equation system (6.5)–(6.6) to assess thermal inertia of the planet, and the reaction of $\Delta T_S(t)$ of the model as a whole to gradual temperature change from T_S^0 to $T_S^0 + \Delta T_{SR}$ is studied to obtain characteristic relaxation time. Here ΔT_{SR} corresponds to the response of an inertialess system to doubled or quadrupled atmospheric CO_2 concentration (see, for example, Schlesinger, 1985). However, since the real CO_2 concentration varies gradually, it is better to study the system's response to time-dependent external perturbation $\Delta F(t)$ for a more precise assessment of future temperature changes, as, for example, Wigley (1984), as well as Wigley and Schlesinger (1985). Their studies demonstrated that the $\Delta T_S(t)/\Delta T_{SR}(t)$ ratio mainly depends on the sensitivity parameter b_0 and the diffusion coefficient K in equation (6.6). We know that incorporation in climatic models of equations that describe the interaction between the atmosphere and oceanic upper layer with depth generally of 50 to 100 m, leads to small surface temperature lags 5 to 10 years behind the $T_{SR}(t)$ value in equation (6.4a), where $\Delta F(t)$ increases continuously with time. With a characteristic scale of $\Delta F(t)$ of several decades, this effect is not significant in forecasting anthropogenic climatic changes.

When the climatic models incorporate equations describing propagation of thermal perturbation in the deep ocean and the heat transfer by currents, sometimes there is a significant discrepancy between calculated changes for T_S and T_{SR}. They depend both on the method of describing heat transfer processes in the ocean and the selection of parameter values for heat transfer intensity in the deep ocean. It is thus often difficult to determine precisely which factor exerts a greater influence on the behavior of function $\Delta T_S(t)$, thermal inertia of the system, i.e., the characteristics describing the ocean, or the change in the relationship between the terms in the thermal balance equation for the oceanic surface. For this reason, thermal inertia of the ocean should be estimated separately, without incorporating the corresponding block into climatic models. This will clarify the influence on function $\Delta T_S(t)$ of the parameters characterizing the ocean itself. The results of modeling current heat transfer in the ocean can be used in this case, however, the oceanic model should contain the fewest possible parameters to be determined empirically.

The model proposed by Kagan et al. (1979) and developed in Kagan, Ryabchenko, 1981, 1982, in which the ocean is divided into three regions, is the most suitable:

1) the upper quasihomogeneous layer interacting with the atmosphere, and occupying 90% of the ice-free World Ocean surface area;
2) the deep layer below containing most of the oceanic water;
3) the area of deep cold water formation, where convective mixing practically down to the bottom occurs due to intensive cooling of the surface ocean layers (see Figure 6.4).

The energy of wind and solar radiation is absorbed by surface water layers. As a result of its conversion into the energy of turbulent and convective motions, the upper quasihomogeneous layer is formed with the characteristics including thickness h_1 that vary throughout the year. These studies obtained values of seasonal variations in temperatures, concentrations of dissolved gases and some other characteristics that generally are in good agreement with full-scale observations. These investigations allow us to construct a simplified scheme of stationary mean annual heat exchange between the ocean and the atmosphere, if the main mixing characteristics are defined from empirical data.

Figure 6.3 provides data from L. A. Strokina on latitudinal dependence of mean annual values of the heat flux absorbed by the ocean needed to construct the model. The corresponding mean annual temperature pattern for the oceans in the Northern and Southern Hemispheres separately and averaged over the World Ocean is found in the recently published atlas by Levitus (1982). These averaged characteristics can be true for the model "World Ocean on the Average," whose scheme is presented in Figure 6.4.

This scheme conditionally divides the ocean into four blocks with various dynamic properties. Block 1 represents the upper well-mixed layer, in which essentially all the shortwave radiation reaching the sea surface is absorbed, the mechanical energy transmitted from the atmosphere dissipates into turbulence energy and ultimately, into heat. Block 2 next to the mixed layer is the major ocean water mass that did not directly contact the atmosphere, but interacts with the upper layer and with Block 3, the region of deep cold water. Block 3 occupies about 10% of the World Ocean's area (S), is situated near the polar caps and is characterized by strong heat emission into the atmosphere and such an intensive mixing of water that there is practically no boundary between the upper mixed and deeper layers.

Circulation exists in the system of these blocks: warm currents carry water of the upper quasihomogeneous layer towards the poles, i.e., to block 3, and thence water enters the deep layers and ascends to the surface via the lower boundary of block 1, thus closing the cycle. Besides circulation, heat diffusion occurs in this system through a closing layer at the interface between blocks 1 and 2, since $T_1 > T_2$.

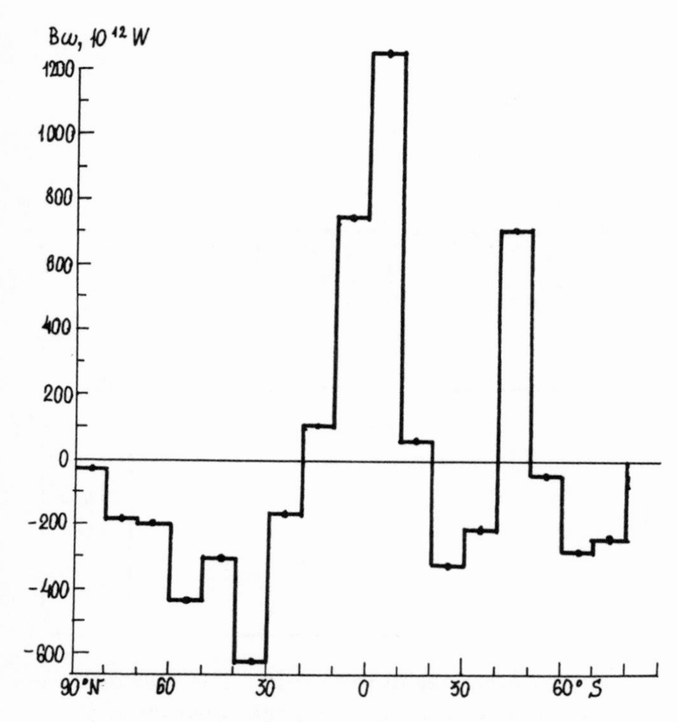

Figure 6.3. Latitudinal distribution of heat flux into ocean B_w from L. A. Strokina (1986).

For a better description of mean temperature patterns in the ocean, we isolate in block 2 the main thermocline (block 2′) of h_2 thickness, with small turbulent time and depth independent diffusion coefficient k, and region 2″ without turbulence. Assume that deep cold water from block 3 enters block 2″ directly, and there is diffusion exchange between blocks 2′ and 2″.

Mean annual values of major parameters W, W_1 and W_2 that determine the intensity of heat transfer in the ocean can be obtained by parameterization of climatic data on ocean-atmosphere energy exchange and on oceanic temperature patterns as applied to the model in Figure 6.4. According to recent calculations by Strokina (1986), the mean annual thermal energy flux $B_W = R - B_a - LE$ (where R is radiation balance of the ocean surface, B_a is turbulent heat

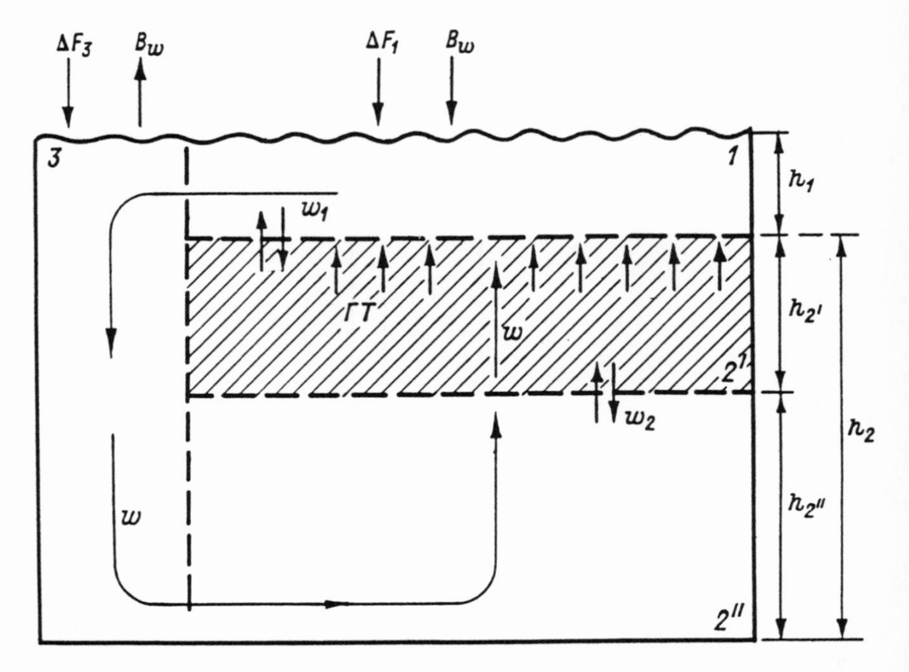

Figure 6.4. Scheme of heat exchange in the model "The World Ocean on the Average." The value B_w designates heat emission of $2.87 \cdot 10^{15}$ W/yr from the area of formation of cold deep waters into the atmosphere and corresponding absorption in block 1; MT is the main thermocline.

exchange with the atmosphere, LE is energy loss for evaporation) into the World Ocean is $\pm 2.87 \cdot 10^{15}$ W. The latitudinal dependence of B_W is rather complex. The distribution of this flux between the hemispheres is uneven, nevertheless, energy absorption prevails at low latitudes and liberation at high levels (see Figure 6.3). As applied to the model in Figure 6.4, this means that block 1 occupying 90% of the ocean surface receives from the atmosphere $2.87 \cdot 10^{15}$ W of energy $\overline{B}_W$, i.e., 9.57 W/m^2, on the average; 86.1 W/m^2 is released into the atmosphere from the surface of block 3, which closes the oceanic energy balance as a whole. Temperature values for different model blocks should be chosen in accordance with data on the temperature pattern averaged over all the oceans. Judging from data presented in the Levitus atlas, the mean temperature in block 2′ should be about 9°C, in block 2″ 2°C and in block 3 0°C. The mean temperature in block 1 is about 17°C. The temperature differential between surface layers and the main thermocline thus equals 8°C. Analogous temperature values were used by Skopintsev (1977) to calculate the biological oxygen expenditure in oxidation of organic matter in the ocean.

The mean oceanic temperature can be expressed in terms of temperatures of individual blocks in the following way:

$$\overline{T} = \frac{1}{V}(V_1 T_1 + V_{2'} T_{2'} + V_{2''} T_{2''} + V_3 T_3).$$

Here $V = V_1 + V_{2'} + V_{2''} + V_3$ is the oceanic volume, where the depth of the region of deep cold water formation is considered to be equal to the mean oceanic depth, i.e., 4000 m. Actually, it is a bit smaller, but this has practically no effect on the value of $\overline{T}$. When the mean thickness of block 1 is 70 m, and blocks 2′ and 2″ are 1000 and 3000 m respectively, the value of $\overline{T}$ equals 3.67°C. According to Stepanov (1974), the mean water temperature of the World Ocean is 3.8°C, and 3.62°C according to the Levitus atlas (4.04°C for the Northern and 3.33°C for the Southern Hemispheres). The model value of $\overline{T}$ consequently can be assumed to agree satisfactorily with empirical data for the aforementioned temperatures of individual blocks.

The system of equations describing heat transfer in the ocean is as follows:

for block 1

$$c(W + W_1)(T_1 - T_{2'}) + B_W = 0, \tag{6.8}$$

for block 2′

$$cW(T_{2''} - T_{2'}) + cW_1(T_1 - T_{2'}) - cW_2(T_{2'} - T_{2''}) = 0, \tag{6.9}$$

for block 2″

$$cW(T_3 - T_{2''}) + cW_2(T_{2'} - T_{2''}) = 0, \tag{6.10}$$

for block 3

$$cW(T_1 - T_3) - B_w = 0. \tag{6.11}$$

Here c is the specific heat of water; parameters W, W_1 and W_2 have the same dimension ($L^3 \times T^{-1}$), but different physical meaning. Parameter W is the quantity of cold deep water formed in block 3 and circulating in the ocean. The $W/0.9S$ ratio is the mean upwelling rate w in the ocean. Parameter W_1 characterizes diffusion properties of the barrier layer at the lower boundary of the upper quasihomogeneous layer, that show strong variations during its annual cycle. We note that this leads to different values of effective mean annual transfer rates for various substances (i): $W_1^{(i)}/0,9S$ between blocks 1 and 2′, which is not always taken into account in calculation of heat transfer intensity in the ocean. Parameter W_2 characterizes thermal and diffusion resistance of block 2′. Since the coefficient of turbulent exchange in water is the same for the transfer of heat and dissolved substances, the values of W_2 (in contrast to W_1) for these processes should also be the same.

Equation (6.11) allows us to evaluate the intensity of oceanic water circulation by data on $\overline{B}_W$ and annual mean temperature differential $T_1 - T_3$. With $\overline{B}_W = 2.87 \cdot 10^{15}$ W, $T_1 = 17°\text{C}$, $T_3 = 0°\text{C}$ and $c = 4.18 \cdot 10^6$ J/(m$^3 \times$°C) we obtain

$$W = \frac{\overline{B}_W}{c(T_1 - T_3)} = 40 \times 10^6 \text{m}^3/\text{sec}, \tag{6.12}$$

which corresponds to the mean upwelling rate of approximately 1.3×10^{-5} cm/sec over 90% of the oceanic surface area.

Values W_1 and W_2 are determined similarly from equations (6.8–6.10) and equal 1.12 W and 0.29 W respectively.

As is shown below, correct selection of the parameter W_1 value is most important for evaluating thermal inertia of the ocean. According to the scheme described here, it mainly depends on the value of W defined from (6.12), in good agreement with the available estimates (e.g., see Munk 1966), as well as on the difference between mean temperatures $T_1 - T_{2'}$.

Analysis of the oceanic response to external thermal perturbation should incorporate functions $\Delta F_1(t) \cdot 0,9S$ and $\Delta F_3(t) \cdot 0,1S$ (see formula (6.4a)) into the system of equations that describes temperature change in all blocks of the model (Figure 6.4).

We will introduce: $T_1 - T_1^0 = x$, $T_{2'} - T_{2'}^0 = y$, $T_{2''} - T^0_{2''} = v$, $T_3 - T_3^0 = z$. The system of equations for determining the dependence of functions x, y, z and v on time is as follows

$$cV_1 \frac{dx}{dt} = -c(W + W_1)(x - y) + 0,9S[\Delta F_1(t) - B_{01}x], \tag{6.13}$$

$$cV_{2'} \frac{dy}{dt} = cW(v - y) + cW_1(x - y) - cW_2(y - v), \tag{6.14'}$$

$$cV_{2''} \frac{dv}{dt} = cW(z - v) + cW_2(y - v), \tag{6.14''}$$

$$cV_3 \frac{dz}{dt} = cW(x - z) + 0,1S[F_3(t) - B_{03}z] \tag{6.15}$$

with initial conditions $x = y = v = z = 0$ when $t = 0$. Values V_1, $V_{2'}$, $V_{2''}$ and V_3 are the volume of blocks in the oceanic model. Values V_1, V_2 and $V_{2''}$ can be regarded as fixed and approximately equal to 0.21×10^{17} m^3, 3×10^{17} m^3 and 9×10^{17} m^3, respectively; the exact value of V_3 is unknown, therefore, it will vary in the range of (0.34–1.36) $\times 10^{17}$ m^3, which corresponds to the effective depth of block 3 of 1000–4000 m.

Since the time for establishing radiation equilibrium in the atmosphere is very short as compared to the heat content change in blocks 1 and 3, temperature changes T_1 and T_3 could occur under the influence of external thermal pulse equal to $[\Delta F_1(t) - B_{01}x]$ and $[\Delta F_3(t) - B_{03}z]$ respectively. This is Cess's idea (see details below), but the suggested approach to solving the problem of oceanic response to changes in the greenhouse effect differs from that developed by Cess and Goldenberg (1979), because the parameters determining the intensity of heat transfer in the ocean are evaluated by empirical data on heat exchange under natural conditions using equations (6.8–6.11).

Functions $\Delta F_1(t)$ and $\Delta F_3(t)$ increase linearly with time in the next century (see Byutner, 1987):

$$\Delta F_1(t) = A_1 t; \qquad \Delta F_3(t) = A_3 t, \tag{6.16}$$

and the value of A_1 should be about 5×10^{-2} W/(m$^2 \times$ yr.), which corresponds to rate A_1/B_{01} of temperature increment in an inertialess system of about 0.03°C/yr., with values of B_{01} varying from 1.4 to 1.8 W/(m$^2 \times$°C).

It is far more complicated to correctly choose the values of parameters A_3 and B_{03} that characterize external perturbation and the reaction to it of high-latitude regions of the climatic system, and its solution requires a more detailed modeling of the whole climatic system. We will therefore regard the A_3/B_{03} ratio as a parameter ranging from A_1/B_{01} to $2A_1/B_{01}$, and will investigate its influence on the calculation results. Equations (6.13–6.15) can be presented in a more convenient form for analyzing the results:

$$\frac{dx}{dt} + x\left(\frac{1}{\tau_1} + \frac{1}{\tau_2} + \frac{1}{\tau_3}\right) - y\left(\frac{1}{\tau_1} + \frac{1}{\tau_2}\right) = At, \tag{6.13a}$$

$$\frac{dy}{dt} + y\left(\frac{h_1}{h_{2'}\tau_1} + \frac{h_1}{h_{2'}\tau_2} + \frac{1}{\tau_4}\right) - v\left(\frac{h_1}{h_{2'}\tau_1} + \frac{1}{\tau_4}\right) = 0, \tag{6.14'a}$$

$$\frac{dv}{dt} + v(\frac{h_1}{h_{2''}\tau_1} + \frac{1}{\tau_5}) - \frac{y}{\tau_5} - z\frac{h_1}{h_{2''}\tau_1} = 0, \quad (6.14''\text{a})$$

$$\frac{dz}{dt} + z(\frac{1}{N\tau_1} + \frac{B_{03}}{B_{01}}\frac{1}{9N\tau_3}) - \frac{x}{N\tau_1} = \frac{A_3}{A_1}\frac{A}{9N}t, \quad (6.15\text{a})$$

where

$$\tau_1 = \frac{V_1}{W} = 17.7 \text{ yrs.}, \qquad \tau_2 = \frac{V_1}{W_1} = 15 \text{ yrs.},$$

$$\tau_3 = \frac{ch_1}{B_{01}} = 5.2 \text{ yrs.}, \qquad \tau_4 = \frac{V_{2'}}{W_2} = 800 \text{ yrs.},$$

$$\tau_5 = \frac{V_{2''}}{W_2} = 2400 \text{ yrs.};$$

$$A = \frac{A_1}{ch_1} = 5 \times 10^{-3}\,^\circ\text{C/yr.}^2; \qquad N = \frac{V_3}{V_1} = \frac{1}{9}\frac{h_3}{h_1}.$$

The results of calculating relative temperature changes in block 1 as compared to these in the inertialess system are presented in Figure 6.5 for $B_{03}/B_{01} = 1$ and $A_3/A_1 = 2$.

Computations showed that surface oceanic temperature changes vary depending on whether a seasonal thermocline exists or it degenerates, as in the region where cold deep water is formed. In general, the presence of a high-latitudinal ocean should diminish the temperature change in high latitudes as compared to the corresponding change above land. However, a precise quantitative estimate of such a "smoothing effect" is impossible to obtain using a simplified model, although it evidently must take place. This essentially does not affect the relative change in $\Delta T_S/\Delta T_{SR}$ of mean global temperature, which should be close to function x/x_R, and consideration for the presence of continents will evidently somewhat increase the actual values of $\Delta T_S/\Delta T_{SR}$ versus those in Figure 6.5.

Selection of τ_2 in the 15– to 25–year range has little influence on calculation results, however small values of τ_2

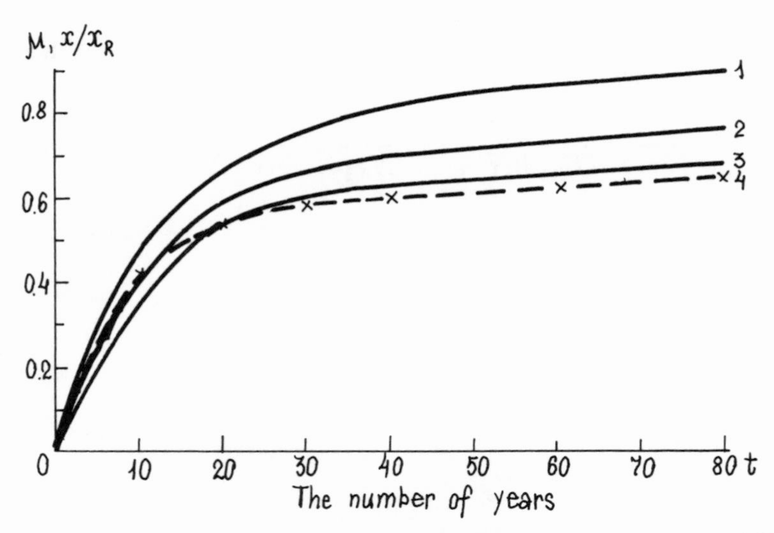

Figure 6.5. Dependence of inertia coefficient μ equal to $\Delta T_S(t)/\Delta T_{SR}(t)$ on time for linear function $\Delta_{SR}(t)$. 1–3, calculation result from equation (6.16′) for the values of parameter ch_1/B_0 equal to 7 years, and B_1 equal to zero, 0.28 and 0.52 W/(m$^2\times$°C), respectively; 4, function x/x_R obtained as a result of solving the full system of equations (6.13)–(6.16) for values $ch_1/B_0 = 5$ yr ($B_1 = 0.6$ W/(m$^2\times$°C). $B_{03}/B_{01} = 1$; $A_3/A_1 = 2$.

(5 years or less) reduce the x/x_R ratio and make it sensitive to changes in τ_2. Similar results are found in studies that model block 2′ as a continuous turbulized inhomogeneous medium with diffusion coefficient k, but without barrier layer at the boundary between blocks 1 and 2. When the values of k are small (0.5–1.7 cm^2/sec), the conclusions drawn by different authors are similar and close to the barrier layer models.

When k is larger (3–5 cm^2/sec), the temporal dependence of function x/x_R changes considerably: it becomes much smoother and reaches 0.7 at $t_1 \approx 50$ years. As pointed out by Munk (1966), the mean temperature profile in the main thermocline can be described correctly with W/S ratio equal to $1.3 \cdot 10^{-5}$ cm^{-1}. This has also been confirmed by more recent empirical data generalized in Levitus, 1982:

for a satisfactory description of the mean temperature profiles, values of W/k should not differ by more than 20% from those of Munk. Therefore, large values of k that were selected to describe heat exchange processes and are usually based on data on the distribution of such tracers as ^{3}H and ^{14}C in the ocean, should correspond to unjustly large values of mean upwelling rates W. The relationship x/x_R illustrated in Figure 6.5 agrees with the results obtained in the work by Harvey and Schneider (1985), that studies oceanic temperature response to a graduated impulse. The function x/x_R depends, of course, not only on the choice of parameters that describe thermal transfer in the ocean, but also on the form of the perturbation function, therefore, no exact comparison can be made. The aforementioned study adopts the intensity of deep cold water formation as 48.4×10^6 m^3/sec, which is close to our value, and the water exchange time between the upper and deep oceanic layers, corresponding to our τ_2 is 10 years, with 30 m effective thickness of the vertical layer. As a result, parameter W_1 is close to our estimate. The function that characterizes the relative change in mean global temperature $\Delta T_S/\Delta T_{SR}$ also varies little within the 10 to 100–year interval after perturbation starts and ranges from 0.6 to 0.8, which is in good agreement with the results given in Figure 6.5. We note that Harvey and Schneider used a complete equation of energy balance at the atmosphere-ocean interface instead of our simple approximation of function ΔF. The form of the relationship between the continent and ocean was significant only for computations for the first five years.

In order to study the influence of parameters B_0 and W_1 on the calculation of function $T_S(t)/T_{SR}(t)$, we will introduce the equation for altered enthalpy of surface layers in the following form:

$$ch_1 \frac{dT_S}{dt} = \{\Delta F(t) - B_0[T_S(t) - T_s^0]\} - B_1[T_S(t) - T_S^0]. \tag{6.16'}$$

The term in the left side of this equation describes altered enthalpy of the upper quasihomogeneous oceanic layer; the first term in the right side describes thermal pulse equal to $\Delta F - B_0[T_S(t) - T_S^0]$ which decreases as temperature $T_S(t)$ approaches T_{SR}, while the second term is heat flux into the deep ocean, that we approximate by $B_1[T_S(t) - T_S^0]$. Insofar as specific heat of the entire troposphere is 30 times lower than that of the upper quasihomogeneous oceanic layer, it is evident that altered tropospheric enthalpy can be ignored.

The mean global value $\overline{h}_1$ (mean annual thickness of the quasihomogeneous layer) can be estimated from empirical data on seasonal variation in heat flux received by the ocean and from surface water temperature. Strokina (1963, 1982) obtained these data.

For the Northern Hemisphere we can thus obtain $ch = 2.1 \cdot 10^8$ J/(m$^2 \times$°C) and for the Southern, 3.15×10^8 J/(m^2 $\times$°C). This corresponds to thickness $\overline{h}_1$ equal to 51 and 77 m, respectively.

Since the influence of continents on variations in surface oceanic temperature is stronger in the Northern Hemisphere, we can adopt $h = 70$ m in calculating $T_S(t)$. The value of the mean annual coefficient of heat exchange B_1 is about 0.6 W/(m$^2 \times$°C), which corresponds to $W_1 = 1.12W$. Byutner (1983) estimated this value using indirect methods as ranging from 0.2 to 0.52 W/(m$^2 \times$°C).

The solution to equation (6.16′) with approximation of the pulse $\Delta F(t)$ by the exponential time-dependent function

$$\Delta F(t) = Ae^{\gamma(t-t_0)} \tag{6.17}$$

has the following form:

$$T_S(t) - T_S^0 = \frac{\frac{A}{B_0}\left[e^{\gamma(t-t_0)} - e^{-\frac{B_0+B_1}{ch_1}(t-t_0)}\right]}{1 + \frac{B_1}{B_0} + \gamma\frac{ch_1}{B_0}}. \tag{6.18}$$

It follows from equation (6.18) that temperature $T_S(t)$ of a real planet varies with time according to the same law that governs temperature $T_{SR}(t)$ of a planet that does not have inertia, but has constant time lag Δt equal to

$$\Delta t = \frac{1}{\gamma} \ln(1 + \frac{B_1}{B_0} + \gamma \frac{ch_1}{B_0}). \tag{6.19}$$

Such a constant value of Δt is characteristic for an exponentially varying pulse. If there is no interaction between the upper and deeper oceanic layers, i.e., $B_1 = 0$, and if $\gamma ch_1/B_0 << 1$, then $\Delta t = ch_1/B_0$ does not depend on the characteristic time of the external factor $1/\gamma$. Hunt and Wells (1979) drew the same conclusion when they solved a similar problem on surface temperature change in the absence of heat exchange at depth $h_1 = 300$ m. We note that they did not use Cess's hypothesis, which was the basis for deriving equation (6.16′), and they considered the actual thermal balance of the oceanic surface.

Generally, Δt depends both on the medium parameters (C/B_0 and B_1/B_0) and on the characteristic perturbation time $1/\gamma$.

In the case of exponential thermal impulse, the ratio

$$\frac{\Delta T_S(t)}{\Delta T_{SR}(t)} = \frac{1}{1 + \gamma \frac{ch_1}{B_0} + \frac{B_1}{B_0}}$$

is also time-independent.

When the thermal pulse increases linearly with time, both "lag-time" and $\Delta T_S/\Delta T_{SR}$ vary with time.

Figure 6.6 illustrates this through function $\Delta T_S(t)$ for the exponential impulse. The impact of specific heat in the upper quasihomogeneous oceanic layer causes surface temperature lag of 5–7 years as compared to the temperature of an inertialess planet. If heat transfer into deeper layers is taken into account, then the approximated surface temperatures never reach the radiation equilibrium of the planet. Higher surface temperature of an inertial climatic system depends considerably on the mean annual coefficient of heat exchange B_1 between the upper and deeper layers. Curves 1, 2 and 3 (Figure 6.5) illustrate the influence of this factor on the inertia coefficient μ, equal to $\Delta T_S/\Delta T_{SR}$.

Comparison of calculations of the coefficient μ for the linear and exponential impulse shows that function $\mu(t)$ depends on the form of the impulse at least up to the values of argument t about $2ch_1/(B_0 + B_1)$. It follows from (6.18) that with $\gamma << ch_1/B_0$, coefficient μ essentially becomes constant and equal to $1/(1 + \frac{B_0}{B_1})$, if t exceeds $2ch_1/(B_0 + B_1)$. Small values of γ or linear function $\Delta F(t)$

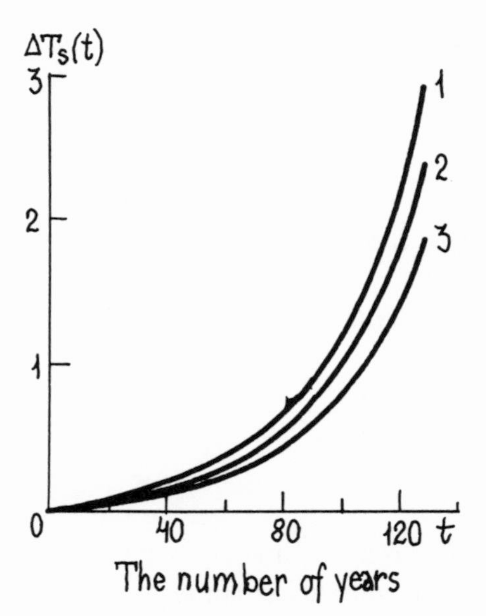

Figure 6.6. Temporal variations of the surface temperature of the planet covered with ocean with different characteristics of its time lag for exponential heat disturbance. 1, inertialess planet; 2, planet with heat capacity of surface layers of 10 W × yr/(m^2×°C) and in the abswence of heat exchange with deeper layers; 3, with the same heat capacity of surface layers and heat exchange coefficient with deeper layers B_1 = 0.52 W/(m^2×°C).

are characteristic of the impulse associated with higher atmospheric CO_2. Chapter 5, therefore, calculates the inertia of the climatic system for other trace gases, such as methane, nitrous oxide, freons and tropospheric ozone for ΔT functions that correspond to altered concentration of these gases (see Table 1.4).

In order to obtain the total value of ΔT_S, for example, for the period 1970–2000, we have to multiply the value of $\Delta T_{SR}^0 = 0.65°\mathrm{C}$ of an inertialess system corresponding to a 50 ppm change in P_a by coefficient μ equal to about 0.7 (see Figure 6.5), and to add the value 0.4°C from Table 1.4 to the result. So we obtain the total forecasted change ΔT_S equal to about 0.9°C, relative to the mean 1970 level (see Chapter 9).

Estimates of thermal inertia of the climatic system using equations (6.5) and (6.6) lead to results that differ slightly from these obtained here, when the effective coefficient of diffusion k for the ocean is approximately 1 $\mathrm{cm}^2/\mathrm{sec}$ and even smaller. A long lag time of surface temperature relative to T_{SR}, as well as considerable lagging of surface oceanic temperature behind air temperature are obtained when large values of coefficient k are used in equation (6.6). These values are usually obtained, when estimates of k are based on data on tracer diffusion in the ocean. When basic parameters are estimated using data on natural heat exchange in the ocean, as we, and Harvey and Schneider have done, then computation results are close to those in Figures 6.5 and 6.6.

The results of modeling seasonal evolution of the upper quasihomogeneous layer of the ocean (Kagan, Ryabchenko, 1981) also show that the mean annual values of heat transfer coefficients between blocks 1 and 2′ should be considerably smaller than mass exchange coefficients, especially when gas is the tracer. This is because mixing of the upper and deeper layers in the ocean occurs at the beginning of spring warming, when temperature of the upper layers is close to the minimum. During this period, concentration of tracers disseminating into the ocean from the atmosphere should be the maximum, since solubility of the majority of gases increases as temperature is lowered.

It is possible to roughly estimate the relationship between mean annual values of parameter $W_1^{(T)}$ that characterize heat transfer, and the analogous parameter $W_1^{(\mathcal{D})}$ that determines the intensity of the transfer of a dissolved gas via the following expressions for mean annual fluxes of heat and dissolved gas through the interface of blocks 1 and 2′:

$$W_1^{(T)}(T_1 - T_{2'}) = \tilde{W}_1(T_1 - a_T - T_{2'}),$$

$$W_1^{(\mathcal{D})}(\mathcal{D}_1 - \mathcal{D}_2) = \tilde{W}_1(\mathcal{D}_1 + a_{\mathcal{D}} - \mathcal{D}_{2'}).$$

Here, a_T and $a_{\mathcal{D}}$ are the amplitudes of seasonal variations in temperature and concentration $\mathcal{D}$ of a dissolved gas, respectively, $\mathcal{D}_1$ and $\mathcal{D}_{2'}$ are mean annual concentrations in blocks 1 and 2′, respectively, while the coefficient $\widetilde{W}_1$ characterizes the rate of water exchange through the interface between blocks 1 and 2′ from February to April. At this time the temperature differential between blocks is the smallest $(T_1 - a_T - T_{2'})$, while the difference between concentrations is the greatest for the year. Estimates of this relationship

$$\frac{W_1^{(\mathcal{D})}}{W_1^{(T)}} = \frac{T_1 - T_{2'}}{\mathcal{D}_1 - \mathcal{D}_{2'}} \cdot \frac{\mathcal{D}_1 + a_{\mathcal{D}} - \mathcal{D}_{2'}}{T_1 - a_T - T_{2'}}$$

made using available empirical data on fields of dissolved oxygen in the ocean and on seasonal variation in its concentrations yield a value of 2.4, while the corresponding estimates for dissolved inorganic carbon yield about 3.0. It is evident that the value of $W_1^{(\mathcal{D})}/W_1^{(T)}$ should depend on the type of tracer disseminating in the ocean (Hoffert, Flannery, 1985).

The survey of climatic models (Section 6.1) used to investigate climatic changes due to a doubling of carbon dioxide shows that in the majority of cases, the results obtained agree with each other. These results include:

1) greater surface air temperature changes in higher than in lower latitudes;
2) greater temperature changes in winter than in summer;
3) more precipitation and runoff in the middle and high latitudes throughout the year, except perhaps for summer.

The estimates given in Section 6.2 indicate that inclusion of thermal inertia of the climatic system in calculations of mean global surface temperature for several decades ahead yields values approximately 30% lower than temperature for an inertialess system.

However, a lot of work still has to be done to obtain reliable quantitative forecasts using numerical models. This work has been organized within the World Climate Program and other international and national scientific programs. It includes research on physical interaction between clouds and radiation, interaction between the ocean (primarily tropical) and the global atmosphere (TOGA project), studies of circulation of the entire World Ocean (at least the upper 1 km layer — Global Experiment on Ocean Circulation), investigation of the cryosphere, in particular, sea ice, and processes on the land surface, in particular, hydrological. Implementation of these and other projects to collect different data will allow us to obtain more reliable parametrizations of various processes and provide the material to verify climatic theory models. This will allow us to narrow the climatic sensitivity limits to changing external factors, including the higher atmospheric concentrations of CO_2 and other "greenhouse" gases, and make more reliable forecasts of the geographical distributions of these variations.

CHAPTER 7

CLIMATIC SENSITIVITY TO CHANGING ATMOSPHERIC GAS COMPOSITION DERIVED FROM DATA ON CURRENT CLIMATIC VARIATIONS

7.1. Current Global Climatic Changes

Climatic changes during the last 100 years are usually called current because this temporal interval is close to the maximum human life expectancy. Since we have data from instrument meteorological observations for the Northern and part of the Southern Hemispheres for this time frame, we can use empirical data to detect many important laws governing global and regional climatic changes that are typical of the modern epoch. The latest advances in this field have been made using theoretical climatic models to study the properties of the global climatic system and the methods of mathematical statistics for empirical analysis of climatic data. The problems covered in this chapter are discussed in greater detail in a recently published monograph (Vinnikov, 1986).

Studies on current climatic changes make extensive use of information on variations in mean hemispheric or global surface air temperature. Although this approach was applied as early as the late nineteenth century (Köppen, 1873), the conclusion that mean air temperature can be used to describe the state of the global climatic system and to study its sensitivity to modifications in the Earth's radiation balance was drawn only in the late 1960s (Budyko, 1968) and advanced in subsequent works on models of climate and its change.

The monograph of Vinnikov (1986) makes a comprehensive survey of data on varying mean air temperature, therefore we will only discuss major sources of information on global climatic change.

In the mid-1970s, several attention-grabbing statements were made that average global temperatures were cooling. These statements were based on incomplete and sometimes unreliable data, which were extrapolated to subsequent decades. It is noteworthy that most of these statements were made at a time when cooling had already been replaced by warming. It was thus obvious that current information was needed on the changing mean global surface air temperature.

Studies in the USSR (Budyko, Vinnikov, 1976; Borzenkova et al., 1976; Vinnikov et al., 1980; and others) have been decisive in establishing the sign of global temperature variations in recent decades. These studies were based on comprehensive analysis of climatological data for surface air temperature of the Northern Hemisphere during the period from 1881 to 1960 prepared in the A. I. Voyeykov Main Geophysical Observatory. The data were published as an atlas (Maps of Air Temperature Deviations from Long-term Averages for the Northern Hemisphere, 1960–1967). The maps were based on data of meteorological observations at approximately 2000 stations, as well as on weather, research and commercial ships, and on drifting Arctic stations.

These and routine maps compiled by the USSR Hydrometeorological Center (Synoptic Bulletin, Northern Hemisphere, 1961–1983) have provided fairly complete information on variations in mean annual surface air temperature for most of the Northern Hemisphere (17.5–87.5°N) for the period from 1881 to the present (Borzenkova et al., 1976; Vinnikov et al., 1980). All the data were carefully checked and if necessary, corrected to ensure homogeneity of the time series. Vinnikov et al. (1980) used a magnetic tape archive of the All-Union Scientific Research Institute of Hydrometeorological Information — World Data Center up-dated with information on mean monthly surface air temperature in the Northern Hemisphere at the $5° \times 10°$ latitudinal-longitudinal grid-points (Gruza, Rankova, 1980).

The main feature of the technique used in these studies to obtain data at the grid-points was that the interpolation was carried out manually, i.e., subjectively. Another feature was the method of data averaging over latitudinal belts based on the following equation:

$$T'(k_1, k_2) = \frac{\sum_{k=k_1}^{k_2} \cos\phi_k \left[\frac{q_k}{N_{Lk}} \sum_{\ell=1}^{N_{Lk}} T'_{Lk\ell} + \frac{(1-q_k)}{N_{0k}} \sum_{\ell=1}^{N_{0k}} T'_{0k\ell}\right]}{\sum_{k=k_1}^{k_2} \cos\phi_k}, \tag{7.1}$$

where N_{Lk} and N_{0k} are the number of grid-points over land and oceans covered by values of air temperature anomalies over the continents ($T'_{Lk\ell}$) and over the oceans ($T'_{0k\ell}$) at latitude ϕ_k, where q_k is the portion of land in the latitudinal belt $\phi_k \pm 2.5°$, and k is the number of the latitudinal belt.

This averaging technique considerably exaggerates the weight of information on air temperature over oceans, where data are fewer and less reliable than over the continents. Consequently, random error of spatial averaging increases, but the probability of obtaining false trends, because of the developing meteorological observational network, decreases. Special data processing methods ensured restoration of the homogeneity of the temporal series that was disrupted when the aforementioned maps of air temperature anomalies were compiled. However, even the corrected series of mean annual surface air temperature in the extratropical Northern Hemisphere (17.5–87.5°N) showed disturbances in homogeneities that can reach 0.03–0.04°C between 1940 and 1941, 0.05°C between 1960 and 1961, and about 0.1°C between 1969 and 1970. This uncertainty is much greater for data on temperature variations in narrower latitudinal zones and for separate seasons or months. This is obviously the most significant shortcoming of this method for processing observational data.

The following conclusions were drawn on the basis of estimated variations in surface air temperature in the Northern Hemisphere for the last 100 years:

— intense warming of the Northern Hemisphere that started in the late nineteenth century ended by the 1940s;

— this warming was followed by cooling in the 1940s that continued to the 1960s;

— the warming was renewed in the Northern Hemisphere starting in the mid-1960s;

— the growth rate of surface air temperature in the Northern Hemisphere over the last 100 years averaged 0.5°C/100 yr.;

— the amplitude of air temperature variations decreases considerably from high and middle to lower latitudes;

— air temperature variations in the cold half-year in high and middle latitudes noticeably exceed those in the warm half-year;

— the mean meridional air temperature gradient decreases as mean temperature rises and increases as the temperature drops.

The mean annual surface air temperature series for the zone 17.5–87.5°N over 100 years obtained in these studies was later extrapolated into the past up to 1579 (Groveman and Landsberg, 1979). The technique was based on the assumption by Landsberg, Groveman and Hakkarinen (1978) that this series for the last century can be reconstructed by multiple regression equations using data from only nine separate stations (the multiple correlation coefficient in this case exceeds 0.8). Similar equations were constructed on the basis of the longest series of measurements of air temperature and a number of indirect climatic characteristics.

Figure 7.1 presents the estimated mean annual surface air temperature in the zone 17.5–87.5°N for 1841–1983 from these studies and compares them with other estimates. We

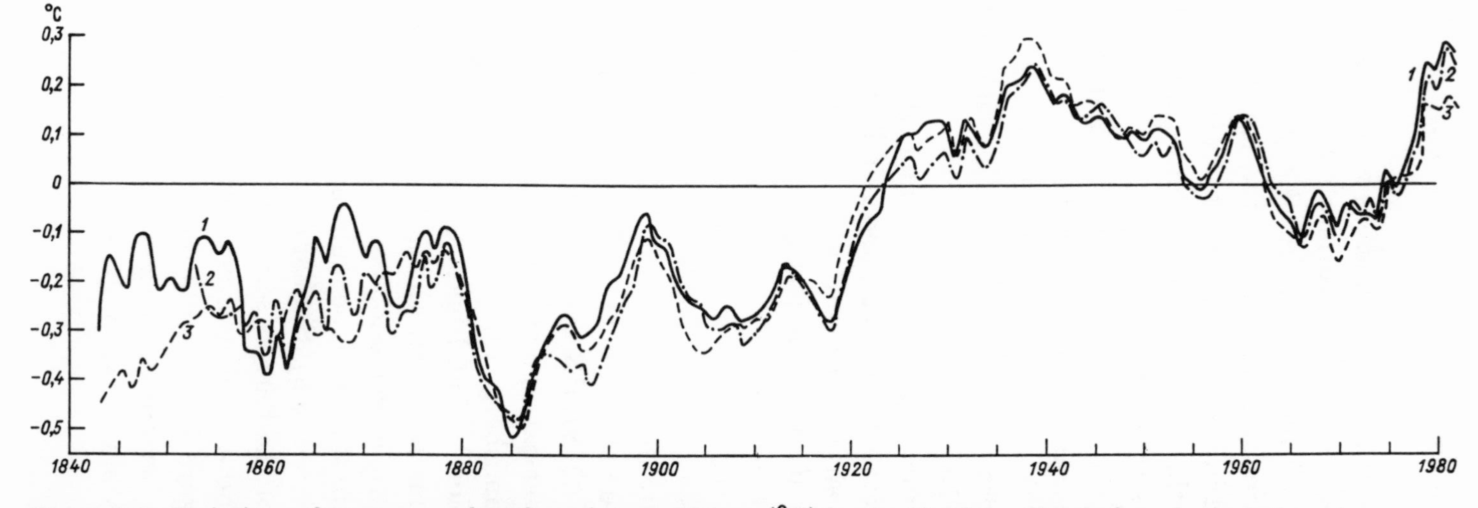

Figure 7.1. Variations of mean annual surface air temperature (°C) in the Northern Hemisphere in deviations from the mean for the period 1951–1975. Estimates are averaged over 5–year intervals. 1, from data of Vinnikov et al., 1987; 2, from Jones et al., 1986; 3, from data of Vinnikov, et al., 1980, and Groveman and Landsberg, 1979.

will show that the estimated mean annual surface air temperature of the extratropical Northern hemisphere (17.5–87.5°N) is fairly close to the mean air temperature variations for the entire hemisphere (0–90°N) and can be used to describe them.

These data on surface air temperature variations in the Northern Hemisphere during meteorological instrument observations were the basis for studies to assess major factors of modern climatic changes, i.e., to make empirical evaluations of mean air temperature sensitivity to altered atmospheric CO_2 concentration and aerosol turbidity, as well as to study regional climatic consequences of mean air temperature variations. In addition to these data, there are other information sources on varying global thermal conditions that pinpoint previous data.

Among these sources, most important is the working monitoring system of current variations in mean monthly surface air temperature of the entire Northern Hemisphere, as well as northern and southern polar regions, developed in England (Jones, Wigley and Kelly, 1982; Kelly et al., 1982; Raper et al., 1984).

This system is based on climatic information published in international (World Weather Records, Monthly Climatic Data for the World) and national reference books. The number of stations used to estimate the Northern Hemispheric mean surface air temperature rose from about 300 in 1881–1900 to approximately 1300 in 1951–1960, and then dropped to 800–900 in subsequent years. Spatial averaging was preceded by interpolation into the 5° × 10° latitudinal-longitudinal grid-points of mean monthly temperature deviations from the average value for the same reference period (1946–1960) for all stations. For the majority of grid-points, the interpolation was made by approximating the pattern of surface temperature deviations in the vicinity of each point by a plane, based on data from the six nearest stations. Subsequent spatial averaging was carried out only for that part of the hemisphere or latitudinal zone, where interpolated values were obtained at the grid-points. This approach distinguishes this method from the method of global spatial averaging based on Equation 7.1.

The time series of mean annual, seasonal, and monthly surface air temperature anomalies for the entire Northern Hemisphere and the latitudinal zones 65–90°N and 65–90°S are available in a digital form and are published regularly in the journal *Climate Monitor*. The analysis by the authors of these studies indicates that the data are high quality and agree well with the aforementioned Soviet studies.

The merits of British climatological studies are the use of the objective data-analysis methods, and consequently, complete reproducibility of all results. However, the experience of using polynomial approximation for an objective analysis of meteorological fields for a numerical weather forecast from the USSR and other countries shows that this method is ineffective and significantly inferior to statistically optimal methods when the station network is sparse.

The most debatable procedure in the British studies is the derivation of mean hemispheric values. It could be incorrect when the density of the meteorological station network in various latitudinal zones changes drastically in time. In spite of support by Jones, Wigley and Kelly (1982) for the correctness of this procedure, it is clear that it can distort the long-term trend components of mean hemispheric values.

The British climatologists confirmed the previous conclusions that warming developed in the Northern Hemisphere in the beginning of the current century, followed by cooling and subsequent warming. This warming began in the second half of the 1960s for winter and spring, and in the mid-1970s for autumn, and later for summer.

The data of the British climatologists are widely used for empirical analysis of the causes of modern global climatic changes and to develop scenarios for altered regional climatic conditions for the epochs with higher atmospheric CO_2 content. The information base was recently expanded and the data analysis technique was noticeably simplified (Kelly et al., 1985; Jones et al., 1986). Jones et al. cite improved estimates of mean annual and mean monthly surface air temperature changes in the Northern Hemisphere for the 1851–1984 period compared with the previously published. These estimates are also presented in Figure 7.1, where they are compared with the evaluations obtained in the USSR. Similar data have recently been published for latitudinal zone 65–90°N for 1851–1984 (Jones, 1985).

The USSR has recently focused on developing statistically optimal methods for global averaging of meteorological fields (Kagan, 1979; Kagan and Lugina, 1981; Vinnikov and Lugina, 1982; Vinnikov et al., 1987; Vinnikov et al., 1989; etc.). These studies are currently underway at the State Hydrological Institute and have obtained new estimated changes in mean air temperature in the Northern Hemisphere for 1841–1985. The estimates obtained by Vinnikov, Groisman, Lugina and Golubev (1987) are also presented in Figure 7.1. They are based on observational data from about 300 long-series stations, that were carefully monitored to rule out random errors and the homogeneity disorders. The use of the statistically optimal technique resulted in fairly high accuracy of spatial averaging by using data from a comparatively small number of stations.

There is a satisfactory agreement among all three smoothed series of estimated altered mean annual surface air temperature in the Northern Hemisphere presented in Figure 7.1. Cross-correlation coefficients of these series for the period after 1880 are about 0.95. Analysis of this series confirms that the mean air temperature in the Northern Hemisphere over the last 100 years rose at an average rate of 0.5°C/100 yr. This overall warming in the Northern Hemisphere was uneven and in some intervals of this period, intensified or even reversed. There are grounds to believe that the warming detected in the Northern Hemisphere is global. This is primarily confirmed by preliminary estimates of varying mean global surface air temperature (Hansen et al., 1981) and data on changing mean surface air and water temperatures over the world oceans (Folland, Parker and Kates, 1984). More correct estimates of changing mean annual and monthly temperature in the Southern Hemisphere for 1851–1984 based on observational data from the continental and island meteorological stations were obtained in Jones, Raper and Wigley, (1986). They showed that the Southern Hemispheric mean temperature has increased by 0.5°C in the last 100 years and this warming was more monotonous than in the Northern Hemisphere.

A similar conclusion is drawn from the simultaneous analysis of data on surface air temperature over the oceans and continents (Jones, Wigley and Wright, 1986). There were no significant variations in the second half of the nineteenth century, then noticeable warming to 1940, relatively stable conditions up to the mid-1970s and subsequent rapid warming This conclusion is also confirmed by preliminary estimates of changing mean annual surface air temperature in the Southern Hemisphere that were obtained in the USSR by statistically optimal methods of spatial averaging.

We will discuss in greater detail the technique for obtaining a time series of mean air temperature based on statistically optimal spatial averaging.

Assume that $f(\phi, \lambda)$ is an averaged meteorological characteristic at a point with geographical coordinates ϕ and λ (latitude and longitude). As the estimated mean value of this characteristic $\Psi\phi_1, \phi_2)$ for the latitudinal belt between ϕ_1 and ϕ_2:

$$\Psi(\phi_1, \phi_2) = \int_0^{2\pi} \int_{\phi_1}^{\phi_2} f(\phi, \lambda) \cos\phi d\phi d\lambda / [2\pi(\sin\phi_2 - \sin\phi_1)] \quad (7.2)$$

we use a linear combination of measurement data of this characteristic $\tilde{f}_i$ $(i = 1, 2, \cdots, n)$ at n separate meteorological stations and with random errors.

Thus:

$$\Psi(\phi_1, \phi_2) \approx \tilde{\Psi}(\phi_1, \phi_2) = \sum_{i=1}^{n} p_i \tilde{f}_i, \tag{7.3}$$

where $i = 1, 2, \cdots, n$ is the number of a station, p_i the weight factors, $\tilde{f}_i$ is measurement data f_i at point i.

Variance of the error from estimating the $\Psi(\phi_1, \phi_2)$ value due to random errors of measurement and spatial averaging procedures equals

$$\mathcal{D}^2 = \overline{(\Psi - \tilde{\Psi})^2}. \tag{7.4}$$

The line above means statistical averaging.

From Equation 7.4 we have

$$\mathcal{D}^2 = M^2 + \sum_{i=1}^{n} \sum_{j=1}^{n} p_i p_j B_{ij} - 2 \sum_{i=1}^{n} p_i \Omega_i + \sum_{i=1}^{n} p_i^2 \delta_i^2, \tag{7.5}$$

where

$$B_{ij} = \overline{f_i' \cdot f_j'} \tag{7.6}$$

is the covariance of the f values at points i and j,

$$f_i' = f_i - \overline{f_i};$$

$$M^2 = \left(\overline{\Psi'(\phi_1, \phi_2)}\right)^2 = \int_0^{2\pi} \int_0^{2\pi} \int_{\phi_1}^{\phi_2} \int_{\phi_1}^{\phi_t} B(\phi, \phi', \lambda, \lambda') \times \cos\phi \cos\phi' d\phi d\phi' d\lambda d\lambda' / 4\pi^2 (\sin\phi_2 - \sin\phi_1)^2 \tag{7.7}$$

is variance of values $f(\phi, \lambda)$ averaged over the latitudinal belt,

$$\Psi'(\phi_1,\phi_2) = \Psi(\phi_1,\phi_2) - \overline{\Psi(\phi_1,\phi_2)};$$

$$\Omega_i = \overline{(\Psi'(\phi_1,\phi_2)f_i')} = \int_0^{2\pi}\int_{\phi_1}^{\phi_2} B(\phi,\phi_i,\lambda,\lambda_i) \times \cos\phi d\phi d\lambda / 2\pi(\sin\phi_2 - \sin\phi_1) \quad (7.8)$$

is mutual covariance of values $\Psi(\phi_1,\phi_2)$ and f_i, $\delta^2 = \overline{(f_i - \tilde{f}_i)^2}$ is the measurement error variance.

The statistically optimal averaging method implies that weights p_i $(i = 1, 2, \cdots, n)$ in Equation 7.3 are determined from the condition of minimum spatial averaging error variance ($\mathcal{D}^2$) and are found by solving the equation system

$$\sum_{i=1}^{n} p_i B_{ij} + p_j \delta_j^2 = \Omega_j, \qquad i,\ j = 1, 2, 3, \cdots, n. \quad (7.9)$$

In practice, when averaged meteorological characteristics contain trend components, for example, if measurement data are processed to reveal hypothetical climatic changes, it is advisable to use the optimal averaging method with normalized weights, that provide the minimum variance in random error of spatial averaging, when an additional condition is fulfilled

$$\sum_{i=1}^{n} p_i = 1. \quad (7.10)$$

In this case, according to Kagan (1979), weights p_i can be calculated by a formula

$$p_i = p_i' + p_i''(1 - \sum_{i=1}^{n} p_i') / \sum_{i=1}^{n} p_i'', \quad (7.11)$$

where p_i' is the weights obtained when equation system (7.9) is solved; p_i'' is the weights obtained by the same system of equations, but assuming that all the values of Ω_i equal one.

The application of statistically optimal methods to spatial averaging is possible, if the spatial autocovariance function of the field of averaged value is specified.

The aforementioned studies (Vinnikov, 1986; Vinnikov et al., 1987) are based on data obtained by K. M. Lugina on mean latitudinal values of standard deviations in mean monthly and mean annual surface air temperature in the Northern Hemisphere presented in Table 7.1. Empirical estimates of the spatial autocorrelation function of mean annual surface air temperature $r(\rho)$ were obtained by Lugina and Speranskaya (1984) and approximated as

$$r(\rho) = \begin{cases} \exp(-0,34\rho^{1,62}) & \text{if } \phi > 60°\text{N} \\ \exp(-0,21\rho^{0,893})J_0(0,852\rho) & \text{if } 30°\text{N} \leq \phi \leq 60°\text{N} \\ \exp(-0,78\rho^{0,63}) & \text{if } 0 \leq \phi < 30°\text{N} \\ \exp(-0,81\rho^{1,11}) & \text{if } 0 < \phi \leq 30°\text{S} \\ \exp(-0,54\rho^{0,88})J_0(0,95\rho) & \text{if } 30°\text{S} < \phi < 60°\text{S} \\ \exp(-0,58\rho^{1,68}) & \text{if } \phi \geq 60°\text{S}, \end{cases} \tag{7.12}$$

where ρ is the distance in thousand kilometers, J_0 is the Bessel function. This approximation is also used to analyze fields of mean monthly surface air temperature.

Thus, the B_{ij} values are calculated by the formula

$$B_{ij} = \sigma_i \sigma_j r(\rho_{ij}), \tag{7.13}$$

where $\sigma_i = \sigma(\phi_i)$, while ρ_{ij} is the distance between points i and j.

Variances in the random error of mean monthly surface air temperature measurements at meteorological stations were assumed to be equal to

$$\delta_i^2 = 0.01(°\text{C})^2. \tag{7.14}$$

Table 7.1. Mean Latitudinal Values of Standard Deviations of Mean Monthly and Mean Annual Surface Air Temperature (°C) in the Northern Hemisphere

ϕ°N	I	II	III	IV	V	VI	VII	VIII	IX	X	XI	XII	Year
90	3.2	2.7	2.4	2.0	1.4	0.8	0.8	0.8	1.2	1.8	2.1	2.7	0.79
85	3.3	2.8	2.6	2.2	1.4	0.8	0.9	0.9	1.2	1.9	2.3	2.9	0.84
80	3.9	3.4	3.1	2.5	1.6	1.0	1.1	1.0	1.2	2.1	2.9	3.4	0.87
75	3.8	3.6	3.2	2.5	1.7	1.2	1.4	1.3	1.4	2.1	3.0	3.6	1.02
70	4.1	3.8	3.4	2.8	1.7	1.7	1.7	1.5	1.7	2.4	3.0	3.7	1.07
65	4.3	4.0	3.3	2.6	1.9	1.6	1.6	1.6	1.6	2.4	3.0	3.9	1.06
60	3.9	3.6	2.8	2.2	1.6	1.4	1.4	1.3	1.3	2.0	2.6	3.6	0.96
55	3.3	3.0	2.4	2.0	1.5	1.4	1.3	1.3	1.3	1.6	2.4	3.0	0.82
50	2.9	2.7	2.3	1.9	1.4	1.3	1.3	1.3	1.3	1.6	2.1	2.7	0.73
45	2.5	2.4	2.0	1.5	1.3	1.2	1.2	1.2	1.3	1.5	1.9	2.3	0.66
40	2.0	1.9	1.7	1.3	1.0	1.2	1.2	1.1	1.2	1.3	1.5	1.9	0.60
35	1.7	1.5	1.4	1.1	1.0	1.0	1.0	0.9	1.0	1.1	1.2	1.6	0.53
30	1.4	1.3	1.1	1.0	0.9	0.9	0.8	0.8	0.9	0.9	1.0	1.2	0.48
25	1.0	1.0	0.9	0.9	0.9	0.8	0.8	0.7	0.8	0.8	0.9	1.0	0.44
20	0.9	0.9	0.8	0.8	0.7	0.7	0.6	0.6	0.7	0.7	0.8	0.8	0.41
15	0.7	0.7	0.8	0.8	0.8	0.7	0.6	0.6	0.7	0.7	0.7	0.7	0.38
10	0.7	0.7	0.8	0.8	0.7	0.6	0.6	0.6	0.6	0.6	0.6	0.7	0.38
5	0.6	0.6	0.8	0.8	0.6	0.5	0.5	0.5	0.6	0.6	0.6	0.6	0.36
0	0.6	0.7	0.8	0.8	0.6	0.5	0.5	0.5	0.6	0.6	0.6	0.6	0.35

Attempts have been made by Vinnikov and Lugina (1982) to estimate the number of evenly distributed stations necessary to obtain global characteristics of the thermal conditions with prescribed accuracy, assuming that the statistically optimal technique for spatial averaging is used. Calculations showed that with 200 to 250 stations distributed over the hemisphere, the variations in mean annual surface air temperature can be tracked with the root-mean-square error of 0.02°C, while half the number of stations ensures a standard error of 0.03°C. These results were considered in the selection of the station network to develop a new system for obtaining current information on modern variations in global temperature conditions based on the optimal averaging method.

The selected network of stations contains 301 stations in the Northern and 265 in the Southern Hemisphere. The deviations in the mean monthly or mean annual temperature from the average over reference periods 1951–1975 for the Northern Hemispheric and 1957–1970 for the Southern Hemispheric stations were spatially averaged.

Mean monthly values in the Northern Hemisphere have been analyzed since 1881. Only mean annual surface air temperatures were analyzed during the preceding period from 1841 to 1880 in the Northern and from 1881 to 1987 in the Southern Hemispheres. Statistically optimal averaging with normalized weights was carried out for the latitudinal belts 60–90°N, 30–60°N, 0–30°N, 0–30°S, 30–60°S, and 60–90°S. However, data could only be obtained for the 60–90°S zone from 1957, when the Antarctic network of meteorological stations was set up.

Summarized calculation results for 1881–1987 are presented in Table 7.2 and Figure 7.2. Statistical characteristics of the series of mean annual surface air temperature for zones 0–90°N, 0–60°S and 90°N–60°S are shown in Table

Table 7.2. Variations in Mean Annual Surface Air Temperature (°C) of the Northern Hemisphere in Deviations from the Mean During 1951–1975

Year	0	1	2	3	4	5	6	7	8	9
1880	—	-0.25	-0.26	-0.34	-0.72	-0.46	-0.48	-0.59	-0.26	-0.10
1890	-0.32	-0.27	-0.37	-0.33	-0.30	-0.24	-0.17	0.04	-0.30	-0.08
1900	0.08	-0.05	-0.27	-0.31	-0.38	-0.17	-0.10	-0.40	-0.33	-0.24
1910	-0.29	-0.22	-0.29	-0.27	-0.07	0.00	-0.21	-0.49	-0.39	-0.21
1920	-0.12	-0.05	-0.12	-0.11	-0.07	0.05	0.27	0.14	0.13	-0.10
1930	0.17	0.28	0.17	-0.19	0.15	0.10	0.12	0.35	0.29	0.14
1940	0.21	0.23	0.19	0.09	0.20	0.03	0.10	0.21	0.14	0.05
1950	-0.05	0.14	0.11	0.28	0.00	-0.09	-0.24	0.03	0.19	0.14
1960	0.10	0.11	0.11	0.11	-0.27	-0.22	-0.03	0.02	-0.13	0.00
1970	0.02	-0.18	-0.21	0.18	-0.14	-0.02	-0.22	0.18	0.08	0.20
1980	0.23	0.49	0.11	0.39	0.08	0.03	0.23	0.35	—	—

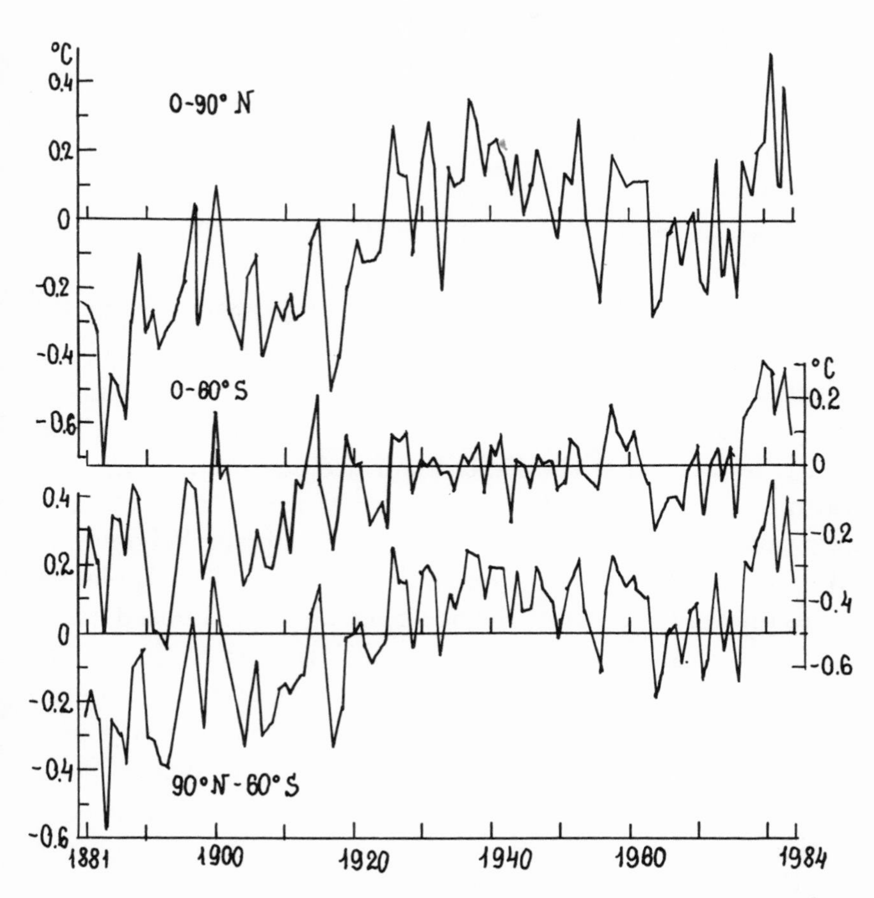

Figure 7.2. Variations in mean annual surface air temperature (°C) in deviations from the mean for the period 1951–1975 for different latitudinal zones.

7.3. Among these characteristics there are elements of the correlation matrix, standard deviations (σ) and linear trend parameters (β), i.e., the mean rates of change in surface air temperature during the 1881–1987 period.

The estimates show the close relationship between the mean air temperature variations in the Northern and Southern Hemispheres. The mean rate of warming in the

Table 7.3. Standard Deviations (σ) and Linear Trend Parameters (β) for Mean Surface Air Temperature During 1881–1987

$\phi°$	$\sigma°C$	$\beta°C/100$ yr.
0–90°N	0.23	0.48
0–60°S	0.19	0.49
90°N–60°S	0.20	0.48

Northern Hemisphere during 1881–1987 was 0.47°C/100 yr., and in the Southern 0.49°C/100 yr. Based on these combined estimates and the potential contribution of the Antarctic polar region (60–90°S), we conclude that since the end of the last century, the Earth's surface air temperature has risen by less than 0.5°C, which agrees well with the aforementioned studies (Hansen et al., 1981; Jones, Wigley and Wright, 1986; Jones, Raper and Wigley, 1986). Table 7.4 gives information about the accuracy of the estimated mean surface air temperatures at different latitudes of the Earth. It presents the root-mean-square errors of spatial averaging computed by Equation 7.5. It is apparent that the estimates are less accurate for the sparser networks in the distant past, however they are fairly reliable for essentially the entire time period.

We emphasize once more that these are new data. Although they have not yet been analyzed in detail, it is already clear that they completely confirm the conclusions drawn from other information sources.

7.2. The Dependence of Mean Air Temperature on Altered Atmospheric CO_2 Content

Chapter 1 noted that the results of empirical analysis of observational data and theoretical investigations show that the main features of current variations in mean air temperature can be explained and described quantitatively by taking into account changes in two major external factors:

Table 7.4. Root-Mean-Square Errors of Mean Annual Surface Air Temperature Spatial Averaging and Standard Deviation (σ) of the Same Series (°C)

ϕ°	1841–1870	1871–1880	1881–1900	1901–1920	1921–1950	1951–1984	σ
60–90°N	—	—	0.26	0.16	0.10	0.04	0.63
30–60°N	0.15–0.20	0.11	0.08	0.07	0.05	0.05	0.25
0–30°N	—	—	0.09	0.06	0.05	0.04	0.20
0–30°S	—	—	0.15	0.09	0.07	0.06	0.22
30–60°S	—	—	0.13	0.10	0.10	0.07	0.21
60–90°S	—	—	—	—	—	0.17	0.48

fluctuations in atmospheric transparency due to changing optical density of the stratospheric aerosol layer, and altered atmospheric CO_2 (and some other gas components) content. Both of these factors, natural and anthropogenic, exert an influence on radiation energy exchange processes that form the global energy balance.

The data on changing global climate and the factors causing them can yield empirical assessments of the parameters of sensitivity of the global climatic system to these factors.

The idea of making empirical estimates of mean air temperature sensitivity to altered atmospheric CO_2 content based on climatic variation data during the period of instrument meteorological observations was proposed by M. I. Budyko (1977b and c), who obtained the value of $\Delta T_c =$ 3.3°C. Similar estimates were published simultaneously by Miles and Gildersleeves (1977), who obtained $\Delta T_c \approx 2.0$–2.2°C.

Gilliland and Schneider (1984) recently made more detailed diagnostic studies. However, their studies were based on poorly substantiated hypotheses regarding the factors of modern global climatic change. Gilliland (1982) found the value of ΔT_c for the Northern Hemisphere to be 2.3°C, while Gilliland and Schneider (1984) found $\Delta T_c \approx 1.6 \pm 0.3$°C. The first study was criticized by Enting, Pittock and Pearman (1984) for the most part, correctly. Schönwiese (1984) quite recently obtained another empirical estimate of this parameter for the Northern Hemisphere ($\Delta T_c \approx 4.1 \pm 0.7$°C).

Vinnikov and Groisman (1981, 1982) showed that the multiple regression method that is used most often in diagnostic studies of the causes of modern climatic changes does not allow asymptotically unbiased empirical estimates to the ΔT_c parameter, and proposed a more accurate analysis technique. They used a transient energy balance model of the Northern Hemisphere to study its ascertained variation in thermal conditions in the last century. They took into consideration the thermal inertia of the climatic system, albedo variations due to transparency fluctuations (turbidity) in the atmosphere, CO_2 concentration effect on longwave outgoing radiation, and the albedo-temperature feedback. The analysis was empirically based on temporal series of mean annual surface air temperature in the Northern Hemisphere, and data on altered atmospheric transparency (turbidity) and atmospheric CO_2 content. The statistical study of the parameters of mean air temperature sensitivity to changes in the main factors that affect the global climatic system was based on the instrumental variable method.

We will consider in more detail the major ideas of these studies, including some later refinements made in Vinnikov (1986).

The equation of the employed diagnostic model looks like:

$$q\frac{dT}{dt} = \frac{S_0}{4}(1 - \alpha(P,T)) - F(T,C) - F_0, \qquad (7.15)$$

where T is the mean annual surface air temperature, t is time, q is the parameter that describes the effective heat capacity of the climatic system relative to comparatively short-term modifications, S_0 is the solar constant, α is the Earth-atmosphere system albedo, $P(t)$ is the characteristic of atmospheric transparency (aerosol turbidity), F is the longwave outgoing radiation, $C(t)$ is atmospheric CO_2 concentration, and F_0 is heat exchange between the upper quasihomogeneous and deeper oceanic layers.

The following parametrizations are used:

$$\alpha(P,T) = \overline{\alpha} + \frac{\partial\alpha}{\partial P}(P - \overline{P}) + \frac{\partial\alpha}{\partial T}(T - \overline{T}), \qquad (7.16)$$

$$F(T,C) = \Gamma(C)(a + BT). \qquad (7.17)$$

Here the bar above means temporal averaging, $\partial\alpha/\partial P$ and $\partial\alpha/\partial T$ are parameters describing albedo sensitivity to variations in the atmospheric transparency (turbidity) and mean air temperature, a and B are empirical coefficients, $\Gamma(C)$ is the function describing the influence of altered atmospheric CO_2 concentration on outgoing radiation. The form of function $\Gamma(C)$ is determined by the hypothesis that the additional heat influx to the Earth due to the CO_2 greenhouse effect is a logarithmic dependence on its concentration. It was also assumed that the term F_0 in the right side of Equation 7.15 can be ignored.

The solution to Equation 7.15 with regard for Equations 7.16 and 7.17 can be written as

$$T'(t) = T'(0)A_1(t) + \frac{\partial\alpha}{\partial P}A_2(t) + \Delta T_c A_3(t), \qquad (7.18)$$

where $A_1(t)$, $A_2(t)$ and $A_3(t)$ are temporal functions, $T'(t) = T(t) - \overline{T}$, and ΔT_c is the parameter of climatic system sensitivity equal to changing mean temperature when the atmospheric CO_2 content doubles.

The first term in the right side of Equation 7.18 describes the influence on altered mean temperature of the initial value of $T'(0)$, and the function $A_1(t)$ alternates sufficiently rapidly. The second term describes mean temperature variations due to atmospheric transparency (turbidity) fluctuations. The third is mean air temperature variations due to altered atmospheric CO_2 concentration. Actually, $A_2(t)$ and $A_3(t)$ represent the result of averaging by exponential smoothing of the temporal series of atmospheric transparency characteristics and CO_2 content. The degree of smoothing is determined by selection of the effective heat capacity values in the climatic system q.

This model is only used for the climatic system of the Northern Hemisphere; it is covered by all the necessary data, including for factors influencing its radiation balance. It was assumed in this case that energy exchange between the hemispheres can be ignored.

Byutner suggested the following approximating expression to describe altered atmospheric CO_2 content during 1883–1977:

$$C(t') = 290 + 1.533 \exp\big(0.0287(t' - 1860)\big), \qquad (7.19)$$

where t' is the number of the year, $C(t')$ is the CO_2 concentration in ppm. This expression can be improved. Two independent series for relative anomalies in direct solar radiation by Pivovarova (1977) and for aerosol optical stratosphere density by Bryson and Goodman (1980) were used as data on $P(t)$.

The Pivovarova series is based on measurement data from 13 actinometric stations within the 40–62°N belt. The mean annual direct radiation intensity in this case was determined from measurements with atmospheric mass of 2 (zenith solar angle equals 60°) at stations in all countries except for the Soviet Union, where the measurements were made at noon. Deviations from the mean were calculated for each station and were derived by the mean for the entire monitoring period. These normalized deviations were averaged for all available stations. Pivovarova's data were repeatedly used in earlier analyses of the influence of atmospheric transparency fluctuations on the mean temperature of the Northern Hemisphere. Studies (Karol', 1977; Pivovarova, 1977; Karol' and Pivovarova, 1978) show that fluctuations in integral atmospheric transparency reflect variations in the stratospheric aerosol layer.

Bryson and Goodman's series is based on data from 42 actinometric stations within the 20–65°N belt. Although data processing and selection of stations are not described by the authors in detail, it is clear that their approach is quite correct for combining monitoring data from different stations, for various seasons and atmospheric masses.

Many studies have used the series of atmospheric volcanic dust content indices by Lamb (1970) and estimated dust content by Mitchell (1975) for this purpose. Both Lamb and Mitchell used the concept that lower atmospheric transparency is caused by solid particles entering the atmosphere during major volcanic eruptions. The concept of stratospheric sulfate aerosol layer was developed in the 1970s. It revealed that climate perturbations were not caused by volcanic dust, but by stratospheric aerosol consisting of tiny sulfuric acid droplets that are mainly produced by gas components of volcanic emissions. Since the emission of solid components does not always seem to correspond to the influx into the stratosphere of gaseous sulfur compounds and to altered optical mass of stratospheric aerosol, the discrepancies between Lamb's volcanic dust content index or Mitchell's estimated volcanic dust mass, on the one hand, and the atmospheric transparency (turbidity) characteristics obtained by measurements of direct solar radiation by the actinometric station network on the other hand, are rather important.

It should be emphasized that Lamb's indices are frequently estimated from mean temperature changes, so they cannot be used to analyze the causes of current variations in global temperature. Lamb and Mitchell's indices are significant only because actinometric data were used to obtain them.

Table 7.5 presents the correlation coefficient matrix from Vinnikov and Groisman (1981) for the four aforementioned atmospheric transparency (turbidity) characteristics.

Table 7.5. Correlation Coefficients of Mean Annual Atmospheric Transparency (Turbidity)

Number of the Characteristic	I	II	III	IV
I	1	–0.68	–0.37	–0.38
II		1	0.43	0.31
III			1	0.87
IV				1

Note: I is the relative anomalies of direct solar radiation from Pivovarova (1977); II is the aerosol optical depth of the stratosphere from Bryson and Goodman (1980); III is the dust veil indices from Lamb (1970); IV is the mass of atmospheric volcanic dust from Mitchell (1975).

The correlation coefficients estimated from data for the 1883–1977 period were fairly low, which shows that the available information is not accurate enough. It differs from that of Bradley and Jones (1985) who did not have the series by Pivovarova in a digital form. Based on all these remarks, we draw the conclusion that the two characteristics of atmospheric transparency (I) and (II) (Figure 7.3), obtained directly by actinometric observation data are preferable.

An independent source of information on past changes in particle concentration in the stratospheric aerosol layer has

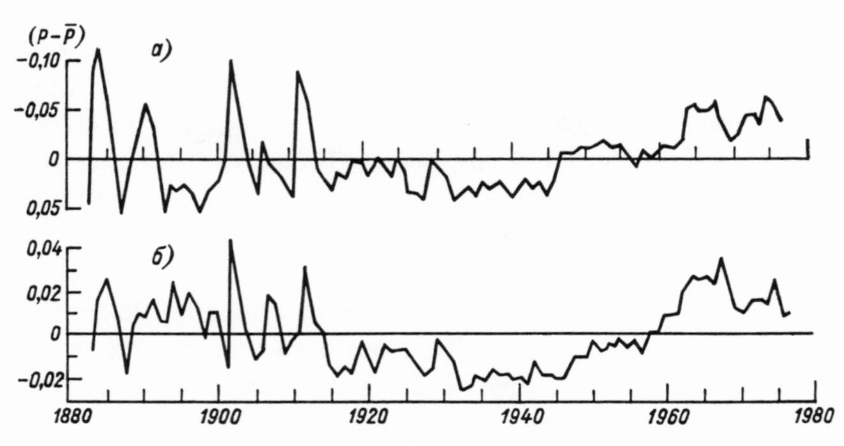

Figure 7.3. Secular variations in mean annual transparency (turbidity) characteristics of the atmosphere. a, from data of Z. I. Pivovarova; b, from data of Bryson and Goodman.

recently been discovered and described in studies by Hammer, Clausen and Dansgaard (1980), Delmas and Boutron (1980), in which ice core samples drilled from Greenland and the Antarctic are investigated for changing sulfate ion concentration that entered the glaciers with precipitation containing settling particles of the sulfate aerosol layer. Since particle concentration in the stratospheric aerosol layer following major explosive volcanic eruptions rises by several orders, these eruptions are easily detected from data on acidity of annual layers in ice cores. These data have been used recently as an objective characteristic for the changing stratospheric aerosol layer. Data from Greenland ice core samples are usually used to characterize the Northern Hemisphere. It is rather complicated to interpret these data, since even a weak eruption of a volcano in North America or Iceland can result in a more noticeable acidity peak than a large eruption far from Greenland.

Without emphasizing the method of statistical evaluation of the value of parameter ΔT_c, we will indicate that the resulting estimates depend on selection of parameter q in Equation 7.15 and on which series of atmospheric transparency (turbidity) characteristics $P(t)$ was used. The method of instrumental variable suggests that together with the basic series, an instrumental variable be used, which excludes the effect on the estimation results of random errors in determining terms of the basic $P(t)$ series. In this case, any of the aforementioned characteristics of atmospheric transparency, including Lamb or Mitchell's indices can be used as the instrumental variable.

The studies of Vinnikov and Groisman (1981, 1982) used the value $q \approx 2W \cdot \text{yr}/(\text{m}^2 \cdot {}^\circ\text{C})$ from data on seasonal variations in radiation balance of the Earth-atmosphere system and mean air temperature of the Northern Hemisphere. The evaluation results for parameter ΔT_c are presented in Table 7.6. This parameter equalled 2.1–4.2°C depending on which data on mean air temperature and atmospheric transparency variations were used. Analysis showed that anthropogenic increment in atmospheric CO_2 concentration led to higher mean annual surface air temperature in the Northern Hemisphere 0.4–0.6°C at the end of the covered period as compared with the mid-1880s. This warming intensified in some time intervals of this period or was obscured by atmospheric transparency fluctuations. Figure 7.4 presents the results of calculating mean air temperature variations and its components that were due to certain factors (Vinnikov and Groisman, 1982).

The study led to the important conclusion that hypothetical absence of CO_2 influences on mean surface air temperature is rejected at the 99% confidence interval. The same conclusion can be easily drawn by comparing the values of ΔT_c and $\sigma_{\Delta T_c}$ from Table 7.6, or by verifying the zero hypothesis ($\Delta T_c \equiv 0$) in the analysis of variance in data on changing mean temperature.

Vinnikov and Groisman later made some improvements in the previous analysis. They showed that calculation of the mean annual surface air temperature in the Northern

Table 7.6. Empirical Estimates of Parameter ΔT_c(°C) and Root-Mean-Square Errors of Estimating It $\sigma_{\Delta T_c}$(°C) Based on Three Series of Mean Annual Air Temperature of the Northern Hemisphere (1–3) and Two Series of Atmospheric Transparency Characteristics (A and B)

	1		2		3	
	ΔT_c	$\sigma_{\Delta T_c}$	ΔT_c	$\sigma_{\Delta T_c}$	ΔT_c	$\sigma_{\Delta T_c}$
A	3.1	0.5	3.7	0.6	4.2	0.7
B	2.1	0.4	2.7	0.5	2.8	0.5

Note: 1 is Vinnikov et al. (1980); 2 is Hansen et al. (1981); 3 is Jones, Wigley and Kelly (1982); A is Pivovarova (1977); B is Bryson and Goodman (1980).

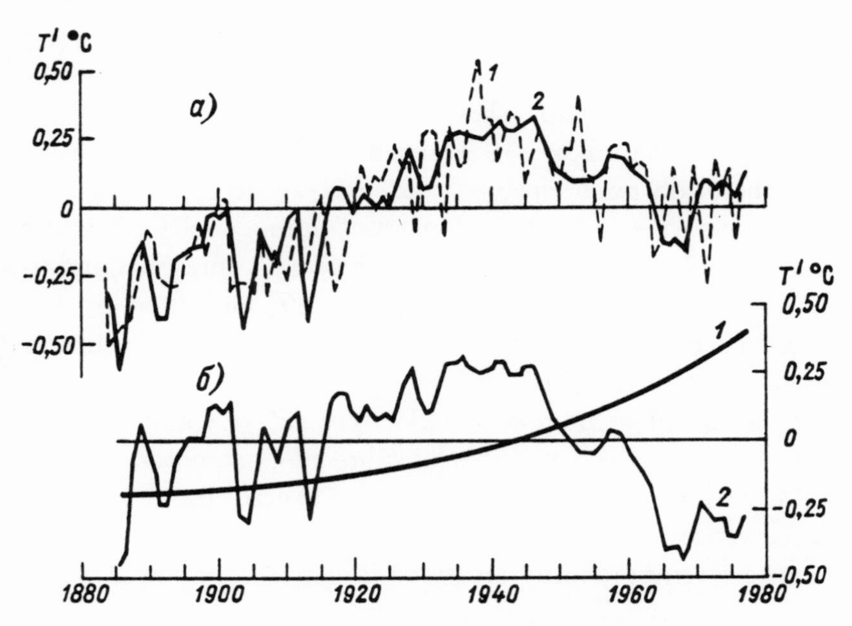

Figure 7.4. Variations in mean annual surface air temperature T' in the Northern Hemisphere (a) and components of air temperature changes due to atmospheric CO_2 variations and transparency fluctuations (b). a: 1, empirical data (Vinnikov et al., 1980); 2, calculation by Equation 7.18. b: 1, with varying CO_2 content; 2, with transparency fluctuations.

Hemisphere over 100 years should use the value of the parameter that describes effective heat capacity $\left(q = 5 \pm 2\text{W} \cdot \text{yr}/(\text{m}^2 \times^\circ \text{C})\right)$. The lower evaluation limit corresponds to the hypothetical absence of heat exchange between the hemispheres, while the upper limit to the hypothetical intense heat exchange. At the same time, corrections were introduced into series $P(t)$ by Pivovarova. Estimates of parameter ΔT_c presented in Table 7.6 differ systematically, and the estimates based on Pivovarova's series slightly exceed those based on Bryson and Goodman's series.

Pivovarova has repeatedly noted that her series of relative anomalies in direct solar radiation starting in the mid-1940s possibly contains a trend of increasing anthropogenic aerosol pollution of the atmosphere near big cities, where essentially all of the long-series actinometric stations are located. This trend for 1959–1982 was assessed at –0.12% per year by comparing it with observational data at 12 mountain stations (Loginov, Pivovarova and Kravchuk, 1983). Budyko and Vinnikov (1973) used slightly different considerations to obtain a close estimate of the corrections needed in Pivovarova's series to reconstruct homogeneity after the mid-1940s (linear trend of 0.1% per year). Analysis demonstrated that estimates of parameter ΔT_c obtained by the corrected Pivovarova series and by Bryson and Goodman's series practically coincide.

The use of value $q = 5 \pm 2\text{W} \cdot \text{yr}/(\text{m}^2 \cdot {}^\circ\text{C})$ and the corrected $P(t)$ series produces a value of parameter $\Delta T_c = 2.6^\circ\text{C}$ with root-mean-square evaluation error $\sigma_{\Delta T_c} = 0.5^\circ\text{C}$. These values do not depend on which information source for mean air temperature variations is used. The signal/noise ratio equal to $\Delta T_c/\sigma_{\Delta T_c}$ is about 5, which corresponds to high statistical confidence that mean air temperature variations due to higher atmospheric CO_2 concentration will be detected.

We note that by ignoring term F_0 in Equation 7.15 that Byutner (1983) believes can be approximately parametrized as

$$F_0 \approx B_1 T', \tag{7.20}$$

we obtain an underestimated value of ΔT_c. The monograph by Vinnikov (1986) shows that empirical estimates of ΔT_c based on the assumption that $b_1 = 0$ are underestimated by about 20–30%.

Estimates of ΔT_c obtained by analysis of empirical data are thus close to the mean value $\Delta T_c \approx 3°\mathrm{C}$ found by climatic theory models.

We will discuss how empirical estimates of sensitivity of latitude-seasonal distribution of surface air temperature are made. For this purpose, Vinnikov and Groisman (1982) have introduced normalized functions of latitudinal-seasonal distribution of surface air temperature response to atmospheric transparency (turbidity) fluctuations $f_P(\phi, m)$ and to altered atmospheric CO_2 content $f_C(\phi, m)$, where ϕ is latitude, $m = 1, 2, \cdots, 12$ is the number of the month (if there is no argument m, the function refers to mean annual conditions):

$$f_{P,C}(\phi) = \frac{1}{12} \sum_{m=1}^{12} f_{P,C}(\phi, m), \tag{7.21}$$

$$\int_0^{\pi/2} f_{P,C}(\phi) \cos\phi d\phi = 1. \tag{7.22}$$

The following assumption was used to describe the determined components of changing mean zonal monthly surface air temperature

$$T'(\phi, m, t) = T'(\phi, m, 0)A_1(t) + f_P(\phi, m)\frac{\partial \alpha}{\partial P}A_2(t) + f_C(\phi, m)\Delta T_c A_3(t). \tag{7.23}$$

This is an approximate expression that is converted into Equation 7.18 for annual and hemispheric averaging.

The same statistical technique that was developed and used to assess parameters ΔT_c and $\partial \alpha / \partial P$ for sensitivity of the entire climatic system was applied to empirical evaluation of $f_{P,C}/(p, m)$. The estimate was based on a temporal series of mean monthly surface air temperature averaged over latitudinal belts ($\phi = 20, 25, \cdots, 85°$N) (Vinnikov et al., 1980).

The estimated normalized functions of the response by mean annual and mean four-season zonally averaged surface air temperature to altered atmospheric CO_2 concentration are presented in Table 7.7. The estimates were made with effective heat capacity of the climatic system $q = 5\text{W} \cdot \text{yr}/(\text{m}^2 \cdot °\text{C})$. These are the mean evaluations obtained using Pivovarova's, and Bryson and Goodman's series of atmospheric transparency characteristics. After Pivovarova's series were corrected by the aforementioned method, the estimates from both series essentially do not differ.

Table 7.7 data distinctly reveal the well-known property of the climatic system that sensitivity of surface air temperature is the greatest in high latitudes in the cold season. These estimates differ slightly from the earlier values with $q = 2\text{W} \cdot \text{yr}/(\text{m}^2 \cdot °\text{C})$ (Vinnikov and Groisman, 1982).

Table 7.8 presents the signal/noise ratios equal to $f_C(\phi, m)/\sigma_{f_C}(\phi, m)$.

Condition $f_C(\phi, m)/\sigma_{f_C}(\phi, m) \geq 2$ was assumed to be a criterion of statistical significance (difference from zero) for estimates of function $f_C(\phi, m)$. If this condition is met, it is possible to detect and reveal air temperature variations due to higher atmospheric CO_2 content. In this case, for high latitudes ($\phi \geq 65°$N) it is possible to detect CO_2 effects from data on mean annual surface air temperature, as well as on winter and summer temperatures, and for low latitudes ($\phi \leq 45°$N) from data on mean annual temperature and temperature of all four seasons. The detection of CO_2

Table 7.7. Empirical Estimate of Normalized Function $f_C(\phi, m)$ for Latitudinal-Seasonal Distribution of Surface Air Temperature Response to Changing Atmospheric CO_2 Content

ϕ°N	Winter	Spring	Summer	Autumn	Year
85	6.5	0.9	0.27	0.2	2.5
80	6.2	1.6	2.0	–0.5	2.3
75	5.4	1.8	1.7	0.5	2.4
70	4.7	1.3	1.1	0.9	2.0
65	3.4	2.0	1.6	1.1	2.0
60	1.2	0.4	0.5	0.2	0.6
55	0.8	0.5	0.3	0.2	0.5
50	1.4	1.1	0.5	0.5	0.9
45	1.8	1.4	0.9	0.7	1.2
40	1.2	0.8	0.6	0.4	0.7
35	0.8	0.8	0.9	0.7	0.8
30	0.9	0.8	0.7	0.7	0.8
25	0.8	0.7	0.6	0.9	0.8
20	0.5	0.6	0.3	0.6	0.5

effect on climate is most difficult for latitudes 50–60°N. The conclusions drawn from Table 7.8 consequently do not agree with the opinion of Madden and Ramanathan (1980) who recommended this zone for detection of the CO_2 influence on climate.

Meteorological observation data from the last 100 years can thus be used as the information base for empirical studies of climatic sensitivity to changing atmospheric CO_2 content.

The empirical estimates presented here are realistic because they agree with the climatic theory models assessments.

The statistical significance of the empirical estimates allows us to consider that climatic changes caused by higher atmospheric CO_2 concentration are detected with a high

Table 7.8. Ratio of the Temperature Change Caused by Higher Atmospheric CO_2 Content to the Root-Mean-Square Error of Estimating This Change

ϕ°N	Winter	Spring	Summer	Autumn	Year
85	3.3	1.4	3.8	1.0	3.3
80	3.6	2.2	3.8	0.6	3.3
75	4.3	2.7	3.7	1.6	4.1
70	4.8	2.3	3.1	2.3	4.3
65	3.9	2.8	4.1	3.0	5.1
60	1.9	1.5	2.4	1.9	2.4
55	1.4	1.8	1.8	1.6	2.4
50	2.3	2.7	2.2	2.4	4.0
45	3.9	3.7	3.6	2.9	6.0
40	3.8	3.2	2.8	2.3	4.0
35	2.8	3.8	4.4	3.8	4.4
30	3.6	3.5	4.2	5.1	4.9
25	4.1	3.0	3.3	5.6	5.5
20	2.4	2.4	1.1	2.9	2.5

degree of confidence. A similar conclusion was drawn in some other studies (Budyko and Vinnikov, 1983; Budyko, Byutner and Vinnikov, 1986; etc.).

7.3. Regional Climatic Changes During Global Warming

Even before the question of unavoidable anthropogenic warming of global climate was raised, study of the relationship between regional variations in climatic conditions and changes in characteristics of global thermal conditions had been started. One of the first attempts of such kind was made by Drozdov (1966), who used regression analysis to link variations in precipitation in the warm and cold seasons to changes in air temperature in the polar basin. He found that the pattern of regional precipitation variations in the Northern Hemisphere corresponds qualitatively to the available physical concepts.

Geographical distribution of the differences in mean values of meteorological elements over quite long periods has been evaluated in many studies (Lysgaard, 1950; Rubinshtein and Polozova, 1966, etc.). Although these estimates described the pattern of climatic changes during that period, at the same time they reflected general features of regional climatic changes due to varying global thermal conditions.

Certain information about these features and their stability was obtained by assessing the geographical distribution of linear trend parameters for meteorological conditions for fairly long periods of time (Van Loon and Williams, 1974; Gruza and Ran'kova, 1980; Jones and Kelly, 1983; etc.).

Similar investigations were made later to build scenarios of possible future regional climatic conditions. They used three simple approaches.

1. The differences in climatic conditions for individual very "warm" years or for the periods of "warm" years from the mean long-term climatic conditions are estimated.
2. The difference in mean climatic conditions of two periods, warm and cold, each covered by observational data for several sequential years is assessed.
3. The difference in climatic conditions for several extremely "warm" and "cold" years during the instrumental meteorological observations is evaluated.

The first approach is represented by studies (Williams, 1980; Jäger and Kellogg, 1983) and conditionally a work by Lamb (1974). The second includes works by Namias (1980); Lough et al. (1983); Potential Climatic Impacts ..., 1984; Bach, Jung and Knottenberg (1985). The third is represented by works of Wigley, Jones and Kelly (1980); Love and Wigley (see Bach et al., 1985), Pittock and Salinger (1982). All these approaches have the same shortcoming; i.e., it is impossible to make full use of all the information available for the period of instrumental meteorological observations.

This noticeably diminishes the statistical significance of the findings.

A somewhat different method was suggested and developed in the works of Vinnikov and Groisman, 1979; Vinnikov and Kovyneva, 1983; Groisman, 1981; Kovyneva, 1984, etc. It studies the relationship between the altered global climatic system described by the mean annual surface air temperature (in these studies, for the Northern Hemisphere) and regional climatic variables, and thus makes greater use of the available climatic information to estimate regional climatic changes that accompany variation in the mean air temperature.

The results of Vinnikov and Groisman (1982) show that empirical estimates of latitudinal-seasonal distribution of surface temperature sensitivity to changing atmospheric CO_2 content and to altered atmospheric transparency have no statistically significant differences.

A similar conclusion was drawn by Manabe and Wetherald (1980) on the basis of numerical experiments with the general atmospheric circulation model. They found that latitudinal distributions of changes in mean annual surface air temperature and hydrological variables, including soil moisture content, with doubled (quadrupled) atmospheric CO_2 concentration essentially coincide with the changes caused by a 2% (or 4%) increase in the solar constant.

Hansen et al. (1984) later found that this conclusion refers both to latitudinal-seasonal and spatial distribution of temperature field changes and thus of other climatic elements.

This capacity of the climatic system to have similar reactions to additional heat fluxes with different latitudinal distribution permits a direct study of the relationship between variations in mean annual surface air temperature in the Northern Hemisphere (a major parameter describing the state of the global climatic system) and changes in regional climatic characteristics without being distracted by the physical mechanisms that are responsible for certain fluctuations in the global thermal conditions.

Vinnikov and Groisman (1979) suggested a simple linear statistical model for this purpose

$$Y_i \approx \alpha_i T + \beta_i, \qquad (7.24)$$

that describes the relationship between different regional climatic characteristics in various seasons Y_i ($i = 1, 2, \cdots$) and mean annual surface air temperature in the Northern Hemisphere T (α_i and β_i are model parameters). The model parameters are assessed by the method of instrumental variable, which produces valid asymptotically unbiased estimates of these parameters when there are random errors in the observational meteorological data. The estimation algorithm is described more comprehensively in Groisman (1979). The influence of global warming or cooling in the Northern Hemisphere on changes in the normal fields of the major meteorological elements: air temperature, atmospheric pressure, precipitation, some characteristics of atmospheric circulation and land hydrology were studied in Vinnikov and Groisman (1979); Vinnikov and Kovyneva (1983); Groisman (1981, 1983); Kovyneva (1982, 1984). We will discuss the main results of this research.

An empirical model by Vinnikov and Groisman (1979) of modern changes in zonal air temperature in the Northern Hemisphere was constructed for mean annual and mean four-season temperature, and in 1982 for mean monthly temperature. The estimated parameters were based on a series of mean monthly surface air temperature averaged for latitudinal belts 85, 80, $\cdots$, 20°N for 1881–1980 (Vinnikov et al., 1980). Improved model parameter estimates are presented in Table 7.9.

Parameters of model α_i are dimensionless in this case and show how many times the air temperature changes at a given latitude and in the corresponding season or year differ from variations in mean annual surface air temperature of extratropical latitudes of the Northern Hemisphere (17.5–87.5°N). These data confirm the extant hypotheses that variations in global thermal conditions are mainly apparent in high latitudes, and particularly in the cold season. Although the 95%-confidence intervals in Table 7.9 are very broad, they show, however, that almost all estimates of the parameters differ significantly from zero. Based on the estimates in this table, we conclude that global warming or cooling for the mean annual and mean seasonal zonal surface air temperature have the same trend at all latitudes north of 20°N. Perhaps, they are the same in lower latitudes of the hemisphere.

Based on a comparison of estimates in Table 7.9 and estimates of function $f_C(\phi, m)$ of similar sense in Table 7.7, it is clear that their differences are not statistically significant, therefore, the reaction of the climatic system to factors that influence global radiation balance is not specific and does not depend on the physical nature of the modifying factor.

This empirical pattern agrees qualitatively well with conclusions based on climatic theory models (Manabe and Stouffer, 1980; Hansen et al., 1984, etc.). However, the sensitivity of the empirical estimates versus climatic models is more dependent on latitude and season. It is possible that these models do not realistically describe the feedback mechanisms between albedo and air temperature. This difference can be partially attributed to the fact that the aforementioned theoretical estimates characterize changes in the temperature field, when atmospheric CO_2 content is very altered (doubled or quadrupled), causing a change in mean air temperature by several degrees, and the empirical model is based on modern climatic change data, when the range of changes in mean air temperature did not exceed ±0.5°C.

Table 7.9. Empirical Model of Changing Zonal Air Temperature in the Northern Hemisphere: Estimated Parameter α and 95%-Confidence Intervals ($\wedge$) and ($\vee$)

		ϕ°N													
Season		85	80	75	70	65	60	55	50	45	40	35	30	25	20
Winter	α	5.6	5.8	5.2	4.5	3.3	1.7	1.2	1.1	1.2	1.0	0.7	0.7	0.7	0.7
	$\wedge$	9.5	9.6	7.8	6.3	4.6	2.9	2.5	2.5	2.3	1.6	1.3	1.1	1.1	1.0
	$\vee$	2.6	3.0	3.1	3.0	2.1	0.6	0.1	0.0	0.2	0.3	0.3	0.3	0.4	0.3
Spring	α	1.7	2.1	2.1	1.9	1.9	0.9	0.7	0.8	0.7	0.6	0.7	0.6	0.4	0.3
	$\wedge$	3.6	4.1	3.7	3.4	3.1	1.6	1.4	1.6	1.3	1.0	1.1	1.0	0.8	—
	$\vee$	-0.1	0.5	0.7	0.7	0.9	0.2	0.1	0.2	0.2	0.2	0.3	0.3	0.1	—
Summer	α	1.4	1.4	1.5	1.5	1.5	1.0	0.8	0.7	1.0	0.7	0.8	0.6	0.2	-0.1
	$\wedge$	2.6	2.4	2.5	2.3	2.4	1.6	1.3	1.2	1.5	1.3	1.4	1.0	—	—
	$\vee$	0.4	0.7	0.7	0.0	0.9	0.5	0.4	0.4	0.5	0.3	0.4	0.2	—	—
Autumn	α	2.6	2.5	2.6	2.5	2.3	1.5	1.2	1.1	1.0	0.8	0.7	0.7	0.6	0.3
	$\wedge$	4.3	4.4	4.1	3.7	3.5	2.4	2.0	1.7	1.5	1.2	1.1	1.0	0.9	0.7
	$\vee$	1.1	0.9	1.5	1.6	1.3	0.8	0.6	0.6	0.6	0.5	0.5	0.5	0.3	0.1
Year	α	2.8	2.9	2.8	2.6	2.2	1.3	1.0	1.0	1.0	0.7	0.7	0.6	0.5	0.3
	$\wedge$	4.5	4.6	4.2	3.5	3.2	1.9	1.5	1.4	1.3	1.1	1.2	0.9	0.8	0.6
	$\vee$	1.5	1.7	1.8	1.8	1.6	0.8	0.5	0.5	0.6	0.5	0.5	0.4	0.3	0.1

Schematic zonal data presentation impairs the practical use of quantitative estimates obtained by this model. The next step, therefore, had to study the geographical distribution of changes in the fields of major meteorological elements during global warming and cooling.

The magnetic tape archives of climatic data prepared in the USSR and United States were the basic information sources to assess the influence of global warming and cooling on the fields of mean annual and mean seasonal values of main meteorological elements:

—The archive of mean monthly anomalies in surface air temperature at regular 5° × 10° grid-points for the Northern Hemisphere during 1891–1983 (USSR, All-Union Scientific-Research Institute of Hydrometeorological Information – World Data Center (VNIIGMI-WDC), Main Geophysical Observatory (MGO), Hydrometeorological Center of the USSR (HMC));

—The archive of mean monthly and mean annual values of air temperature and precipitation for 148 stations of the USSR from 1891 through 1981 (USSR, VNIIGMI-WDC);

—The archive of mean monthly and annual temperature values of 1135 foreign stations (USSR, VNIIGMI-WDC-MGO);

—The archive of mean monthly and annual precipitation totals at 980 foreign stations (USSR, VNIIGMI-WDC-MGO);

—The archive of mean monthly climatic surface data at 2500 global stations (USA, NCAR);

—The archive of mean monthly pressure at the regular 5° × 5° grid-points during 1899–1972 (USA, NCAR);

—The archive of monthly precipitation totals in percentage of the norm for 619 stations of the USSR during 1891–1977 (USSR, VNIIGMI-WDC);

—The archive of mean monthly characteristics of

coordinates and intensity of eight "centers of atmospheric action" of the Northern Hemisphere during 1891–1975 (USSR, VNIIGMI-WDC, HMC);

—The archive of mean monthly characteristics of the frequency of different types of atmospheric circulation in the Northern Hemisphere (USSR, HMC-Arctic and Antarctic Scientific-Research Institute - Institute of Geography, USSR Academy of Sciences).

These archives contain a considerable portion of the available climatic information.

Estimates of parameter α that describe the linear relationship between the mean annual and mean four-season temperatures on the one hand, and mean annual surface air temperature of the hemisphere, on the other hand, have been obtained for individual stations and regular grid-points based on long-term series of mean monthly surface air temperature from these archives.

Figure 7.5 shows the estimated model parameters Equation 7.24 for annual, winter and summer mean surface air temperature. The estimates of α-parameters are dimensionless and show how many times regional changes in surface air temperature in a given season differ from the changes in mean annual surface air temperature in the Northern Hemisphere.

There is a new feature in the altered surface air temperature field compared to the zonal model. Hemispheric warming and cooling are no longer of the same sign. There are areas (hatched areas on maps), where global hemispheric warming is followed by lower air temperature.

The greatest regional changes in surface air temperature that accompany global warming or cooling in the hemisphere are observed in winter. They are particularly drastic in the Arctic and circumpolar regions. During hemispheric warming, winter becomes colder in some regions of Eurasia and North America.

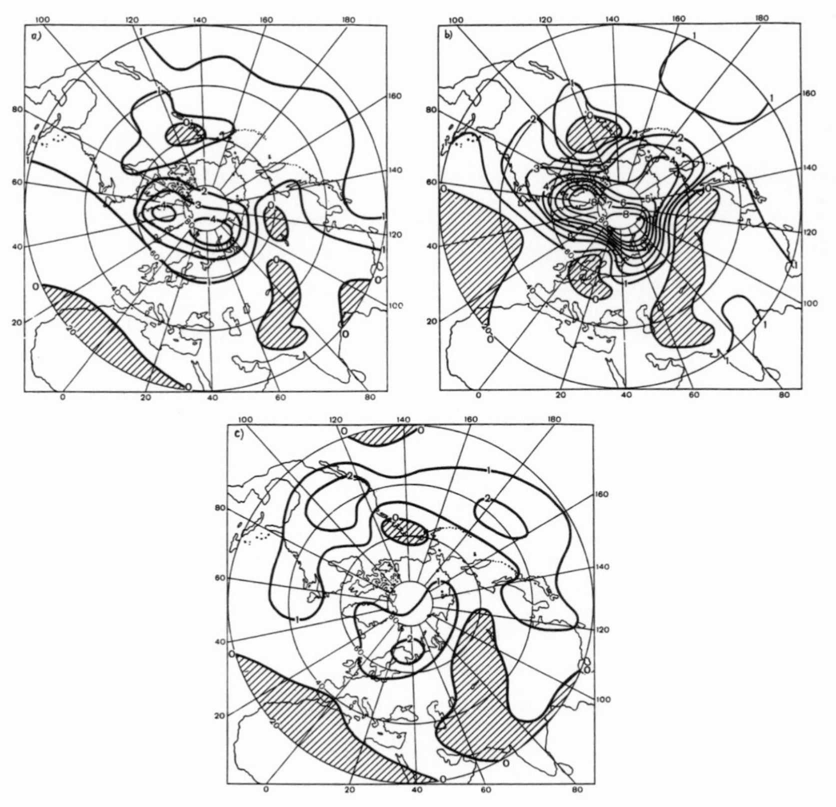

Figure 7.5. Estimates of parameters α of linear dependence of mean annual (a), average for winter (b) and summer (c) seasons of surface air temperature from the mean annual surface air temperature in the Northern Hemisphere.

In summer, warming is stronger in high latitudes of the Eastern Hemisphere and in temperate latitudes of the Western. The scale of summer changes is considerably smaller than in other seasons. Kovyneva (1982, 1984) discusses the estimates for all seasons.

The maps of estimates for α_i for the surface temperature field confirm previous conclusions that high latitudes are more sensitive to variations in global thermal conditions, and at the same time indicate the great importance of circulation mechanisms responsible for nonzonal components of climatic response by the temperature field to global warming or cooling.

Analysis of the altered surface pressure field shows that with varying mean air temperature, the position and intensity of the main pressure centers changes.

Kovyneva (1982) reveals the following important features of higher mean air temperature in the Northern Hemisphere: In winter, the Siberian anticyclone intensifies and its center shifts to the northwest. General attenuation of cyclonic activity over the Atlantic and intensification over the Pacific occur. The Azorean anticyclone center shifts to the northeast; in the 55–70°N zone, on the whole, atmospheric pressure rises, which should result in overall attenuation in cyclonic activity of high and middle latitudes of the Northern Hemisphere.

In spring, atmospheric pressure drops everywhere except for Western Europe, Canada and the western coast of the United States. Cyclonic activity increases over the Pacific. The Icelandic low and Azorean high decrease somewhat, and their centers shift slightly to the south.

In summer, cyclonic activity over the Atlantic increases; the Azorean pressure peak shifts to the northeast; atmospheric pressure rises over Europe, drops over Asia, and the Asian pressure minimum intensifies; the Honolulu high shifts to the north, while the California low to the south.

In autumn, cyclonic activity increases over the Atlantic. The Aleutian low deepens and the Honolulu high attenuates. The Asian high attenuates.

The same but weaker pattern that occurs in winter is revealed for the mean annual pressure field.

Based on data by Sorkina (1972) Groisman confirmed the conclusions on changing position and intensity of major atmospheric "action centers" in the Northern Hemisphere.

Systems of synoptic classification of circulation processes in the Northern Hemisphere of Dzerdzeevsky (1976) and

Vangengeym-Girs (Girs, 1974) are frequently used to analyze atmospheric circulation to make long-range weather forecasts. Groisman (1983) showed that warming of the Northern Hemisphere is accompanied with fewer days with polar invasions (elementary circulation mechanisms 3–12 according to Dzerdzeevsky) and the type W circulation (zonal according to Vangengeym-Girs) in all seasons. In spring and autumn, the frequency of type C circulation increases during warming, and in winter, the number of days with type E circulation (C and E are meridional circulation types according to Vangengeym-Girs).

The main results of studying modern changes in the precipitation field for most of the Northern Hemisphere were published in studies (Vinnikov and Groisman, 1979; Groisman, 1981; Vinnikov and Kovyneva, 1983; Kovyneva, 1984; Vinnikov, 1986). Maps were plotted by Groisman for the USSR and by Kovyneva for the rest of the hemisphere.

Estimates for the oceans were based on precipitation measurements from island stations, however they were insufficient to construct a system of isolines. Therefore, only the sign of precipitation changes was assessed and zero isolines were drawn for the oceans. In this case, it was assumed that the sign of precipitation change with varying mean temperature coincides with the sign of changing vertical component of the geostrophic eddy.

Figure 7.6 depicts a schematic map reflecting the major pattern of changes in mean annual precipitation during warming (or cooling) of the hemisphere. Parameters α_i on the map show the percentage change in precipitation from the mean with a 0.1°C rise in temperature. Similar maps for four seasons are published in a work by Kovyneva (1984).

Analysis has shown that the fields of precipitation norms are very sensitive to global warming and cooling, and in various regions and seasons, precipitation changes often have a different sign.

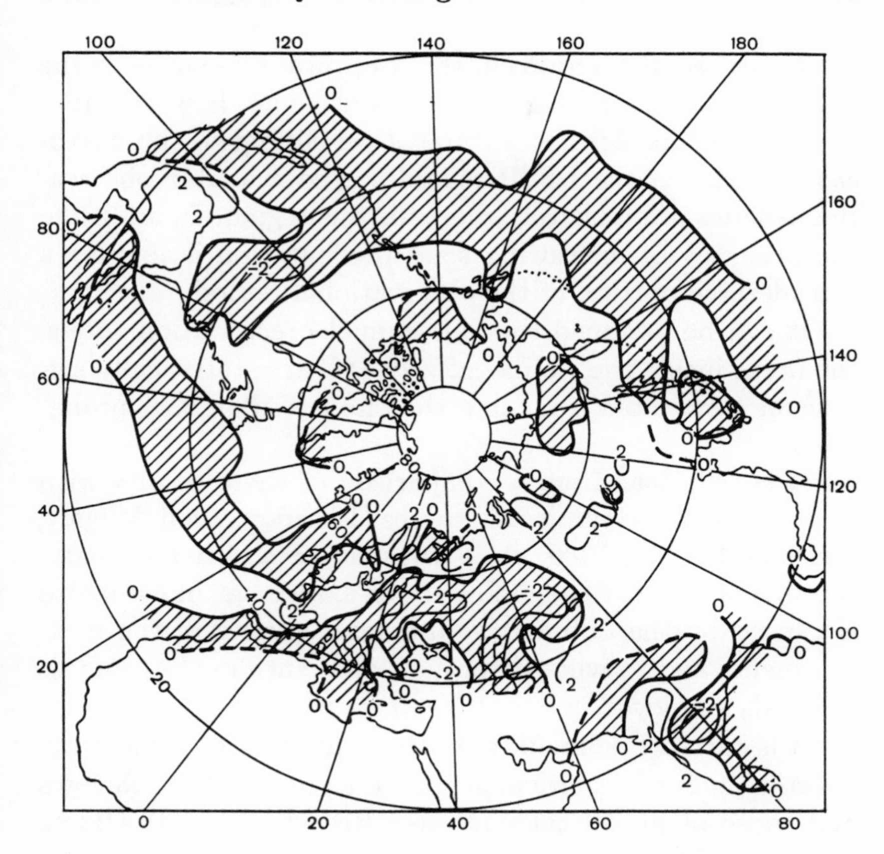

Figure 7.6. Estimates of dependence of variations in mean annual precipitation on changes in mean annual surface air temperature in the Northern Hemisphere. Estimates of parameter α are expressed in % of the norm with a 0.1 °C change in temperature.

For the year as a whole and throughout all seasons, during warming there is a belt of considerably lower precipitation in the middle latitudes on the hemispheric maps. On the average for the year, this belt is situated on the continents between 35 and 50°N, including steppe and forest-steppe zones of Eurasia and North America.

The greatest scale of precipitation changes is observed in winter and autumn, and the smallest, in the warm season. A zone of lower precipitation in Eurasia on the average is slightly more northward than in America.

There are distinctions in the precipitation change in the western and eastern parts of the continents, however, it is difficult to consider the estimates pertaining to the northeastern part of the USSR reliable. The available observational series are relatively short for this region. In addition, the accuracy of measuring solid precipitation is low for a considerable portion of the observational period.

It can be assumed that for annual precipitation totals, the areas inside the isoline $\pm 2\%/0.1^{\circ}C$ are statistically significant (separated from zero with no less than 95% probability).

It is not clear from a comparison of these results with the data of other empirical studies (Drozdov, 1966; Wigley, Jones and Kelly, 1980, etc.) without an additional analysis, to what extent the difference or agreement in estimates is caused by the different volume and quality of the measurements and to what extent, by difference in the analysis methods employed.

It is currently difficult to ensure complete agreement between results of empirical studies on precipitation changes and those of model calculations. Recent attempts (Bach, Jung and Knottenberg, 1985) are not reassuring.

The following conclusions can be drawn from the available materials on modern climatic change. The state of the global climatic system has changed considerably over the last 100 years due to global warming. The surface air temperature in the Northern hemisphere rose during the century at an average rate of about 0.5°C/100 yr. This warming was interrupted by cooling in the 1940s and was renewed in the mid-1960s. Mean air temperature in the Southern Hemisphere varied similarly. There is no doubt that the climatic warming observed during the last 100 years is global.

The modern methods of global averaging of measurement data from meteorological stations provide satisfactory accuracy and statistical significance of the conclusions regarding modern changes in surface air temperature in the Northern Hemisphere. The data and conclusions obtained for the Southern Hemisphere need to be improved.

It has been found that the anthropogenic rise in atmospheric CO_2 has caused an increase in surface air temperature in the Northern Hemisphere by 0.4–0.6°C since the 1880s. In some parts of this period, the warming intensified or was obscured by atmospheric transparency fluctuations. The contribution of variations in atmospheric CO_2 concentration to the variance of mean annual surface air temperature is comparable to the influence of atmospheric transparency fluctuations.

Empirical estimates of the sensitivity of mean annual surface air temperature in the Northern Hemisphere to varying atmospheric CO_2 content agree with the most valid theoretical estimates, confirming that they are realistic. Empirical estimates of parameter ΔT_c that describe mean air temperature change when the atmospheric CO_2 concentration doubles are $\Delta T_c \approx 2.6 \pm 0.5$°C for modern climatic changes. This estimate is probably underestimated by 20–30% because heat exchange between the upper and deeper oceanic layers is ignored.

We note that the results of studying climatic sensitivity by climatic theory models and the analysis of empirical data show that variations in regional climatic conditions accompanying variations in mean air temperature do not depend on the physical nature of the factor that influences the global radiation distribution.

Warmings or coolings in the Northern Hemisphere within the ±0.5°C range typical of the last 100 years are accompanied by considerable variations in precipitation and other meteorological elements. In particular, the greatest air temperature changes occur in high latitudes in winter. The scale of the observed temperature changes somewhat exceeds that predicted by climatic theory models. When the

Northern Hemisphere is warmed within the indicated limits, precipitation decreases in steppe and forest-steppe zones of North America and Eurasia, and the scale of these changes is greater in the cold season than in the warm. This decrease in precipitation in the middle latitudes is mainly caused by changes in atmospheric circulation and is observed not only on the continents, but also over the oceans.

The most common features of the changing mean annual precipitation field revealed by analyzing data on modern climatic changes are confirmed by climatic theory models and the results of climatic reconstructions of the Holocene climatic optimum epoch. This allows the empirical estimates to be used approximately for a greater scale of global warming, possibly, up to 1°C.

The resulting empirical estimates of climatic sensitivity can serve as the empirical basis for developing a forecast for climatic changes caused by current anthropogenic variations in the Earth's atmospheric chemical composition.

CHAPTER 8

PALEOCLIMATIC EVIDENCE OF CLIMATIC SENSITIVITY TO CHANGING ATMOSPHERIC GAS COMPOSITION

8.1. Introduction

Study of the history of altered atmospheric gas composition showed that practically all the warmings in the distant past occurred when the CO_2 concentration was higher (Budyko and Ronov, 1979; Budyko, Ronov and Yanshin, 1985). Fossil fuel burning today as it were restores the atmospheric gas composition that is characteristic of warm Cenozoic periods. Since the anthropogenic trend of CO_2 concentration will be preserved in the near future, climatic conditions in the beginning and mid-twentieth century will be close to climates of warm periods in the Pleistocene (interglacial) and Pliocene. Consequently, detailed climatic reconstructions for these epochs had to be made.

Paleoclimatic data can be used along with analysis of empirical data for the last hundred years and model computations as an independent approach to estimating future climatic conditions. The idea of using paleoclimatic data as analogs of future climate was proposed in the mid-1970s (Soviet-American Workshop..., 1977). The first estimates of possible climatic changes with double atmospheric CO_2 concentration were made at approximately that time on the basis of maps constructed by V. M. Sinitsyn for the Pliocene (Budyko et al., 1978).

The series of paleoclimatic maps compiled by Sinitsyn in the 1950s and 1960s made a great contribution to paleoclimatology of the mid-twentieth century. Great progress in Earth science after that substantially altered the situation in paleoclimatology. The amount of information on climates of the past has recently increased considerably. Rapid progress in paleoclimatic research was stimulated by advances both in physical climatology and geology, and was associated with development and improvement in new methods, including isotope and paleomagnetic, as well as application of statistics to analysis of paleontological and palynological evidence.

The development of deep-sea drilling technology produced unique facts on world ocean climate in the last 100–150 Ma.

The application of statistical methods to analyze microfaunal evidence obtained using an improved technique of deep-sea drilling permitted reconstruction of the global world ocean surface temperature for narrow temporal intervals of the Pleistocene (Imbrie et al., 1973; CLIMAP, 1976, 1984; Barash, 1985; Safarova, Nikolayev and Blyum, 1984).

The data needed to reconstruct past climatic conditions on the continents have been obtained using paleobotanical (spore-pollen) and faunal evidence, as well as data on level fluctuations in lakes and reservoirs, mountain glaciers and ice sheets, etc.

The USSR has been most successful in developing methods for studying paleoclimates using spore-pollen data. These studies resulted in a series of air temperature and precipitation reconstructions for the territory of the USSR and Western Europe for several temporal intervals of the Pleistocene and Holocene (Paleogeography of Europe during the last Hundred Thousand Years, 1982; Velichko et al., 1982, 1983, 1984; Klimanov, 1982; Burashnikova, Muratova and Suyetova, 1982).

Oxygen isotope data on fresh-water lake carbonate (Mörner, 1980) cave deposits (stalactites and stalagmites), travertines and continental ice (Kotlyakov and Gordiyenko, 1982) proved very promising. Ice-sheet core samples from Greenland and Antarctica were the richest sources of new paleoclimatic information on atmospheric CO_2 concentration and the intensity of explosive volcanic eruptions (Hammer, Clausen and Dansgaard, 1980; Neftel et al., 1982; Hammer, 1984; Lorius et al., 1985).

8.2. Methods of Paleoclimatic Reconstructions

The extant methods for studying past climates can be divided into four groups:

1) lithological-genetic,
2) ecological-palaeontological,
3) isotope,
4) geochemical.

Lithological-genetic methods incorporate two basically different subgroups. The first describes lithogenetic processes reflecting types of global climatic conditions existing over a long period of time. This methodology is the basis for isolation of long-term variations in climate from 2–3 Ma to hundreds of millions of years (Sinitsyn, 1966, 1967, 1970). The second subgroup, on the contrary, permits detection of relatively brief climatic variations related to the change in structural deposits. For example, the alteration of loesses with fossil soils or lake silts with salt levels can serve as an excellent indicator of deposits in the regions of moisture deficit, where there is essentially no other type of information. Cryogenic textures that are related to phase transitions of water into ice and vice versa not only allow us to obtain permafrost area data, but also to assess the amplitude of seasonal temperature variations. These data are important, in particular, because they indicate climatic conditions of cold epochs, for which there is considerably less information than for warm ones.

The study of different weathering crusts are considered in detail in the works by Sinitsyn (1966, 1967) and other authors as a very valuable source of quantitative climatic information, especially for the pre-Quaternary climates. Gerasimov (1979) studied the possibilities of obtaining quantitative information on this group of formations as well as on fossil soils of the Pleistocene, although areal and not specific points. The progress in quantitative and qualitative assessment of the climate in the past was summarized in *Methods for Palaeoclimatic Reconstructions* (1985).

Ecological-palaeontological methods are used most widely to estimate thermal and moisture conditions. These methods are discussed in detail in the work of Velichko (1985), therefore, we shall consider only basic peculiarities of their application.

Methods of paleoclimatic reconstructions based on studies of modern benthic sediments and land deposits differ significantly. Paleoclimatic evidence varies more for the continents than the oceans. The most detailed information, including that for determining temperature and precipitation, can be obtained from Paleobotanical data (mainly, spore-pollen) processed by the arealographic method suggested by Iversen and elaborated by Grichuk (1985). This method can determine air temperature for different seasons with accuracy to 1°C and annual rainfall with accuracy to 25 mm, providing quite a few fossil pollen species are analyzed.

Information-statistical methods have recently been used extensively to process paleobotanical (spore-pollen) information. Different transfer functions used in these methods are proposed in the studies by Webb, Street and Howe (1980), Klimanov (1982), Burashnikova, Muratova and Suyetova (1982), Liberman, Muratova and Suyetova (1985).

Reliable data on air temperature and precipitation for individual points can be obtained within the Holocene by the zonal method that uses the correlation between modern plant associations (steppe, forest, tundra, etc.) and climatic indications to reconstruct their past value (Khotinskiy and Savina, 1985).

It seems that the arealographic method of reconstructing temperature and precipitation by combining climatic characteristics for the majority of definite plant species is the most universal for the entire Quaternary. The results of experiments on the use of this method for even earlier periods of Cenozoic are encouraging.

The method of arealograms has recently been successfully applied to small mammals (rodents, first of all) as well as to fossil insects (Velichko, 1985).

For paleoclimatic investigations of land, as a whole, especially when studying temporal variations of climate, various methods and their combinations should be applied, but in this case we have to take into account their reliability, precision and limits of information provided by each of them.

Study of benthic oceanic sediments is a valuable constituent of global paleoclimatic investigations. Paleoecological analysis of marine microfauna is based on the isolation of ecologically representative species dwelling at a certain depth and depending most of all on summer or winter water temperature. The use of factor analysis for microfaunistic associations makes it possible to determine the function of transition from the current correlation between water temperature and species composition of plankton to water temperature in the past.

The method of factor analysis of marine microfauna was developed by Imbrie et al. (Imbrie et al., 1973; CLIMAP, 1976, 1984) and Barash (1985), and is widely used to reconstruct water surface temperature (CLIMAP, 1976, 1984; Blyum, 1984; Safarova, Nikolayev and Blyum, 1985).

Isotopic Methods. The isotopic paleothermometry is based on the assumption that there is equilibrium exchange between oxygen isotopes (^{16}O and ^{18}O) of water and calcium carbonate precipitating from it. When the isotope composition of sea water is constant, data on oxygen-isotope composition of carbonate marine fauna (planktonic and benthic) provide reliable information about surface and bottom water temperature, and a 1‰ change in isotope composition corresponds to a 4°C temperature change.

The problem of interpreting isotopic data pertaining to the period of continental ice sheet formation, when the natural process of oxygen isotopic fractionation occurred is the most difficult one. Duplessy (1980), Savin (1982), Dansgaard (1971), and some other scientists believe that during the changing volumes of continental ice, the oxygen isotopic data obtained on planktonic and benthic foraminifera reflect the change in isotopic composition of sea water to a larger extent (practically by 2/3) than the temperature change in surface or bottom water.

Consequently, many scientists propose that the post-Miocene (the last 15 Ma) paleotemperature data should be regarded not as quantitative, but as a qualitative characteristic of temperature.

Interesting data of oxygen isotope analysis of lake (fresh-water) carbonates has been obtained in recent years. Similar investigations were carried out on lakes in Switzerland, Sweden and Estonia (Mörner, 1980; Eicher Siegenthaler and Wegmüller, 1981; Punning and Raukas, 1985). These data combined with palynological information produce detailed characteristics of temperature conditions for the Late Quaternary.

One of the most important problems in summarizing various types of paleoclimatic data is dating of the evidence. Scales based on paleomagnetic and absolute methods of dating are widely used, in addition to biochronological (biostratigraphic) temporal scales.

Very reliable dating is ensured by some physical methods that are based on radioactive isotopic decay. The most popular is the carbon isotopic method, although its range is limited to the last 30,000–40,000 years. Another method, potassiumargon, allows dating in a broader temporal interval (from a thousand to a billion years), but its application is restricted because there are volcanic lavas or volcanic tufas in the geological sections. The fission-track and so-called non-equilibrium uranium methods (Geochronology of the USSR, 1973) have been widely used in recent years. The latter yields the most reliable results only for information from shells and corals. Basic characteristics of these methods are presented in Table 8.1.

8.3. Global Temperature Trend During the Cenozoic

As early as the middle of our century, based mainly on continental data (Markov, 1941; Sinitsyn, 1967), it was thought that climatic change in the Cenozoic was directed towards cooling, and that this was an uneven process with rhythmical fluctuations that were the strongest during the Quaternary.

Using data on the oxygen isotope of deep-sea sediments excavated by the Swedish expedition in 1947–1949 in the Caribbean sea, Emiliani was the first to establish that bottom temperature of the world ocean declines throughout the Cenozoic. He hypothesized that this decrease reflected cooling at high latitudes. This is quite correct, since oceanic bottom waters are formed as a result of sinking and horizontal spreading of cold and heavy high-latitude (polar) surface waters. Data on benthic microfauna are currently widely used as an indicator for the temperature change in high-latitude surface waters (Emiliani, 1966).

Figure 8.1 presents summarized data on paleotemperatures in surface and deep waters in the central Pacific Ocean derived from oxygen isotope analysis of planktonic and benthic microfauna of deep-sea core samples DSDP 44, 47, 48, 49, 50, 167, 171, 305 and 306 taken during the 29th and 32nd trips of the research ship Glomar Challenger (Douglas and Savin, 1973; Shackleton and Kennett, 1973; Savin, Douglas and Stehli, 1975; Savin, 1982). Based on current assumptions, we can consider that data on benthic fauna reflect

Table 8.1. Basic Characteristics of Absolute dating Methods.

Method	Objects of Dating	Period of of Dating
Dendrochronological	Wood	0–7000 yrs
Radiological methods		
— carbon isotope (^{14}C)	Organic remnants (wood, bones, shells)	0–30,000 yrs
— potassium-argon	Volcanic lavas	1000 yrs–1 billion yrs
— tracks of spontaneous fission of uranium nucleus	Glass, mica, apatites, zircons	From several thousands to 1 billion yrs
— nonequilibrium uranium (U^{234}), Io (Th^{230}), Pa^{231}	Natural water, fossil bones, stalactites and stalagmites, travertines, soils, peat, shells of fresh-water molluscs, corals, marine silts	100 thousand yrs for Pa^{231}, 300 thousand yrs for Io, 1 million yrs for U^{234}
Physicochemical methods	Ceramics, stalagmites, loesses (quartz)	From 0 to several thousand yrs
thermoluminescent	sands, granite	
— amino acid*	Bones	

*At the development stage.

surface water temperature variations at high latitudes over the last 120 Ma, while planktonic data (from Figure 8.1) show the surface water temperature change in the equatorial and tropical latitudes (20°N–20°S). The difference between these curves represents the pole-equator temperature gradient.

Climatic data for continental regions are mainly obtained from paleobotanical evidence for North America (Wolfe, 1980), Siberia (Gol'bert et al., 1977) and the Russian Plain. Velichko used the data of Krishtofovich (1957), Girchuk

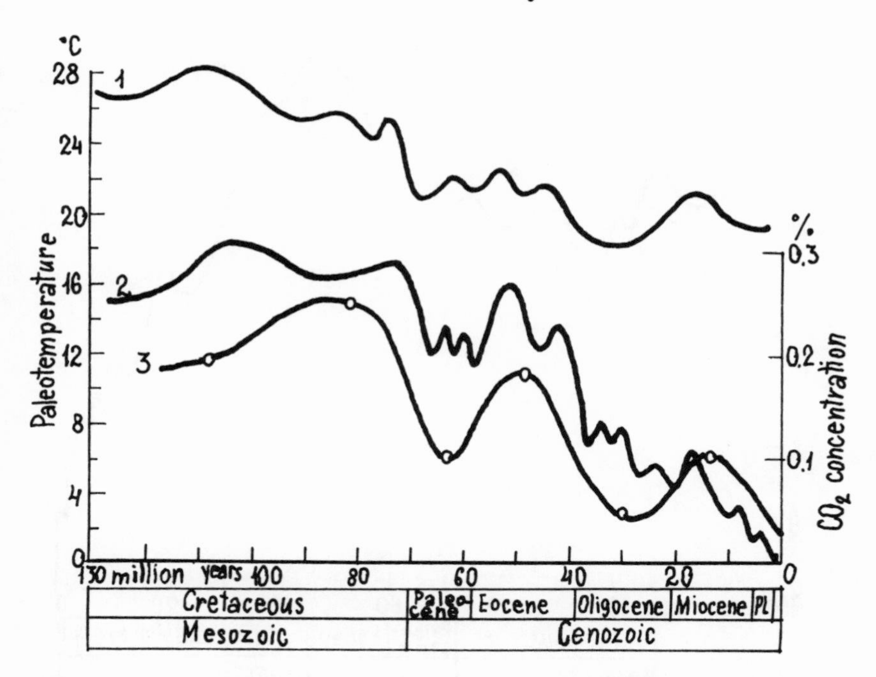

Figure 8.1. Water temperature change in high (2) and low (1) latitudes (Savin, et al., 1975; Savin, 1982; Douglas and Savin, 1973) and atmospheric CO_2 concentration (3) (Budyko et al., 1985) over the last 120 Ma.

(1961), Yasamanov (1978), Sinitsyn (1980), Chepalyga (1980) and others (Figure 8.2) to reconstruct the thermal conditions in the southern Russian Plain.

Joint analysis of continental and marine data (Figures 8.1 and 8.2) allows us to conclude that a relatively stable and warm climate prevailed on Earth for more than 200 Ma (from the Early Permian to the Paleogene). Air temperature in the high latitudes did not fall below 10–15°C, reaching 20–25°C in the middle latitudes. The equator-pole temperature gradient was 2–2.5 times lower than the current one. Climatic cooling began in the Late Cretaceous at the Campanian-Maastrichian boundary, about 70 million years ago, when high-latitude temperature decreased by 5–6°C and tropical by 2–3°C.

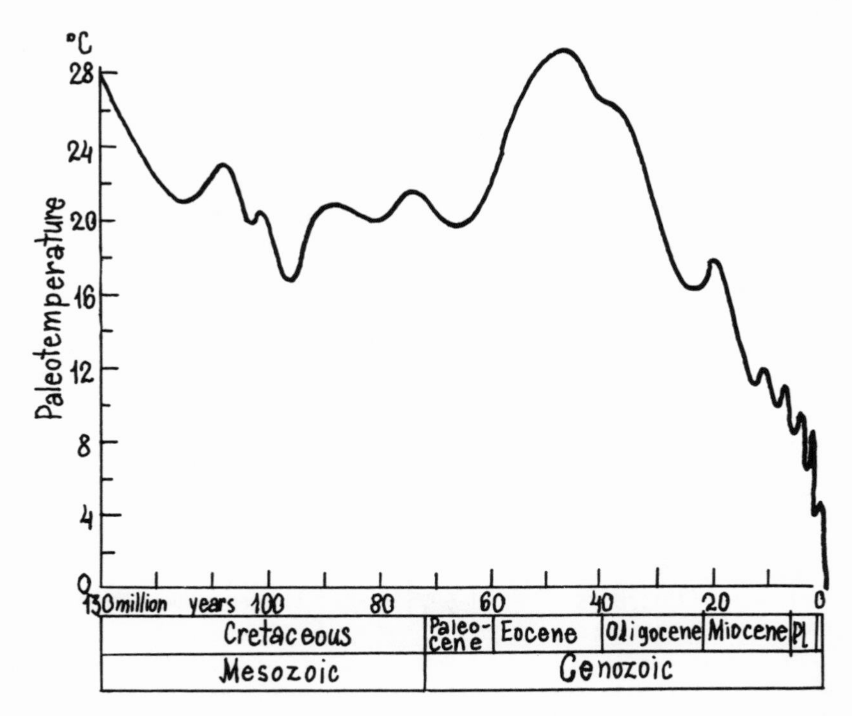

Figure 8.2. Air temperature change in the southern part of the Russian Plain according to A. A. Velichko.

However, warming started again during the Paleocene. It peaked in the middle of the Eocene ("the Eocene optimum"), that was apparently caused by a rapid increase in CO_2 concentration in the atmosphere (see Figure 8.1).

Further history of the Cenozoic climate represents a lengthy transition (for 40–45 million years) from nonglacial to modern climatic conditions, with development of continental and marine glaciation, first in the Southern Hemisphere and then in the Northern. This transition was uneven and most likely consisted of several stages: drastic and relatively brief coolings alternated with periods of relatively stable and comparatively warm climate.

Several major stages in the evolution of the Cenozoic climate can be distinguished from currently available data:

1. Rapid warming at the Paleocene-Eocene boundary that reached its maximum in the middle of the Eocene and

was accompanied by maximum advance of thermophilic flora and faunas to the north. Mean annual air and water temperatures in the high latitudes (north of 70°) did not fall below 10–12°C (Gol'bert et al., 1977; McKenna, 1980; Biske, 1981).

2. Cooling that started in the middle of the Eocene and ended in a sharp temperature decrease of 5–6°C (10–12° in northeast Asia and Siberia), perhaps only during one million years at the Eocene-Oligocene boundary (37–38 million years ago). At that time, sea ice and the first continental glaciation appeared in Antarctica.
3. The climate generally became cooler and more arid during the Oligocene. Distinct cooling in the tropical latitudes is a characteristic feature of this period. It was the only period during the Cenozoic when water temperature in the equatorial latitudes was lower than the current one, while the temperate latitudes were still fairly warm (Savin, Douglas and Stehli, 1975; Borzenkova, 1981).
4. Climatic warming with a 4–5°C temperature rise in the high latitudes in the first half of the Miocene versus the Oligocene. Water temperature in the tropical latitudes increases close to modern values and remains practically unchanged up to now (except for cooler intervals in the Quaternary).
5. Sharp drop in air temperature in the high and temperate latitudes in the mid-Miocene (about 15 million years ago). Expansion of the East Antarctic ice sheet, whose size about 12–11 million years ago is close to the present.
6. The Late Miocene cooling due to the appearance of the West Antarctic ice sheet about 10.3–9.5 million years ago. In the Northern Hemisphere, the first mountain glaciation appears in Alaska, and broad-leaved and taxodiaeae species disappear from high-latitude forests.
7. The Messinian salinity crisis about 5.5 million years ago, accompanied by a drop in the World Ocean level and a

1.5–1.8-fold enlargement in the Antarctic ice sheet compared with the modern. In the Northern Hemisphere, the first elements of tundra vegetation appear in northeast Asia, and taiga-type forests in Iceland.

8. Pleistocene-type climatic variations begin about 3.0 million years ago. The first ice sheets appeared in the Northern Hemisphere in Greenland and Iceland.
9. Events during the last 2 Ma onset the development of ice sheets and the formation of a permafrost area in the Northern Hemisphere. Appearance of pack-ice in the Arctic ocean.

In the last 50 Ma, a lengthy transition thus occurred from nonglacial climatic conditions dominating on the Earth for more than 200 Ma (Mesozoic and Early Cenozoic) to glacial modern conditions. This transition proceeded in several stages (15–12, 10–9, 7–6 and 1.0–0 Ma) during which continental and marine glaciation appeared first in the Southern Hemisphere, and then in the Northern.

As seen in Figure 8.1, there is a definite correspondence between climatic change in the Cenozoic and the overall decrease in CO_2 concentration from values close to 0.2% in the Mid-Eocene down to approximately 0.06% in the Pliocene.

The highest atmospheric CO_2 content of 0.05–0.06% was observed in the last, warmest Cenozoic stage, the so-called Pliocene climatic optimum (Budyko, Ronov and Yanshin, 1985).

According to paleobotanical data, mesophytic forests with a large amount of taxodiaeae and thermophilic exotics existed in the northwest part of the European USSR and near the Urals during that time. In southwestern Siberia and northern Kazakhstan, a fir, hemlock and broad-leaved forest zone was extended. In Mongolia, desert landscapes were almost completely displaced by Manchurian type forests, and precipitation almost doubled versus the present. Continental data are thus indicative not only of considerable warming, but also of a wetter climate in the high and middle latitudes. Warming evidence is even more pronounced in the World Ocean. At that time, sea level was 20–25 m higher than the present. According to data obtained by Ciesielski and Weaver (1974), water temperature along the Antarctic coast was 7–10°C higher than now, the East Antarctic glacier was considerably degraded and the West Antarctic melted completely.

Cooling, that began about 3.2–3.1 million years ago in the high latitudes (Greenland, Ellesmere Island, Baffin Island) took place during a rapid decrease in atmospheric CO_2 concentration. About 3.0 million years ago, the first ice sheets appeared in Greenland and Iceland, some thermophiles disappeared in the east Russia Plain, and in northeastern Eurasia, representatives of tundra phyto- and zoocenoses appeared.

Stronger cooling about 2.5 million years ago resulted in the formation of typical taiga landscapes on most of Eurasia, and in the disappearance of thermophilic elements even in the Mediterranean. According to paleobotanical data, the mean annual temperatures in the east Russian Plain dropped 6–8°C compared to current levels; we can therefore hypothesize that glaciation started in Scandinavia at that time (Grichuk, 1981).

Thus, beginning in the second half of the Pliocene, not only the frequency, but also the amplitude of climatic oscillations increase considerably.

In the Late-Neogene–early Quaternary (2.1–0.8 million years ago) and Pleistocene proper (0.75 million years ago), several coolings occurred. Beginning 0.7–0.8 million years ago, the Earth's climate acquired well-pronounced features of a glacial climate. Landscape and climatic changes were complicated and unique at that time, judging both from continental and deep-sea drilling data (Velichko, 1973; Zubakov and Borzenkova, 1983).

These changes consisted of alternating ice and interglacial epochs, where the mean annual temperature in the high latitudes was 6–7°C higher in the interglacial age than now, and 8–10°C lower in the ice age than at present. During glaciations, a substantial degradation of forests in all latitude zones was observed, as well as an increase in the areas covered by deserts, continental, sea ice and permafrost in the high and middle latitudes. During the interglacials, on the contrary, the forest area expanded considerably, mainly, due to their extension into the high latitudes, where maximum rise in temperature occurred.

Paleobotanical data indicate that climate during all interglacials was considerably warmer than now and any of them could be an analog of future climate. However, we have sufficient paleoclimatic information for only two warm intervals that are the closest to us, Likhvin-Holstein (about 275,000–350,000 years ago) and Mikulino-Eemian-Sangamon with warming optimum of 122,000–125,000 years ago. The last warming is also clear from deep-sea sections where it corresponds to the substage 5e (about 122,000 years ago). The CLIMAP group (1984) reconstructed water surface temperature for the entire World Ocean for this interval. In continental northwestern Europe and the European USSR, it is customary to regard the Eemian (Mikulino) interglacial epoch as an analog to isotope substage 5e. In western Siberia, this warming is referred to as the Kazantsev interglacial, during which landscape zones shifted to the north by 500–700 km, while summer temperatures were 4–5°C higher than the modern. In eastern Siberia and northeastern USSR, light-coniferous forests with spruce and fir dominated; they extended to the north for 500–600 km and occupied the territory of the present-day forest-tundra and tundra. In the basin of the middle and upper reaches of Yenisey, dark-coniferous forests with fir and cedar developed, while in the basin of the middle and lower reaches of the Irtysh, mixed forests with birch, linden and elm (Volkova, 1977). A new stage of global cooling started in the late Eemian interglacial (about 75,000–72,000 years ago) and ended with the greatest drop in temperature for the whole Quaternary, about 20,000–18,000 years ago. Using data on surface water temperature changes 18,000 years ago that were obtained by the CLIMAP project (CLIMAP, 1976) and paleobotanical data for the USSR territory (Velichko, (ed.), 1984), North America (Wright (ed.), 1983) and western Europe (Shotton, ed., 1978; Lowe, Gray and Robinson, eds., 1980; Woillard and Mook, 1982), it was possible to estimate the drop in summer (July–August) surface air temperature for the entire Northern Hemisphere as 4.6°C (Borzenkova, 1987).

Recent studies by Soviet and foreign authors (Borzenkova and Zubakov, 1984; Kerr, 1984b; Ruddiman and Duplessy, 1985; and others), have focused on climatic investigation of the final stage of the Würm glaciation that ended in a noticeable warming known as "the Holocene climatic optimum." The most complete chronology of climatic events during the last 16,000–17,000 years, based on the entire set of paleoclimatic (paleobotanical, microfaunal, deep-sea etc.) radiocarbon dated evidence is presented by Borzenkova and Zubakov (1984) and recently in *Paleoclimates in the Lateglacial and Holocene* (1989). Analysis of these data allows us to conclude that the major climatic variations in the well-pronounced climatic variations throughout the Late Glacial-Holocene were global.

The application of statistical methods to paleobotanical and microfaunal data made it possible to obtain quantitative estimates of surface air temperature and surface oceanic water for the temperate and high latitudes of the Northern Hemisphere (Figures 8.3, 8.4).

As seen from Figure 8.4, the climatic history of the last 16,000 years can be divided into three natural periods:

1) the late glacial age (16,000–10,000 years ago) characterized by warming against the background of rapid (lasting from 200 to 600 years) and severe global temperature variations (2–3°C);

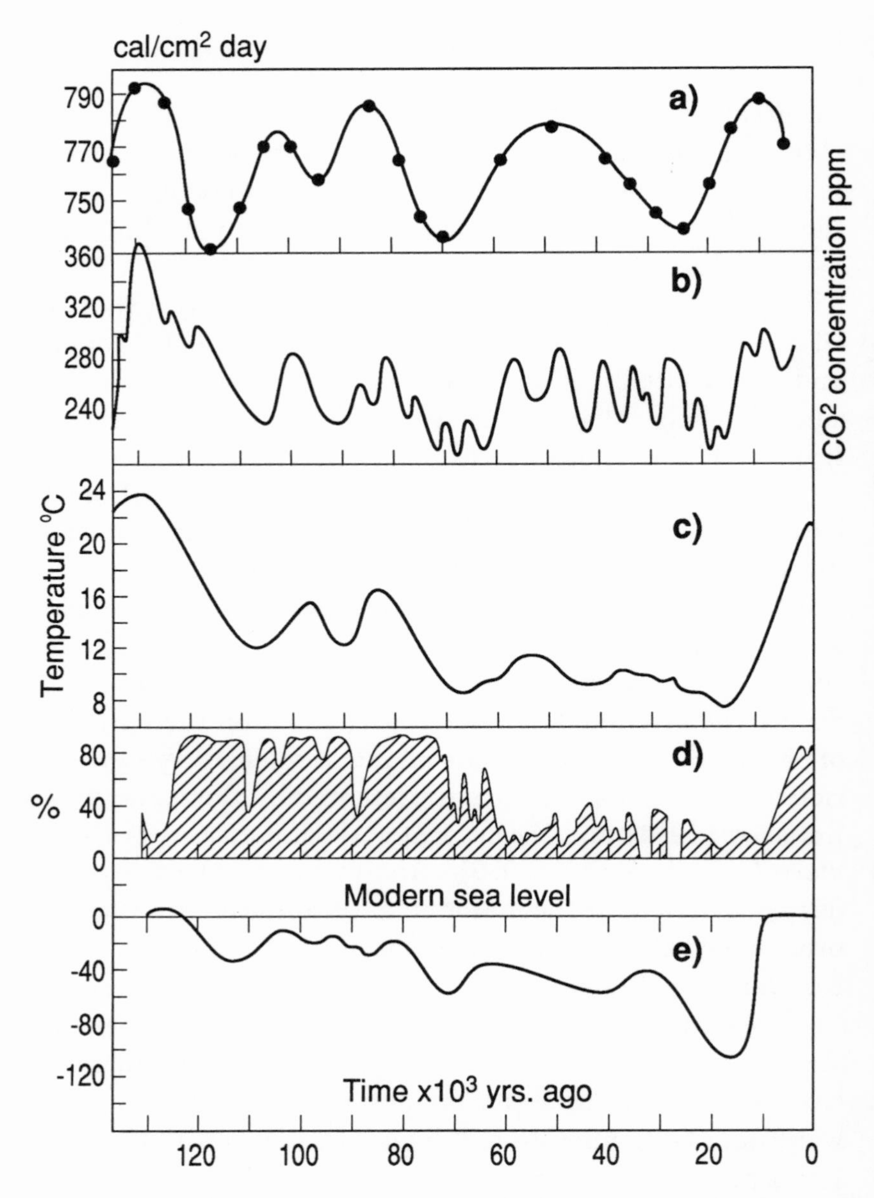

Figure 8.3. (a) The summer radiation at 65°N (Berger, 1978); (b) Variations of the CO_2 concentration (using $^{12}C/^{13}C$) (Shackleton and Pisias, 1985); (c) Sea surface temperatures in the North Atlantic (Imbrie et al., 1973); (d) Relations (%) between arboreal and nonarboreal pollen (site in the central part of Europe) (Woillard and Mook, 1982); (e) Sea-level fluctuations (CLIMAP, 1976).

2) the Early and Middle Holocene (10,000–5,500 years ago), a time of warm and relatively stable climate with high-latitude temperatures 2–3°C higher than now;
3) the Late Holocene (the last 5,000 years) characterized by cooling and greater climatic variability.

The Early-Middle Holocene warming is the most interesting for reconstruction of future climatic conditions. In the period 9,000–5,000 years ago, several warmings are dated: the Early Boreal (8,900–8,800 years ago) and three Atlantic (7,800–7,500; 6,900–6,500 and 6,200–5,300 years ago). The latter (the main thermic optimum of the Holocene) corresponds to the highest World Ocean level for the Holocene (0.5–1.0 m higher than the present).

It is interesting to note that all the significant warmings and coolings of the Late Glacial-Holocene are reflected in changes of the amount of precipitation and lake levels in different latitudinal zones. During coolings, a decrease in precipitation and lowering of lake levels in the subtropical and tropical regions (Street and Grove, 1979; Borzenkova, 1980; Varushchenko, 1984; Kutzbach and Street-Perrott, 1985) are observed, and vice versa, well-pronounced transgressions of the Caspian Sea (Varushchenko, Varushchenko, and Klige, 1980), Lake Geneva and lakes of the Great Basin in the United States (Smith and Street-Perrott, 1983). Thus, three transgressions of the Caspian Sea — the Dagestan (11,000–10,500 years ago), Gousan I (8,500–8,300 years ago), and Gousan II (6,400–6,200 years ago) correspond to the Late Glacial-Holocene, and three regression — Kulaly (9,000–8,700 years ago), Zhelandy (7,100–6,700 years ago) and Izberbash (5,800–5,300 years ago) correspond to Boreal warming and two Atlantic optima (Geochronology of the USSR, 1974). During warmings, the tropical and subtropical latitudes receive sufficient precipitation, while some regions suffer a shortage. The relationship between changes in global temperature and precipitation pattern in different latitudinal zones is discussed in more detail in the works by Zubakov

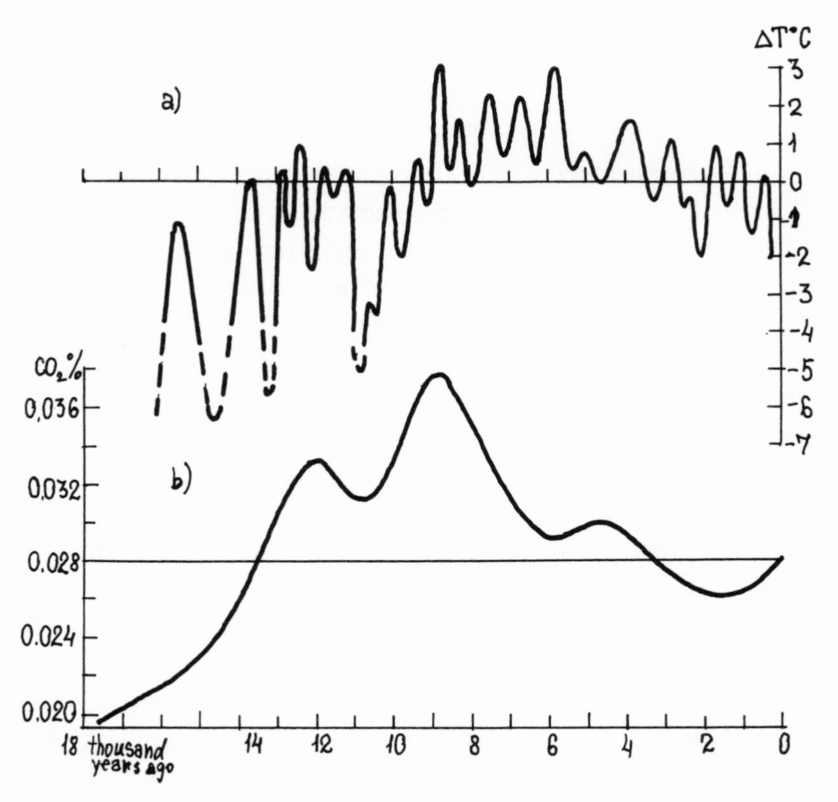

Figure 8.4. Changes in summer air temperature in high and middle latitudes (a) from Borzenkova and Zubakov (1984) and of CO_2 concentration (b) averaged from data of W. Berger (1978).

and Borzenkova (1983), Borzenkova and Zubakov (1984), Velichnev (1987), Borzenkova (1987). These studies showed that the relationship between global temperature and precipitation in different latitudes, according to date for the last 100 years, is confirmed by paleoclimatic data, and as should be expected, these changes were significantly greater in the past.

8.4. Climatic Changes During the Pleistocene and Holocene

When paleoclimatic data are used for climatological forecasts, we have to visualize which factors caused warmings in the past and to what extent they are comparable to the causes of anthropogenic warming.

The data in the report of the Soviet-American meeting to study the impact of higher atmospheric CO_2 content on climate (The Impact of Increased CO_2 ..., 1982) allow us to conclude that there is close conformity between basic regularities of climatic variations during warm epochs in the past and the modern warming caused by increasing CO_2 content. Model projections also allow us to assume that the patterns of surface temperature variations caused by such different factors as fluctuations in solar radiation and higher CO_2 content are quite similar (see Chapters 1, 7 and 9).

Paleoclimatic data indicate that the overall drop in atmospheric CO_2 concentration is the main factor for climatic change during the Cenozoic. Some other factors, in particular, change in the solar constant and the Earth's albedo (Budyko, 1984) also contributed to climatic change. The appearance of large glaciers in the Late Neogene (2.5–2.0 million years ago) and ice sheets in the Early Pleistocene (1.7–1.8 million years ago) resulted in a considerably lower global temperature and greater climatic sensitivity to comparatively slight changes in such climate-forming factors as fluctuations in solar radiation due to astronomical factors. Comparatively small changes in seasonal distribution of solar radiation caused by orbital factors (eccentricity of the Earth's orbit, the angle of inclination of rotation of the Earth's axis to the orbital plane and the time of the equinoxes) were like a "trigger" for the fluctuations in climate during the Quaternary. The impact of orbital factors increased, to a large extent, because of the change in the Earth's albedo caused by enormous fluctuations in the areas of snow and ice cover, deserts and forests during glaciation and interglacials.

However, recent data on natural fluctuations in atmospheric CO_2 content in the Late Pleistocene-Holocene showed that practically all warmings (interglacials) took place with CO_2 concentration higher than now, while during glaciations, CO_2 concentration dropped to 180–200 ppm.

The problem of the existence of comparatively rapid ($10^3 - 10^4$ yr) natural fluctuations in atmospheric CO_2 concentration and their influence on global thermal conditions arose comparatively recently. The first results were obtained in 1980 from analysis of the gas composition of air bubbles preserved in ancient ice (Delmas, Ascencio and Legrand, 1980). Analysis of ice cores from the Antarctic well Dome C, that was made by French scientists showed that atmospheric CO_2 concentration during the glaciation peak (18,000–20,000 years ago) dropped to 180–200 ppm, and then rose (about 10,000–9,000 years ago) to the values exceeding pre-industrial levels. These first indications of such rapid fluctuations in CO_2 concentration gave rise to skepticism and distrust on the part of many scientists. However, this result was confirmed two years later by Swiss scientists from Bern University on the basis of ice core evidence from Byrd (Antarctica) and Camp Century (Greenland) cores. These data indicated fluctuations in CO_2 concentration during the last 40,000 years (Neftel et al., 1982). Although independent estimates of CO_2 concentration made using ice cores are in good agreement, it is very arduous to obtain them. It was necessary to find a simpler method to determine atmospheric CO_2 concentration in the past. Such data were obtained by Shackleton and Pisias (1985) from the results of carbon isotope analysis of benthic and planktonic foraminifera. Figure 8.3 illustrates a part of this record spanning the last 340,000 years. Spectral analysis of these data allowed them to isolate fluctuations with periods of 100,000, 40,000 and 19,000–23,000 years that coincide with periods of solar radiation variation due to astronomical factors. The comparison of data on CO_2 content fluctuations with the oxygen isotope curve that characterizes changes in continental glaciation showed that solar radiation fluctuations are followed by changes in CO_2 concentration which, in turn, precede changes in volumes of glaciation (Kerr, 1984a).

There are currently several hypotheses that link natural fluctuations in CO_2 concentration with altered biological productivity of the World Ocean (Broecker, 1984; Weber and Flohn, 1984). Living marine organisms are known to play the role of a biological pump for carbon from the upper to lower oceanic layers. In this case, the more powerful this biological pump is, the lower the CO_2 partial pressure in the upper oceanic layers and the atmosphere; and vice versa, the lower biological productivity is, the higher the CO_2 concentration in the upper oceanic layers and the atmosphere. Prime factors that limit biological productivity are nitrogen and phosphorous from the deep oceanic layers that are plentiful in the regions of intensive upwelling, as well as in the regions of bottom water formation along the Antarctic coast. All the models that the authors used to explain brief oscillations in atmospheric CO_2 concentration are related to altered biological productivity due to changes in the deep-water circulation rate, broadening or narrowing of active upwelling zones, and change in the surface circulation rate in Antarctic waters (Seigenthaler and Wenk, 1984; Broecker, 1984; Weber and Flohn, 1984). However, whatever the reasons for CO_2 content fluctuations, we can hypothesize that CO_2 intensifies the impact on the surface air temperature produced by comparatively small changes in solar radiation due to astronomical factors.

The question of this influence of astronomical factors was investigated in a series of studies by Berger (Berger, 1978; Berger et al. (eds.), 1984). On the basis of these materials, Pisias computed the amount of solar radiation entering the upper boundary of the atmosphere on the first day of each two-month interval for the last 30,000 years. These data are analyzed by Davis (1984). Computations showed a more complicated seasonal redistribution of solar radiation under the influence of astronomical factors than previously supposed. During the last 30,000 years, several peaks of solar radiation influx were observed in the warm six-month period: in its beginning (May-June) about 13,500 years ago, in the middle (July-August) about 11,000–10,000 years ago, and in the end of the warm period (September-October) about 6,000–5,000 years ago. The peak summer solar radiation influxes precede three peaks of atmospheric CO_2 concentration (about 13,000, 9,000–8,000 and 6,000–5,000 years). These changes in CO_2 concentration and solar radiation influx gradually agree with the peculiarities of altered surface air temperature in the high and middle latitudes during the Late Glaciation-Holocene (Figure 8.4a).

The first peak (13,500–13,000 years ago) coincides with the Raunis interstadial in the northwestern European USSR, appearance of arboreals in Western Europe and higher surface water temperature in the North Atlantic to values close to the modern (Duplessy et al., 1981).

The warming in the Early Holocene-Early Boreal (8,800–8,900 years) was caused not only by an increase in summer solar radiation influx to 7% for the entire Northern Hemisphere (Kutzbach, 1981), but also by a maximum CO_2 concentration of 360–380 ppm for the last 18,000 years (Berger, 1982). It is known from paleobotanical data that Early Boreal warming was most distinct in the Northern and Southern Hemispheres, where there were no continental ice sheets (Taymyr, northern Yakut Republic, Chukchi Peninsula, Alaska, southern regions of Chili and New Zealand (Khotinskiy, 1977; Kaplina and Lozhkin, 1982; Klimanov and Nikol'skaya, 1983; Ritchie, Cwynar and Spear, 1983; Heusser, 1984 and others)).

Model calculations made by Kutzbach (Kutzbach and Otto-Bliesner, 1982) showed a 2–4°C increase in summer temperature about 9,000 years ago in the high latitudes, taking into account that summer radiation for the entire hemisphere was 7% higher than the current. Considering that this atmospheric CO_2 concentration was much higher than the pre-industrial level, the rise in summer temperature in the high latitudes could reach 5–6°C and even more, as confirmed by paleobotanical data.

Figures 8.3 and 8.4 show that all of the most significant warmings in the Late Pleistocene and Holocene took place while summer solar radiation increased as a result of astronomical factors and higher atmospheric CO_2 concentration. This allows us to regard warm intervals of Pleistocene (interglacials) as analogs of future climate.

8.5. Paleoclimatic Analogs in the Twenty-first Century

According to predicted global climatic changes from anthropogenic factors, we can expect an increase of 1.3°C in global temperature by the year 2000 over the pre-industrial epoch, of 2.5°C by 2025, and of 3–4°C by the mid-twenty-first century (see Chapter 9).

If we consider different levels of higher global temperature, then we can find close analogs for some of them in the geological past. We will discuss three warm epochs for this purpose:

1) Holocene optimum (6,200–5,300 years ago),
2) Mikulino (Eem) interglacial optimum (about 125,000 years ago)
3) Pliocene optimum (4.3–3.3 million years ago).

These epochs are characterized by different degrees of detail and accuracy of paleoclimatic information. It seems that the available data should become less detailed and accurate in sequence from the Holocene to Pliocene. However, although it may seem paradoxical, greater difficulties were encountered in reconstructing the youngest Holocene epoch, than the earlier Mikulino interglacial optimum.

Holocene optimum (6.0–5.5 thousand years ago). It is primarily difficult to make global climatic reconstructions for this interval because the paleobotanical data that provide the most reliable information on land climate can not be effectively processed by the arealographic range method, since most of the time there are no specific analyses. Therefore, the most reliable specifics about temperatures and precipitation for individual points have been obtained mostly for Eastern Europe, and partially for Siberia. For other regions of Eurasia, as well as North America, a zonal method was used for data processing (Savina and Khotinskiy, 1982). The estimates of climatic parameters were less accurate than the estimates by the arealographic method. We should note that the global chronological correlation for the Holocene optimum is conditional because there are very few oceanic data.

Paleobotanical, mainly radioactive carbon-dated spore and pollen data were the basis for reconstructing the landscape and climatic conditions in the Northern Hemispheric continents. For central regions of North America, paleotanical data were used from Webb, Street and Howe (1980), Wright (ed.), (1983), Delcourt, P. and Delcourt, H. (1983); for western regions (Washington State), data from Heusser, C., Heusser L. and Streeter, (1980); for southern regions and Florida, from Watts (1983); for the southwest and Great Lakes basin, from Smith and Street-Perrott (1983); for California, from Adam and West (1983).

Archaeological evidence, information about lake level fluctuations in the Rift Valley and Lake Chad, generalizations in a number of works (Street and Grove, 1979; Borzenkova, 1980; Williams and Faure (eds.), 1980; Street-Perrott and Roberts, 1981; Petit-Maire and Riser (eds.), 1983; Zubakov and Borzenkova, 1983), results of deep-sea studies along the western coasts of Africa, data on changes in tropical forests and desert territories (Diester-Haas, 1976; Hamilton, 1976) were the basic sources of information for North Africa (the Sahara and adjacent regions).

Palaeobotanical data, generalized in some research (Shotton, 1978; Lowe, Gray and Robinson, eds. 1980) were used to reconstruct western European landscapes.

Quantitative climatic characteristics (air temperature, precipitation) were obtained from landscape and physiological data. A transition function was used to switch to past climatic parameters based on the relationship between modern plant associations (tundra, forest, steppe, etc.) and modern climate (Savina and Khotinsky, 1982). We also used previous data on climate reconstructions for the 6,000–5,500 time interval by Klimanov (1982), Burashnikova, Muratova and Suyetova (1982), Khotinskiy and Savina (1985) for the USSR territory.

Air temperature above the ocean was estimated from factor analysis data of planktonic fauna in the North Atlantic (Duplessy et al., 1981), at the eastern coast of North America (Balsam, 1981), as well as from data on coral propagation along the coasts of Japan and Taiwan (Taira, 1979).

Figure 8.5 presents differences between summer (July-August) temperatures of the Holocene optimum (6,200–5,300 years ago) and the modern epoch, while Table 8.2 contains the mean latitudinal values.

Table 8.2. Mean Latitudinal Distances Between Summer (July-August) Temperatures of the Late Atlantic Optimum and the Modern Epoch

ϕ°N	80–70	70–60	60–50	50–40	40–30	30–20
ΔT°C	4.4	3.0	1.9	1.0	0.3	–0.2

The greatest changes in July temperature (up to 4°C) were observed in the high latitudes, north of 70°N. In the temperature latitudes, temperature differences drop to 1–2°C, and in the south they become negative. The greatest temperature drop is observed in the central regions of the Sahara, in Central Asia and Arabia. Such changes are apparently associated with higher precipitations in these

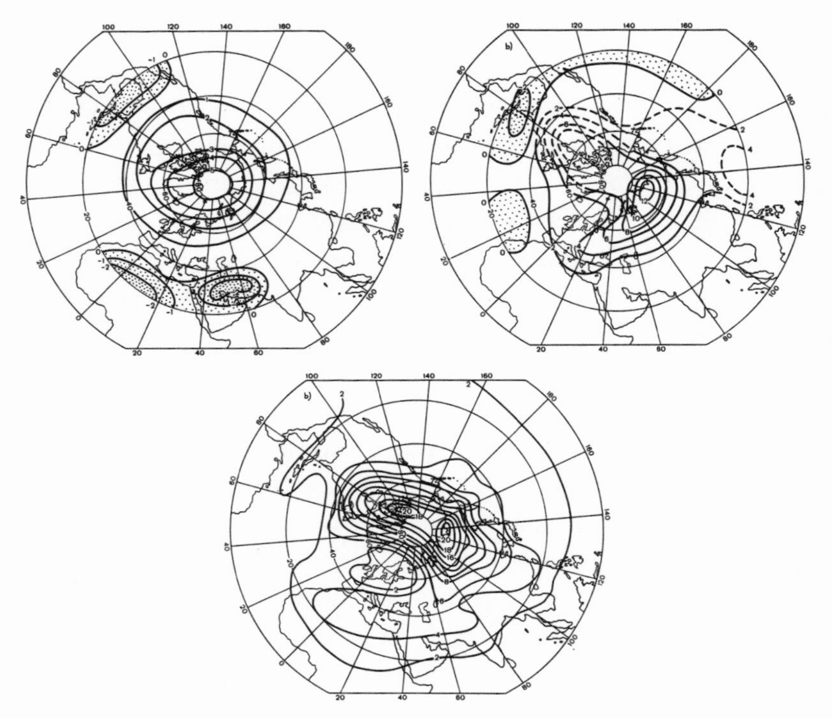

Figure 8.5. Deviation in summer (July-August) air temperature (°C) from modern for the late Atlantic optimum of the Holocene.

regions than the modern level.

Computation of changes in mean air temperature in the Northern Hemisphere should take into consideration that the most detailed data on temperature changes are available for the 25–70°C N belt. Consequently, for higher and lower latitudes, it is expedient to use theoretical estimates of temperature changes along with empirical data. Such estimates can be obtained by multiplying temperature changes within the 25–70°N zone by the ratio of temperature change in the entire Northern Hemisphere to its change within the indicated zone, based on general atmospheric circulation models (Manabe and Stouffer, 1980; Spelman and Manabe, 1984; Manabe and Bryan, 1985).

We can thus estimate the change in mean annual air temperature in the Northern Hemisphere for the Holocene optimum as 1.2°C.

Reconstructions by Klimanov (1982), by Muratova (Burashnikova, Muratova and Suyetova, 1982), and by Khotinskiy and Savina (1985) for the whole USSR territory, supplemented by paleobotanical evidence and data on lake level fluctuations for the Great Basin in the United States (Street and Grove, 1979; Smith and Street-Perrott, 1983), for North Africa (Williams and Faure (eds.), 1980; Street-Perrott and Roberts, 1981; Petit-Maire and Riser (eds.), 1983 and others) were used to restore the precipitation field.

It is seen from Figure 8.6 that during the climatic optimum (with global warming of about 1°C), the amount of precipitation in the northern regions (northward of 55–60°N) was 5–10 cm higher than now. To the south, between 55–50°N, in some regions of the European USSR and Western Siberia, the amount of precipitation decreased approximately by 5 cm.

The greatest changes in moisture content were observed in arid subtropical regions of Africa and Asia where annual rainfall (mainly, due to summer monsoons) increased 2–3 times and more. In the southern and central regions of present-day Sahara, the annual precipitation was 30–40 cm versus the current 10–25 mm. In monsoon regions of India, in the region of the present-day Tar desert, annual precipitation reached 40–50 cm (Bryson and Swain, 1981).

Moisture conditions improved considerably (annual precipitation increased by 10–15 cm) in the regions of central Asia, that are now deserts (Ustyurt, Mangyshlak, Kyzylkumy). Extensive archaeological evidence found here and ^{14}C-dated at 9,000–5,000 years indicate that this area was populated in the early and middle Holocene (Mamedov, 1982; Varuschenko, 1984; Murzayeva et al., 1984).

In North America, south of 55°N, especially in the central regions, precipitation dropped significantly, 20–25 cm

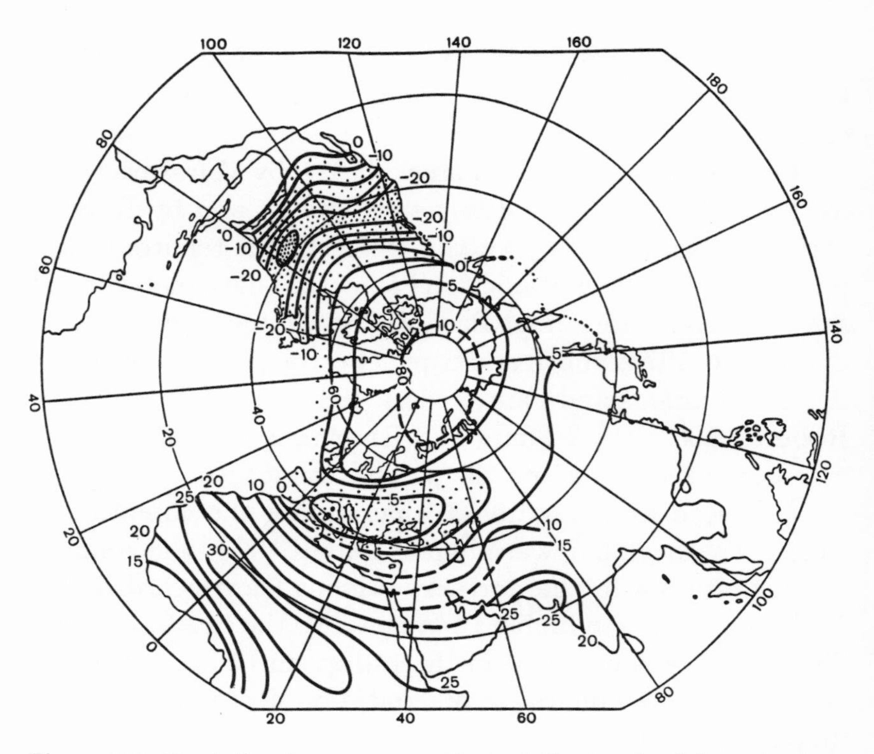

Figure 8.6. Deviation in mean annual precipitation (cm) from modern for the late Atlantic optimum of the Holocene.

lower than the present (Bartlein, Webb and Fleri, 1984). Extremely low lake levels in western regions of the United States (the Great Basin) were observed throughout the whole Holocene warming (Smith and Street-Perrott, 1983). At that time, the area of prairies increased the most and advanced eastward towards the Great Lakes. Reconstruction of temperature and precipitation by Bartlein and others (Bartlein, Webb and Fleri, 1984) for the U.S. Midwest for two time intervals (9,000 and 6,000 years ago) showed that during the Holocene warming there was 25% less precipitation in this region than the present level, while temperature was 0.5–1.0°C higher.

Analysis of paleoclimatic data on altered moisture conditions in different latitude zones throughout the Holocene that was made in a series of works (Street and Grove, 1979;

Borzenkova, 1980; Zubakov, Borzenkova, 1983; Borzenkova and Zubakov, 1984; Velichno, 1987) showed that the relationship between altered global temperature and precipitation in different latitudinal zones determined from modern meteorological data (the last 100 years) (Drozdov, 1983; Vinnikov and Kovyneva, 1983) was the same even in the past with global warming of 1°C. Climatic warming during the Holocene optimum was accompanied with considerably improved moisture conditions in the subtropical regions (between 10–20°N) and the northern regions (to the north of 60°N). The lower amount of precipitation in the temperate and southern latitudes (between 50 and 30°N) was the most obvious in North America. In the European USSR, there was a comparatively slight drop in precipitation at latitudes 50–30°N, while it was close to current levels in some regions. In Western Siberia, moisture conditions differed only slightly from the modern, while precipitation increased somewhat in Eastern Siberia (Khotinskiy and Savina, 1985).

The simplest mechanism to explain the relationship between altered global temperature and precipitation is the lower meridional temperature gradient during global warmings that leads to a certain weakening in zonal circulation and a northward shift in circulation and climatic zones.

This scheme apparently occurs by means of corresponding displacement of the main atmospheric pressure centers and the change in their intensity (Kovyneva, 1984). During global warmings, the Azorean pressure peak moves to the northeast, the Icelandic minimum is slightly weakened, whereas the Aleutian minimum is deepened. Cyclonic activity intensifies over the Pacific and is weakened somewhat over the Atlantic. The drop in precipitation in the central and eastern regions of North America is hypothetically related to a certain weakening in cyclonic activity at latitudes 55–60°N and northward shift of basic cyclonic trajectories.

This contributed to an improvement of moisture conditions in northern Eurasia and Canada. Moisture conditions in subtropical (monsoon) regions depended on seasonal and secular changes in the location of the Intratropical Convergence Zone. Its northward shift during global warming promoted good moisture conditions in these regions.

The Mikulino Interglacial Optimum. The nearest interglacial optimum (about 122,000–125,000 years ago) is very important as an analog of future climate. Paleogeographical data show that the climate at that time was somewhat warmer than during the Late Atlantic Holocene warming. The most efficient methods for processing pollen spectra containing species analyses could be used to determine temperature and precipitation because there are enough paleobotancial data for this interval (Grichuk, 1981, 1985; Velichko et al., 1982, 1983, 1984).

CLIMAP information (CLIMAP, 1984) and data of Barash (1985) for the North Atlantic and Safarova et al. (1984) for the Central Pacific were used to reconstruct water surface temperature.

Figure 8.7 presents the reconstructions of summer (July-August) and winter (January– February) temperatures of the Northern Hemisphere as their deviations from modern values, while Table 8.3 presents mean latitudinal differences in these temperatures. The calculations were made by V. Grichuk, E. Gurtovaya, E. Zelikson, O. Borisova, and A. Velichko.

The strongest warming in summer (up to 6–8°C) is observed in the northeastern regions of the USSR, north of 70°N, in northern Canada and in Greenland. Over most of the USSR and Western Europe, north of 50–55°N, temperature changes do not exceed 1–3°C and are near 0°C to the south. A region of negative values is observed above the Arabian Peninsula, Central Asia, the Sahara, South Atlantic and Indian Ocean.

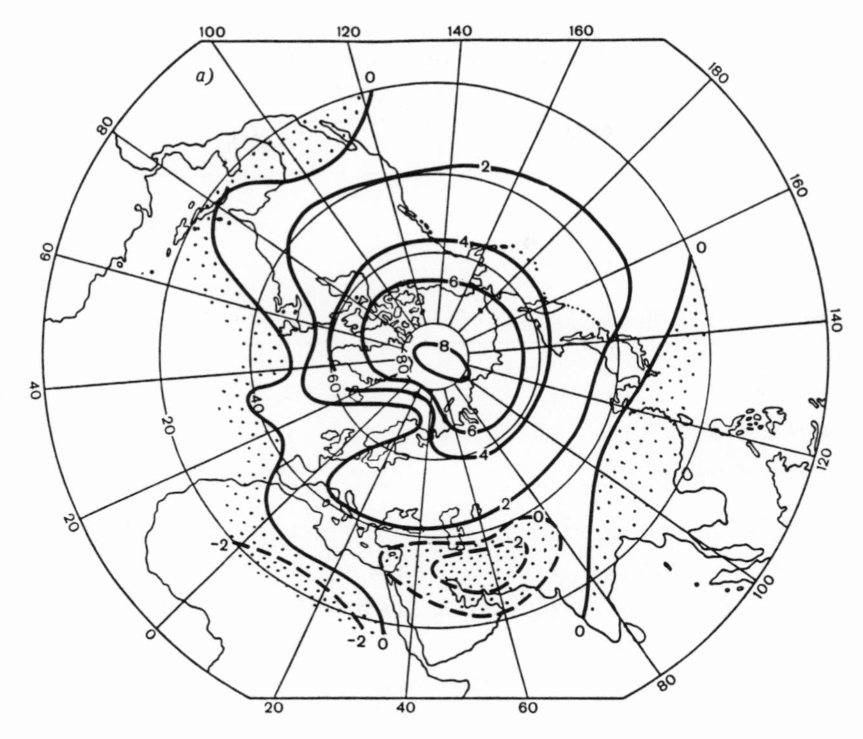

Figure 8.7. Deviation in summer (a) and winter (b) air temperature (°C) from modern for the Mikulino Interglacial.

The situation is much more complicated in winter. The strongest warmings are observed in northern regions of Western and Eastern Siberia (up to 10–12°C). Approximately the same temperature increase takes place in northern Canada, Baffin Island, and in Greenland. In Western Europe, air temperature during the last interglacial exceeded the current level by 2–4°C, and in the central regions of the European USSR it was 4–8°C higher.

To the south of 50°N, there exists a zone of comparatively small (not greater than 1–2°C) negative temperature deviations. It is characteristic that this area is observed both in the Atlantic and the Pacific.

Figure 8.8 presents mean annual deviations in the amount of precipitation from modern levels for the Mikulino

Table 8.3. Mean Latitudinal Temperature Differences (°C) Between Eemian (Mikulino) Warming and Modern Epoch.

	ϕ°N lat.						
Season	80–70	70–60	60–50	50–40	40–30	30–20	Mean annual global temperature
January-February	7.4	6.5	4.7	2.4	1.2	0.2	
July-August	6.0	4.8	2.8	1.6	0.3	–0.2	
Mean annual	6.7	5.6	3.8	2.0	0.8	0.0	2.0

interglacial obtained from paleobotanical data for Eurasia and some regions of northern Africa (Velichko et al., 1982, 1983, 1984). The precipitation distribution is more complicated than air temperature. It is known that the present precipitation pattern is also very complex, since it is not only affected by atmospheric circulation, but also by a number of other factors, primarily topography. Consequently, reconstruction of precipitation as a system of paleoisohyets seems to be very conditional, and we should preferably consider large regions where the amount of precipitation was lower or higher than at present. Such an approach is advisable even for the extratropical territory of Eurasia, where at a definite number of sites, precipitation is computed by the most efficient arealographic method from pollen-spore data with species analysis (computations carried out by E. E. Gurtovaya, E. M. Zelikson, and O. K. Borisova).

As seen from Figure 8.8, the greatest increase in rainfall is observed in coastal regions of Western Europe, on the British Isles and in Central and Southern Europe (30–50 cm/yr more than at present).

In northern regions of Asia (Taymyr, Yakut Republic), precipitation was 20 cm/yr higher than the modern level. In regions of the European part of the interior USSR and

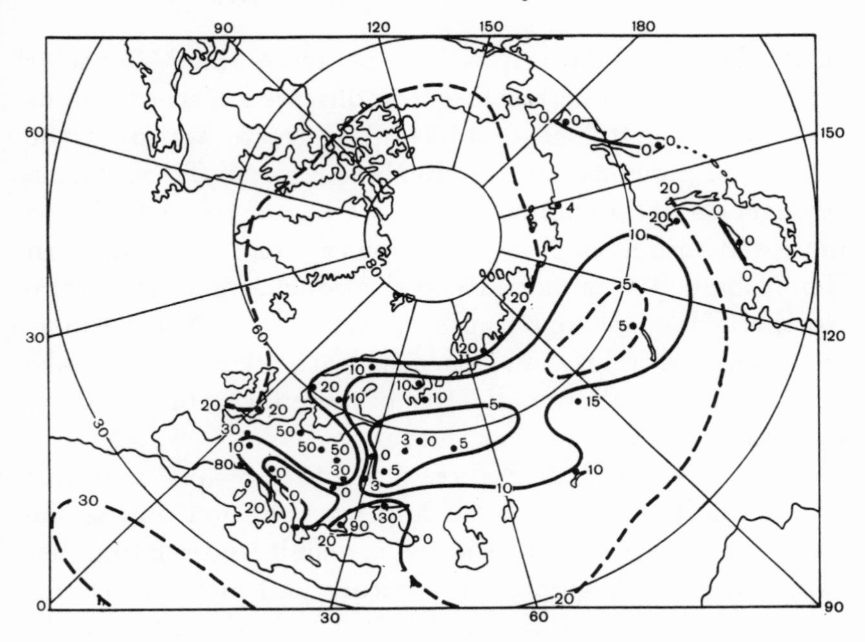

Figure 8.8. Deviation in annual precipitation (cm) from modern for the Mikulino Interglacial.

in the middle latitudes of Western and Eastern Siberia, the precipitation increment was less spectacular and did not exceed 5 cm/yr. Moisture conditions improved in steppe regions of the Russian Plain, Central Asia and in Kazakhstan.

Overall, in most of the Eurasian extratropical zone, precipitation increased throughout the Mikulino (Eem) interglacial (apparently except for the Pacific coast where no deviations have been revealed).

Combined analysis of two maps (see Figures 8.6 and 8.8) for the Holocene optimum and Mikulino interglacial showed that there are some common trends in the precipitation variation. The most considerable change in moisture content during global warming is observed in the high (north of 60–65°N) and subtropical latitudes. A comparatively slight precipitation variation is observed over the greater part of the USSR European territory and in Western Siberia. In this case, it is interesting to note that a slight decrease in precipitation, no more than 50 cm/yr compared with modern values, was observed in these regions throughout the Holocene, while during greater warming, the amount of precipitation rose in these regions. Thus, during stronger global warming (about 2°C), moisture conditions improved throughout the extratropical zone of the Eastern Hemisphere. We cannot extend this conclusion to the entire Northern Hemisphere because we lack data on North America. However, indirect evidence (in particular, data on lake level fluctuations) are indicative of more favorable moisture conditions in these regions than now.

Data presented in Figures 8.6 and 8.8 and materials on fluctuations in lake levels and precipitation during the Late Glaciation-Holocene show a definite correlation between changes in global temperature and moisture conditions over vast territories in different latitudinal zones. At the same time, local relationships between the thermal mode and precipitation are rather complicated and to a certain degree ambiguous, mainly because of meso- and microclimatic conditions. Such indirect indices as lake levels, soil types and landscapes characterize peculiarities of moisture conditions over vast territories and provide an idea of macrocirculation processes within a definite climatic or latitudinal zone.

There are pronounced relationships between thermal and moisture conditions in tropical and subtropical regions, where precipitation depends on the intensity of monsoon circulation and position of the Intratropical Convergence Zone. During the entire Pleistocene interglacial monsoon circulation intensified due to peculiarities in seasonal redistribution of solar radiation caused by astronomical factors (Kutzbach and Street-Perrott, 1985). The warmest intervals (isotope stages 5e, 7, 9 and 11) were accompanied by the most favorable moisture conditions in the Sahara, Arabia, in monsoon regions of India, Asia and Australia. These concepts were the basis for reconstruction of precipitation in monsoon regions of Africa, Asia and America. Investigations showed (see, for example, Rossignol-Strick, 1985) that monsoon circulation intensified during Riss-Würm (5e) more than during the Holocene. This made it possible to assume that there was slightly more precipitation in these regions than during the Holocene warming (see Figure 8.6).

Pliocene Optimum. The most significant Pliocene warming, the so-called climatic optimum of this interval could be a possible analog of future climate (see Chapter 9). The main features of the landscape and climate for that period were considered in section 8.2. Here, reconstructions of the summer and winter temperatures and annual rainfall during the Pliocene warming (4.3–3.3 million years ago, zone N 19 Blow in deep-sea section), are presented (Borzenkova, Zubakov, 1985).

A working landscape model made by Zubakov using paleobotanical and paleontological data served as the basis for paleoclimatic reconstruction for this interval. Factor and oxygen isotopic data obtained from deep-sea drilling by Glomar Challenger were the main source of quantitative information.

The upper oceanic temperature field was reconstructed on the basis of the relationship between modern planktonic foraminifera associations and water surface temperature (Bé, 1977). Several latitudinal zones are distinguished via present-day species of planktonic foraminifera. Their boundaries are defined by water and air temperature. Some research (Bandy, Casey and Wright, 1971; Barash, 1985) has demonstrated that the location of these zones of tropical, temperate and subarctic waters changed considerably with latitude throughout the last 10 Ma. During warmings, boundaries of plankton zones shifted markedly to the north by approximately 10°, and during coolings, by just as much to the equator.

Figure 8.9 shows maps of temperature differences between the Pliocene optimum and the modern epoch.

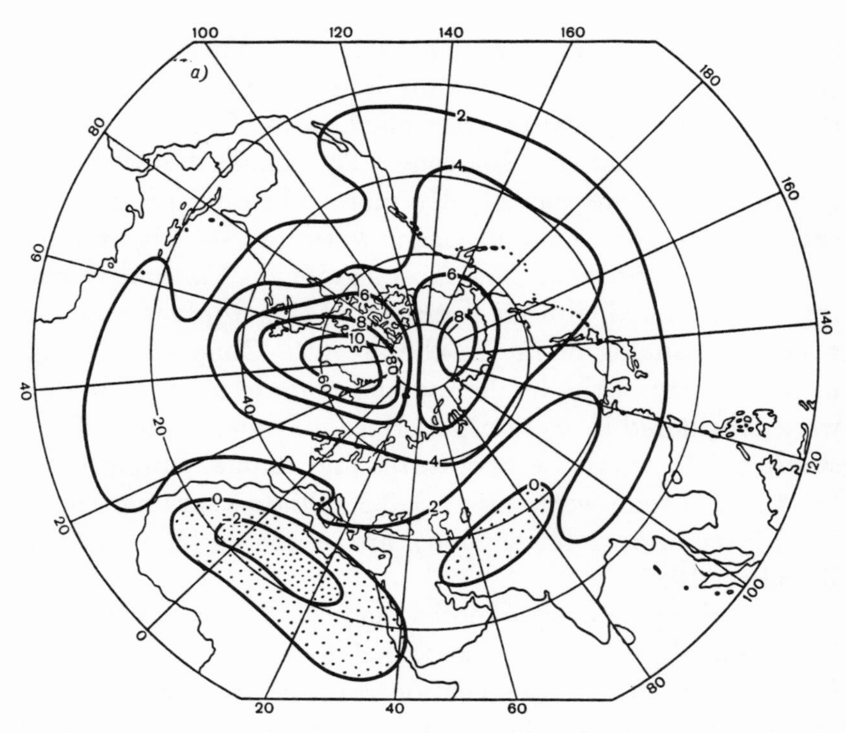

Figure 8.9. Deviation in summer (a) and winter (b) air temperature (°C) from modern for the Pliocene optimum.

Table 8.4 presents mean latitudinal values of these differences, compared with the calculations of Manabe and Bryan (1985) obtained from the general atmospheric circulation model with doubled CO_2 concentration.

Both paleoclimatic data and the general atmospheric circulation model calculations showed much higher temperature changes in the high and temperate latitudes than in the low latitudes. The greatest changes in winter were observed north of 70°N, in the region of Canada, Baffin Island, northeastern USSR, where the temperature was 20–22°C higher than now. The summer map shows high-latitude temperature to be 6–8°C higher than at present. There are also comparatively small areas with negative values in the regions where precipitation considerably exceeds the modern

Table 8.4. Mean Latitudinal Temperature Differences (°C) Between the Pliocene Climatic Optimum and the Modern Epoch

	ϕ°N									
Season	90–80	80–70	70–60	60–50	50–40	40–30	30–20	20–10	10–0	0–90
July	11.9	10.3	6.5	5.4	4.1	2.8	1.1			
January	14.2	13.2	11.3	6.4	4.8	3.8	2.8			
Annual Mean	13.0/10.5	11.8/8.0	8.9/6.0	5.9/4.9	4.4/2.7	3.0/2.6	2.0/2.6	—/2.5	—/2.4	3.6/3.3

Note: The denominator shows the temperature difference from Manabe and Bryan (1985) for comparison.

level (Kazakhstan, the regions of the modern Gobi and Sahara Deserts).

Yefimova (Figure 8.10) reconstructed the spatial distribution of annual rainfall for the Pliocene.

The mean annual air temperatures for the Pliocene optimum were determined from data for July and January (see Figure 8.9). The annual evaporation levels for geographical grid points ($5 \times 5°$) on Northern Hemispheric continents were defined on the basis of the existing relationship between mean annual air temperature and annual total evaporativity for different landscape and vegetation zones both in the modern epoch and the Pliocene optimum. The annual precipitation was analyzed by using the correlation between the mean total evaporativity at the boundaries and within different landscape and vegetation zones as defined by Zubenok (1976). For example, the difference between evaporativity and precipitation in the forest zone is 0–30 cm/yr. Near the boundary between the forest and forest-steppe zones (savanna-steppes in the Pliocene), the total annual precipitation approximately equals evaporativity. In the forest-steppe or savanna-steppe zones, evaporativity can

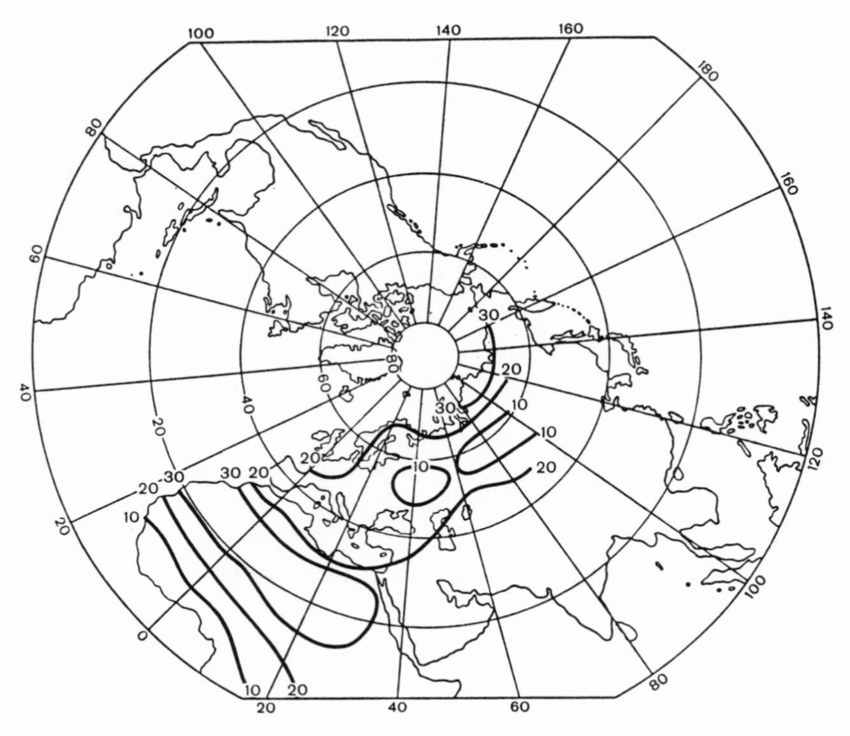

Figure 8.10. Deviation in annual precipitation (cm) from modern for the Pliocene optimum.

exceed precipitation by 20 cm/yr, while in the steppe zone by 20–40 cm/yr. At the boundary between steppe and semidesert zones, the amount of precipitation can be half of evaporativity. Correspondingly, a schematic map of precipitation for the Pliocene climatic optimum and then a map of differences between total annual precipitation in the Pliocene optimum and the modern epochs (see Figure 8.10) were plotted. As seen from Figure 8.10, the amount of precipitation in the Pliocene was greater than at present over almost all of Eurasia. The maximum precipitation increase (up to 30 cm/yr) is characteristic for the northeastern coast of the Asian USSR. Near the Baltic Sea, in the north of Western Siberia, in the Yakut Republic and basins of the Yana and Indigirka, mean annual precipitation was 20 cm higher than the modern. Near the Caspian Sea, the south of Western Siberia, Northern Kazakhstan and Central Asia, it was 20–25 cm higher.

Analysis of the climatic history of the Cenozoic and climatic reconstruction for three warm intervals within this time has thus shown that changes in surface air temperature caused by such different physical phenomena as redistribution of solar radiation or fluctuations in CO_2 concentration were fairly similar (Budyko, 1980; Manabe and Wetherald, 1980; Vinnikov and Groysman, 1982). Paleoclimatic reconstructions can therefore be used for warm intervals of the Holocene, Pleistocene and Pliocene as analogs of future climate when atmospheric CO_2 concentration will exceed the modern level due to human activities.

It should be taken into consideration that we obtained the aforementioned estimates of mean annual temperature changes for the entire Northern Hemisphere of about 1°C for the Holocene optimum, about 2.0°C for the Mikulino interglacial and about 3°–4°C for the Pliocene optimum by comparison with the climatic conditions of the late nineteenth and first half of the twentieth centuries. A certain adjustment should be made for the higher CO_2 concentration. Considering that the increase in mean temperature of the Northern Hemisphere was 0.2°C at that time, temperature increases indicated above were 1.4°C (Holocene), 2.2°C (interglacial) and 3.8°C (Pliocene) versus the climate of the end of the pre-industrial epoch.

It should also be noted that paleobotanic and paleontological data provide sufficiently reliable reconstructions of mean global and local surface air temperatures for different levels of global warming in the Quaternary and in more remote epochs. It is considerably more difficult to reconstruct atmospheric precipitation fields due to their greater temporal and spatial variability. However, preliminary reconstructions of total annual atmospheric precipitation for the Pleistocene warm intervals (the Holocene optimum, Riss-Würm interglacial optimum) and the Pliocene optimum corroborate the previous hypothesis (Budyko, 1980; Drozdov, 1983) that the dependence of atmospheric precipitation conditions on global warming level is nonlinear. As the level of global warming increases, moisture conditions on the continents of the Northern Hemisphere improve, and the area of precipitation deficit in the temperate latitudes is considerably reduced.

It can be assumed that creation of a series of paleoclimatic reconstructions for different levels of global warming will be of practical value in making climatic forecasts during developing global warming.

CHAPTER 9

EXPECTED CLIMATIC CHANGES

9.1. Mean Global Air Temperature

Estimates of mean surface air temperature changes to be expected in the next century due to the rise in CO_2 concentration were first realistically attempted in the early 1970s (Budyko, 1972). The results of this calculation presented in Figure 9.1 as line 2 project that in 120 years, the air temperature will rise by almost 2.5°C, this warming corresponding approximately to a doubling of CO_2 concentration. The same study concluded that future warming will probably reduce precipitation over some regions with insufficient moisture in the middle latitudes and that the higher temperature will diminish the area of polar sea ice.

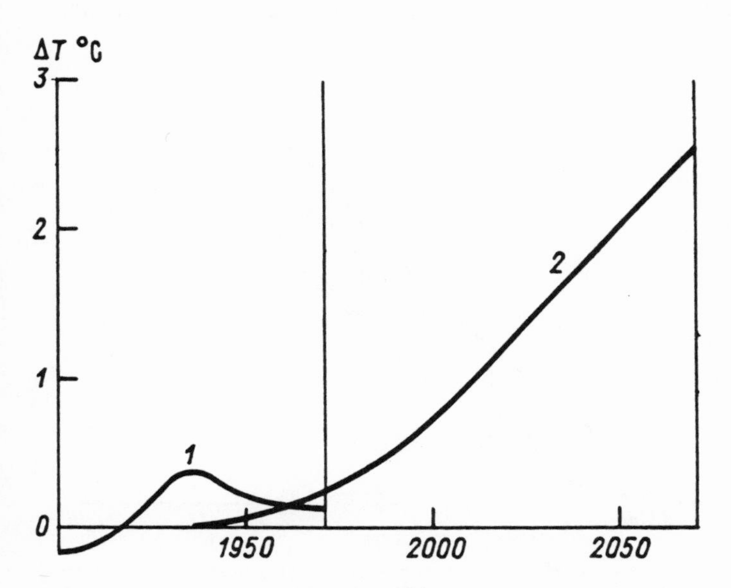

Figure 9.1. Secular variations in smoothed anomalies of mean air temperature in the Northern Hemisphere (1) and expected change in mean surface air temperature ΔT due to higher CO_2 concentration (2).

Many calculations of expected climatic changes were subsequently published, some by individual scientists and others in reports of national and international scientific agencies. Figure 9.2 presents a sample prediction of the forthcoming rise in mean global surface air temperature (curve 2) proposed at a meeting of Soviet-American scientists (Climatic Effects of Increased Atmospheric CO_2 ..., 1982). The figure also shows that the lowest most likely higher temperatures almost coincide with the aforementioned forecast (Budyko, 1972). This coincidence occurs because two additional factors included in the prediction of the meeting of experts, i.e., thermal inertia of the climatic system and intensified atmospheric greenhouse effect due to elevated trace gas concentrations compensate for the influence on higher temperature. This compensation had been discussed even before the meeting of experts (see Budyko, 1980, etc.) took place.

These scientists associate the maximum temperature increase (curve 3) mainly with the more significant climatic

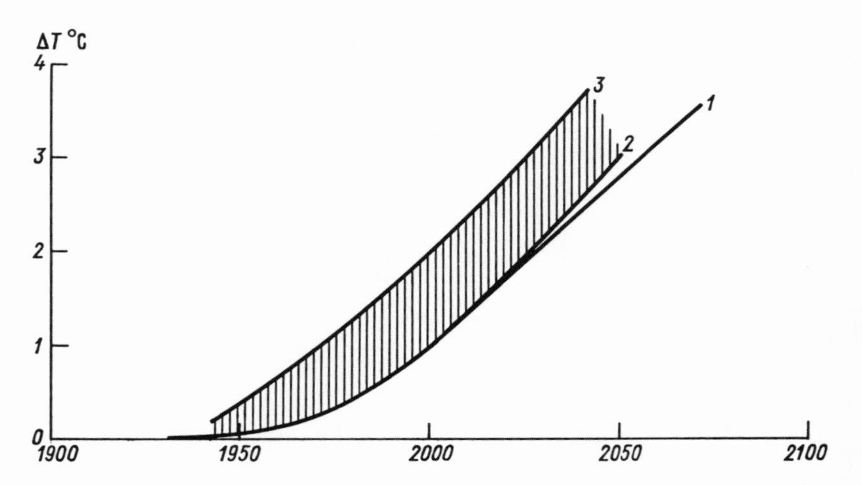

Figure 9.2. Projected surface air temperature changes: 1, according to Budyko, 1972; 2, 3, according to Climatic Effects of Increased Atmospheric Carbon Dioxide..., 1982.

Table 9.1. Increase in Mean Air Temperature (°C) Due to Changing Atmospheric Chemical Composition Compared With the End of the Nineteenth Century

Author	1975	2000	2025	2050
Budyko, 1972*	–	0.7	1.4	2.0
Kellogg, 1977-1978	0.5	1.2	–	4.0
Budyko, 1979	0.5	0.9-1.3	1.8-2.5	–
Flohn, 1981	–	1.0	1.5	3.3
Budyko et al., 1981	–	0.9	1.8	2.8
Climatic Effects of Increased Atmospheric CO_2 ... 1982	–	1–2	2–3	3–5
Anthropogenic Climatic Change 1987	0.5	1.4	2.2	3–4

*As compared with 1970, i.e., 0.5°C lower than the values determined from comparison with the end of the nineteenth century.

effect of higher atmospheric trace gas level. The majority of other predictions for increasing mean global surface temperature published over the last ten years also satisfactorily agree with the first forecast. This is particularly apparent from some estimates of expected variations in mean global surface air temperature in Table 9.1.

The last line of this table presents temperature differences obtained by the authors of this book (see Chapters 1, 3, 5 and 6). This table indicates that all estimates of higher mean temperature by the year 2000 differ slightly. This difference in the estimates increases somewhat for the more distant future, however, they are relatively insignificant and are comparable to the calculation error.

It is interesting to compare the estimates of expected

temperature change in Table 9.1 with the statement of the Villach conference in Austria (Bolin et al., 1986). This statement cites the most likely value of the "equivalent" mass increase in atmospheric CO_2 (including the influence of trace gases on the greenhouse effect), that corresponds to a 100% increase in equivalent CO_2 concentration in 2030 versus the pre-industrial level. The report assumes that the values of parameters ΔT_C range from 1.5 to 4.5°C. Using the mean value, we find the temperature increase for this moment in time to be equal to 3°C, which should be somewhat decreased to make allowances for climatic system thermal inertia. According to estimates of different authors, this decrease is from 20 to 40% of the initial temperature increment. Using the mean value of the indicated correction in this case, we find that the mean temperature increase in 2030 will be 2.1°C. This value agrees satisfactorily with data in Table 9.1.

The estimates of subsequent calculations as well as the first prediction of temperature variation can be compared with observational data on the Northern Hemispheric mean temperature in the 1970s–1980s. The results of this comparison (Chapter 11) confirm that warming developed during this period, and the higher mean temperature agrees well with the predicted value.

Noting that the first prediction of the expected change in mean global temperature was confirmed by subsequent calculations and observational data, we recall that global warming was forecasted at a time when it was believed that the climate was cooling. This is particularly clear from Figure 9.1 where curve 1 represents the temperature variation trend based on observational data. In the early 1970s, more than 20 predictions of expected climatic change were made and they all assumed that cooling would continue. These predictions had a poor scientific basis and were essentially reduced only to extrapolation of the observed temperature trend.

The prediction in Figure 9.1 that cooling would stop

in the near future was confirmed in the mid–1970s, when it became possible to compare observational air temperature data for the early 1970s with prior years (Budyko and Vinnikov, 1976; Borzenkova et al., 1976). As indicated in Chapter 7, it was discovered at that time that cooling had ended in the late 1960s. Later air temperature data showed that warming continued in subsequent years, and the mean temperature for the first half of the 1980s was not lower than that in the warmest years of the 1930s–1940s.

We should focus on the reasons why the future climatic change prediction was so accurate, particularly since long-term weather forecasts are not often correct.

According to the concept set forth in Chapter 1, the changing mean global temperature smoothed over several years can be attributed at present to the anthropogenic component associated with warming that is now occurring faster, and to natural variations mainly due to instability of atmospheric transparency. These fluctuations can result both in warming and cooling.

Human economic activity should have raised mean temperature by 0.5°C by the late 1960s. This is close to the absolute critical value of cooling that was reached during the last 100 years due to lowered atmospheric transparency. Since the mean global temperature, averaged for 5- or 10-year time intervals was close to the norm at that time, it was clear that the probability of subsequent decrease in these smoothed temperature values became extremely low. It was thus possible to reliably forecast the future warming. This was confirmed by subsequent observational data.

It is quite natural that right after this prediction was made, it seemed very paradoxical. This can be explained by the fact that new descriptions for the mechanism of physical climatic changes were not widely published at that time.

9.2. Regional Climatic Changes

Analogs of Anthropogenic Climate Change. Observational data show that anthropogenic warming of the early

1980s is equivalent to a more than 0.5°C increase in mean global temperature as compared to the pre-industrial period. This is quite close to the range of variations in temperature anomalies averaged for 5- or 10-year intervals and caused by natural fluctuations in atmospheric transparency. Although the current warming is noticeably disguised by such fluctuations, it exerts considerable influence on climatic conditions in various regions of the Earth. Chapter 7 indicated that mean global temperature rose on the same scale in the 1930s.

The calculations in this book show that in the early 2000s, anthropogenic warming will raise the mean global temperature by 1.4°C compared with the climate in the late nineteenth century. This warming, as described in Chapter 8, corresponds to climatic conditions of the Holocene optimum, 5,000 or 6,000 years ago. By 2025, the mean temperature will increase by 2.2°C, i.e., it will reach the characteristic level for the last interglacial epoch (Mikulino interglacial, about 125,000 years ago). By the middle of the next century, the mean temperature anomaly can reach 3 or 4°C, which is close to the Pliocene climatic optimum, 3 or 4 million years ago.

It is not clear whether evidence on climatic conditions of the aforementioned past warm epochs can be used to clarify future climatic conditions.

There is reason to believe that each of these warmings is a consequence of the effect of different climate-forming factors. We have found that the warm epoch of the 1930s can mainly be attributed to unusually high atmospheric transparency, and to a lesser extent, to the higher CO_2 concentration that was comparatively insignificant at that time. The Holocene climatic optimum depended in part on variations in the position of the Earth's surface relative to the Sun, i.e., on astronomical factors. The same factors affected climate in the Mikulino interglacial, although climatic change at that time was probably significantly affected by higher

CO_2 concentration. There is no doubt that the Pliocene climatic optimum was caused mainly by higher atmospheric CO_2 content.

Not only do the causes of the aforementioned climate variations not coincide, but all these warmings occurred while the structure of the Earth's surface was changing. Some differences in the Earth's relief were caused entirely by climatic changes (fluctuations in the area of sea ice and snow cover on continents), and some depended also on the other factors. Thus, for example, the continental forest area that affected the Earth's albedo changed in the nineteenth to twentieth centuries because of man's economic activities.

Certain difficulties in using data on past climates to estimate future climatic conditions are created by differences in the temporal intervals covered by the data on the past warm climates. It is clear that for the geological past (up to the Holocene climatic optimum inclusively), the average climatic data can characterize the steady-state climatic system. This question is not as clear for the data on meteorological conditions during the past warm epochs when the world network of meteorological stations was operating, i.e., the last 100 years. The influence of thermal inertia on warming should be greater for brief positive anomalies (on the order of a year) and smaller for warm periods of ten or more years.

Not only is it difficult to choose analogs of anthropogenic climatic change, but the accuracy of the available data on spatial distribution of meteorological elements for warmings of the past is limited. In most cases the errors in these data are thought to grow as the age of the epochs under consideration increases. As this takes place, reliable determination of mean anomalies in meteorological elements for warm epochs is very important, but it is difficult both because it is necessary to take into account climatic changes during these epochs and inaccurate dating of their beginning and end.

When data on past climates are used as analogs of future climatic conditions, the most difficult question is how the anomalies in meteorological elements in different geographical regions depend on the rise in mean global air temperature. When these anomalies are directly proportional to the mean temperature anomalies, the problem of using climatic analog is significantly simplified, since in this case data on different warm epochs can be easily applied for their mutual verification. If this relationship is nonlinear, then the possibility of making this check is limited.

These difficulties in using climatic analogs to forecast anthropogenic climatic change do not at all mean that the empirical approach to solving this problem is less promising than calculating future climatic changes by climatic theory models. At the same time, it is obvious that the aforementioned questions have to be clarified in order to obtain valid data on future climate based on information about past warmings.

Here, we mention the results of the first attempt to use climatic analogs together with model calculations to estimate future climatic conditions (Budyko et al., 1978). It employed data on climatic changes with an 0.5°C increase in mean temperature subsequently published by Vinnikov and Groisman (1979). Paleoclimatic maps of the Pliocene by Sinitsyn (1965, 1967) were used for the later warming that was associated with doubling of atmospheric CO_2 concentration.

The resulting information about regional climatic changes for the western part of the USSR was compared in the indicated study and in a later publication (Budyko, 1980) with climatic model results obtained by Manabe and Wetherald (1975, 1980). Since the model calculations were based on a very schematized arrangement of the continents and oceans, only a very approximate comparison of the results of using empirical analogs and model calculations to estimate future climatic conditions was possible. This comparison indicated a certain similarity in the findings; however, in this case agreement was qualitative rather than quantitative.

It became clear later that for estimates of future regional climatic changes, the Sinitsyn maps were insufficiently accurate, therefore, new paleoclimatic maps of the Pliocene had to be made (see Climatic Effects of Increased Atmospheric CO_2 ..., 1982; and Chapter 8 of this book). The problem was thus only partially resolved.

Turning to the possible use of climatic theory models to estimate expected anthropogenic climatic changes, we note that the accuracy of model calculations is a priori unknown. It is limited because it is difficult to define many parameters of the problem to be solved and for other reasons mentioned in Chapters 1 and 6. Consequently, in order to solve the question of using theoretical calculations of regional climatic changes during warmings in the prediction of anthropogenic climatic change, it is necessary to determine their accuracy.

Estimation of Regional Climatic Changes During Warming. We can use various data to solve the aforementioned problems. Some studies mention, in particular, that regional temperature variations at the Earth's surface can be quite similar even when there are different reasons for climatic change towards warming. For instance, calculations by climatic theory models established that an increase in the solar constant and higher CO_2 concentration lead to similar spatial distributions of surface air temperature deviations from its initial values (Manabe and Wetherald, 1980). Empirical studies (Budyko and Yefimova, 1984, etc.) have shown close similarity in the distribution of the differences in mean latitudinal surface air temperatures and other climatic elements caused both by increased annual radiation influx and higher atmospheric CO_2 concentration.

The most effective method for clarifying regional climatic change patterns during warmings is to compare empirical data on climatic conditions during different warm epochs with climatic model results. Such an analysis was made by Budyko, Vinnikov and Yefimova (1983) for mean latitudinal surface air temperature variations and annual total precipitation on the continents. They (see Chapter 1) compared the deviations from the modern climatic conditions obtained empirically on the basis of temperature and precipitation for climatic warming in the past century and the Pliocene climatic optimum with the results of theoretical calculation for the doubling and quadrupling of the CO_2 concentration above its conventional value of 0.03%. Since during all these warmings, the mean surface air temperature varied, the mean latitudinal temperature and precipitation variations were reduced to one value of higher mean global temperature equal to 1°C.

This comparison has shown that the temperature changes associated with different scales of warming and obtained by independent methods (by empirical data and by climatic theory models) agree well in the latitudinal range from the equator to 60°N. The temperature variations obtained by empirical data for higher latitudes noticeably exceed the model calculation results. This discrepancy is probably attributed to insufficient consideration by atmospheric general circulation models of temperature-snow-ice cover feedback in high latitudes.

In order to compare regional climatic changes during warmings of different scales in more detail, it is necessary to use maps of changes in climatic conditions for large territories. For this purpose, maps of air temperature increment in winter and summer were constructed for four warmings: in the twentieth century, the Holocene optimum, the Mikulino interglacial and the Pliocene optimum for comparison with the late–nineteenth century climate.

These maps, presented in Chapters 7 and 8, cover most of the continents of the Northern Hemisphere. Comparison of these maps with those of temperature changes caused by higher atmospheric CO_2 concentration based on atmospheric general circulation models allow us to clarify the nature of climatic changes during global warmings.

This comparison should take into account that the mean global temperature rose during the greatest warming of the twentieth century by 0.5°C, during the warming of the Holocene climatic optimum by about 1.4°C, during the Mikulino interglacial by 2.2°C, and during the Pliocene climatic optimum by almost 4.0°C. If the temperature anomalies on these maps are divided by changes in mean global temperatures for these epochs (i.e., they are reduced to the standard increase in mean global temperature of 1°C), we can conclude that all the maps are fairly similar. In all cases, winter temperature anomalies over most of the territory in question exceed the mean global temperature increase. In summer, these anomalies average the same 1°C.

At the same time, all maps show an explicit spatial variability of temperature anomalies that increases both in winter and summer as latitude is higher. Greater anomalies are seen on winter maps in regions with a more continental climate.

The following conclusions can be drawn from the great similarity in the temperature anomaly maps compiled by the four independent methods.

Regional changes in temperature anomalies associated with different warming scales in the first approximation are directly proportional to elevated mean global temperature. This means that the regional distributions of surface air temperature anomalies that depend on the aforementioned noncoincident causes of climatic change are similar to each other.

Another conclusion is that the empirical methods used in this investigation to estimate spatial temperature distribution during global warming are sufficiently valid. It is therefore possible to average the four winter and four summer maps to produce a more accurate pattern of temperature distribution during warming as compared with each map based on one of the four independent empirical methods for estimating temperature change.

These averaged maps are depicted in Figure 9.3. Figure 9.3a shows that in a marine climate of the middle latitudes, there are no noticeable changes in winter temperature during global warming. Winters will be much milder in the interior regions of the middle latitudes and in high latitudes. Warmer summers will mainly occur in high latitudes (Figure 9.3b).

In some southern continental regions, the temperature will be probably somewhat lower, because of greater heat expenditure for evaporation of precipitation falling over these regions.

Global warming has a more complicated influence on precipitation falling on the continents than on temperature. Figure 9.4a shows the map of altered annual total precipitation over the USSR territory that was compiled from data on modern climatic change with a 0.5°C increase in mean global surface air temperature.

Data in this figure depict the trend towards a certain increase in precipitation in the northern European part of the USSR and in Siberia. The zone of lower precipitation is in the central European part of the USSR, some of Kazakhstan and of the Far East.

During the warming of the Holocene climatic optimum (Figure 9.4b) the region of higher precipitation covers almost all the Asian and some of the European USSR. Precipitation decreases only in the southern European part of the USSR.

A different pattern of precipitation change is discovered in the maps for the epochs of considerable warmings, the

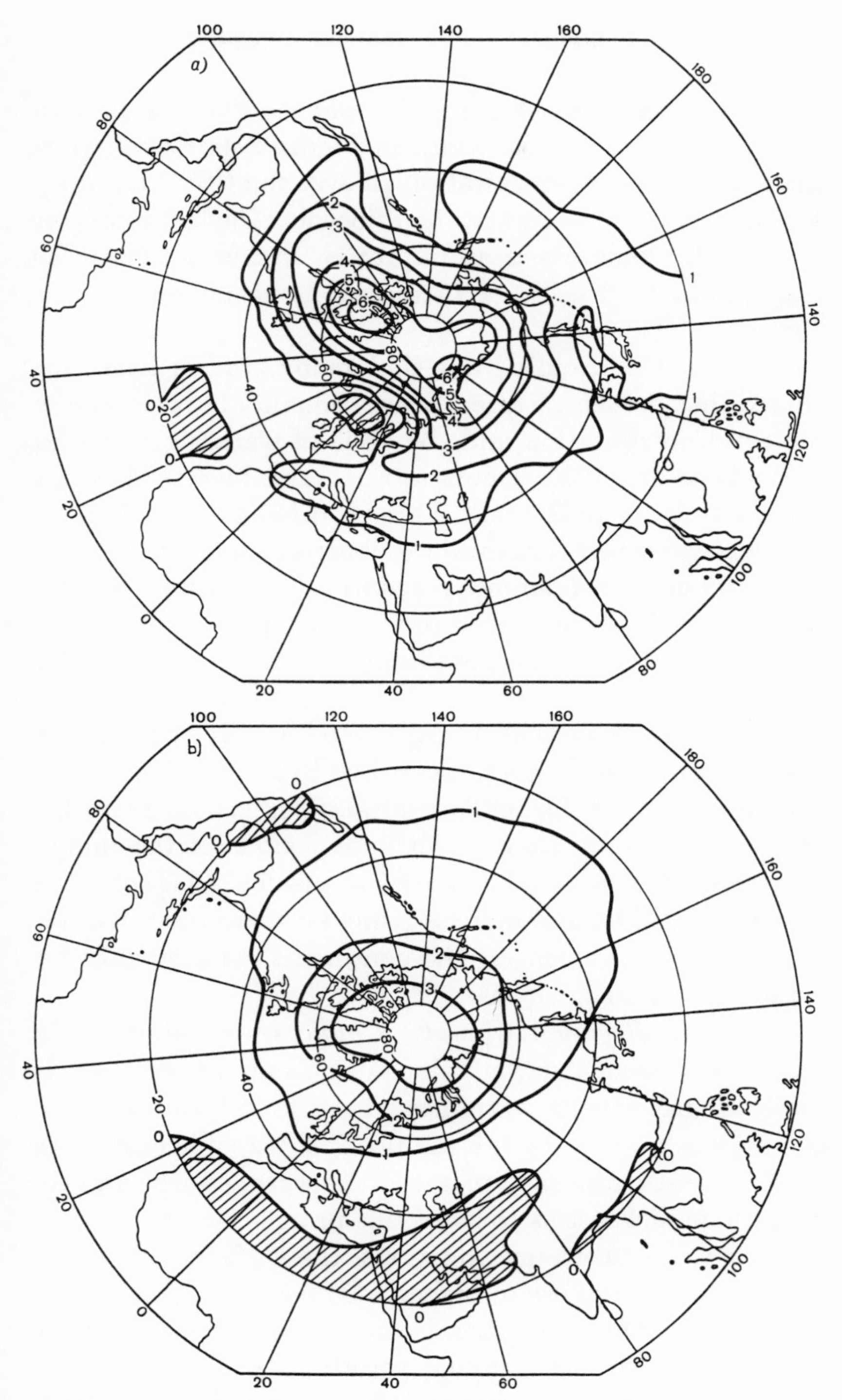

Figure 9.3. Variations in surface air temperature (°C) in winter (a) and summer (b).

Mikulino interglacial (Figure 9.4c) and the Pliocene climatic optimum (Figure 9.4d). Data in Figure 9.4c demonstrate that there is more precipitation all over the USSR territory. A similar pattern occurs in the Pliocene. The last map constructed by Yefimova is an improved version of her earlier map (Climatic Effects of Increased Atmospheric CO_2 ..., 1982).

It should be pointed out that many studies have concluded that there is no direct proportionality between changing total precipitation and the scale of warming (Budyko, 1980; Drozdov, 1981; Climatic Effects of Increased Atmospheric CO_2 ..., 1982, etc.). This hypothesis is based on the rather obvious fact that during global warming, the equator-pole temperature difference is reduced. This results in modification of the atmospheric pressure field and atmospheric general circulation, and essentially affects the precipitation pattern.

It is conceivable that during comparatively slight warming, the subtropical high pressure belt in the Northern Hemisphere spreads to higher latitudes. This causes more frequent droughts in some continental regions of the middle latitudes. The noticeable warming produces higher absolute air humidity, and induces rising air-mass precipitation. The moisture pattern on the continents consequently becomes more homogeneous.

Expected Regional Climatic Changes. Passing to the question of using the available materials about regional climatic change during warmings to estimate future climatic conditions, we will discuss the possible use of the previous empirical estimates of climates in the warm epochs and the climatic model results to solve this problem.

In several recent investigations, the surface air temperature changes because of higher CO_2 concentration growth were calculated by the general atmospheric circulation models, incorporating the annual variations in meteorological elements and real topography (Manabe and Stouffer, 1980;

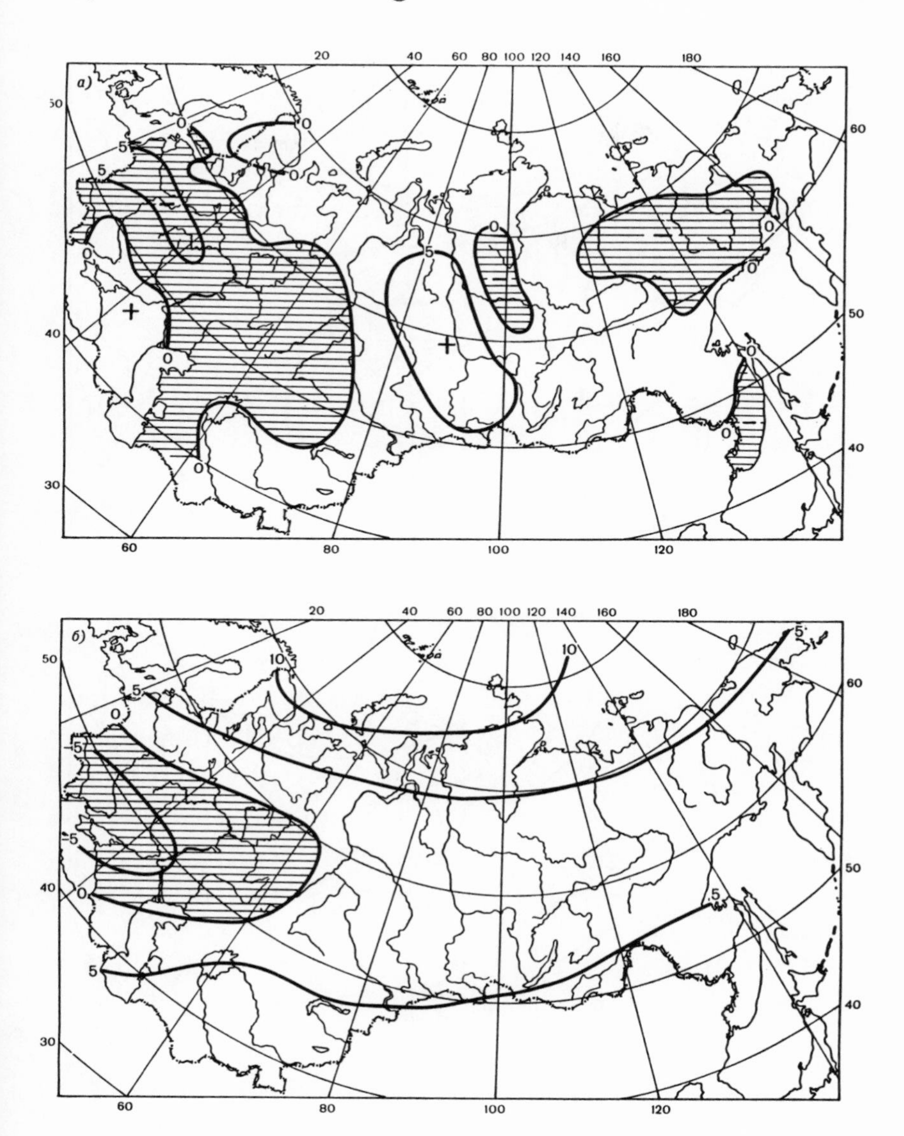

Figure 9.4a and b.

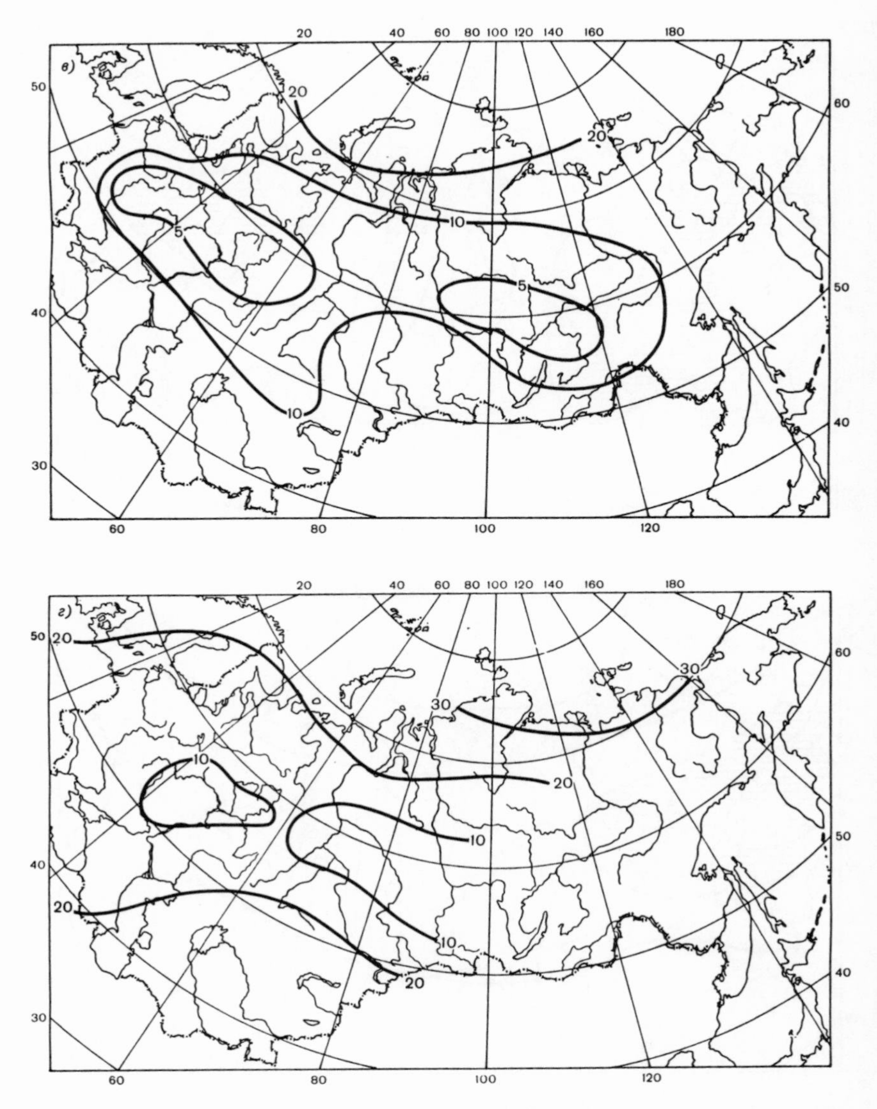

Figure 9.4. Variations in total annual precipitation (cm/yr) in the territory of the USSR. a, with 0.5°C increase in mean global air temperature; b, for the Holocene climatic optimum warming; c, for the Mikulino interglacial warming; d, for the Pliocene climatic optimum warming.

Schlesinger, 1984; Hansen et al., 1984, etc.). Some of these studies also tried to calculate precipitation variations over the continents by this method.

We have already mentioned that changes in mean latitudinal temperatures calculated by this method agree rather well with the conclusions of empirical studies (except for the high latitude zone). However, comparison of model spatial distributions of temperature variations with the results of empirical calculations reveals a certain discrepancy. For instance, a number of climatic model calculations show that the summer temperature increase relaxes somewhat with increasing latitude, whereas empirical data demonstrate that the temperature differences rise with increasing latitude. It might be thought that in this case empirical data yield more valid results. This is based on agreement between the conclusions of four independent empirical investigations. As mentioned above, a number of model calculations underestimate the influence of the temperature–ice-cover feedback which probably causes the indicated discrepancy between the theoretical results and empirical data.

It should be noted however, that there are minor discrepancies between the regional temperature patterns during warmings obtained by models and from empirical data. Bearing all this in mind, we can view the maps of temperature variations in Figure 9.3 as estimates of expected climatic change in the next century. It is more difficult to solve the problem of future precipitation variations.

As mentioned above, the maps of precipitation variations plotted by empirical methods with different increases in mean global temperature do not coincide with each other.

At the same time, the pattern of precipitation anomalies on the maps pertaining to a moderate warming (see Figures 9.4a and b) differs rather slightly. The maps for considerable warmings are also fairly similar to each other (Figures 9.4c and d). Similarity of these maps, on the one hand, shows their reliability, and on the other hand, means that two different kinds of precipitation pattern on the continents appear during global warming.

The first occurs during less significant warmings, that are characterized by 1–1.5°C rise in mean temperature. Under these conditions, in some areas where there is insufficient moisture in the middle latitudes, precipitation declines. During a greater warming, when mean temperature increases by 2°C or more, there is more precipitation over almost all the continents.

It is indicated in Chapter 8 that Figures 9.4b, c and d describe variations in precipitation as mean temperature rises by a certain value compared with the period, whose data were used to compile paleoclimatic maps, rather than with pre-industrial epoch. In this period (end of the nineteenth and the first half of the twentieth century), the rise in mean air temperature due to higher CO_2 concentration averages 0.2°C.

Taking this into account, we should remember that the map depicted in Figure 9.4b shows variations in precipitation during a 1.2°C warming, in Figure 9.4c by 2.0°C and in Figure 9.4d, by 3.6°C. Such warmings will occur at the beginning of 2000, by about 2025 and in the middle of the next century, respectively.

It is difficult to explain the nature of regional precipitation variations over the continents because the results obtained in studying this problem by climatic theory methods are limited. The detailed study by Manabe and Stouffer (1980) omits data on precipitation changes with increasing CO_2 concentration, since these authors considered these data to be unreliable. Calculations of changes in precipitation in other modern theoretical studies do not seem to be more accurate.

It is therefore more rational to use the maps shown in Figure 9.4 to determine precipitation changes at different stages of global warming, taking into account that these maps should be checked by some other independent method and that they can be viewed only as a tentative and approximate estimate of future climatic conditions.

At the same time, there is a possibility of using data of meteorological observations for the 1970s and the early 1980s, to verify the assumption that moisture conditions probably deteriorate during global warming in a number of regions in the middle latitudes which was proposed back in the early 1970s (Budyko, 1972). Assessment of the geographical distribution of atmospheric precipitation anomalies for 1975–1984 over the greater part of the forest-steppe and steppe zones of the USSR reveals a tendency towards lower total precipitation. Droughts were more frequent in the area in this time interval, and the 1975 and 1981 droughts were among the severest in the last 100 years. This agreement indicates that it is possible to use the calculations of expected climatic changes, which is discussed in more detail in section 10.5.

One may think that the cited schematic information about future climatic conditions gives a more realistic idea of climate at the end of the twentieth and early twenty-first century compared to data about climate of the recent past. However, the available information about future regional climatic change is still very limited in detail and accuracy. It is thus extremely important to accelerate preparation of more detailed maps that describe climatic conditions of the next century.

9.3. Influence of Climatic Change on the Hydrosphere

Surface Water. Although the impact of climatic changes on surface water is of great practical importance, it is still not clear how anthropogenic global warming affects the continental hydrological cycle.

All climatic models of general atmospheric circulation show that when the increase in mean air temperature influences the global energy balance, mean precipitation rises mainly due to greater evaporation from the oceanic surface.

Budyko and Drozdov (1976) demonstrated that greater evaporation from the oceanic surface during warming is a simple consequence of the physical dependence of specific air humidity saturated with water vapor on temperature of the evaporating surface.

The relationship between the change in mean temperature and average precipitation is nonlinear, however, by analyzing estimates in the review by Schlesinger (1984), we can conclude that a 1°C global warming augments average precipitation by approximately 1.6–2.6%, or by 2–3 cm/yr in absolute units.

The difficulties in calculating changes that occur in the general atmospheric circulation system have not yet been completely overcome. As indicated above, the climatic theory models cannot be used yet as a sufficient basis for assessing future changes in atmospheric precipitation patterns. We should thus currently use the prediction scenarios for altered surface air temperature and mean annual precipitation based on modern climatic change studies and paleoclimatic reconstructions of past epochs that were warmer than the modern in order to evaluate probable variations in the hydrological cycle of continental water.

The first attempt to assess variations in potential evaporation and mean run-off in the 2020s was made about ten years ago (Budyko et al., 1978). Current data on future climatic conditions allow for improvement of the materials of this article.

It is easier to estimate possible variations in annual river run-off for the epochs, when mean global temperature rose to 1°C above the pre-industrial level (or to 0.5°C above the modern temperature). In this case, the response of river run-off can be estimated by current hydrological observation data.

There are now few rivers in the world whose hydrological cycle has not been affected by man's intensified activities in catchment areas, by artificial hydraulic structures and reservoirs, and irreversible water losses for agricultural and industrial needs. Therefore, only a relatively small part of the available long series of annual run-off data can be assumed to be homogeneous, which refers primarily to some small rivers. In addition, the methods of modern hydrology in some cases can restore the lost homogeneity of the run-off data series. N. A. Speranskaya estimated the parameters of the linear relationship between annual river run-off in the USSR and variations in the mean surface air temperature of the Northern Hemisphere using the data archive based on the Hydrological Reference Book (Main Hydrological Characteristics, 1966–1980) and supplemented by V. I. Babkin with a reconstructed run-off data series. The corrections made in the observational data are based on detailed information about the development of economic and water-management activities in river basins and catchment areas. The most homogeneous observational series lasting about 60 to 90 years that were selected for analysis usually refer to small rivers or the upper reaches of big rivers.

The results of statistical evaluation of the relationship between annual run-off and mean air temperature led to the conclusions that with a 0.5°C warming of the Northern Hemisphere versus the modern climate, the climatic change is probably accompanied by variations in annual run-off:

1) river run-off decreases in the Central and Volga-Vyatka regions of the European part of the USSR; the range of the Volga run-off variations at Volgograd is about 5%;
2) the upper Ob run-off increases; the range of variations is 2 to 7%;
3) the upper Yenisey run-off increases; the range of variations is 7 to 10%;
4) the Amur run-off increases, variations in its middle reaches is the greatest; the range of variations is about 10%.

These estimates are consistent with calculations (Budyko, 1971, and others) of possible variations in mean annual run-off using empirical estimates of changing annual precipitation with a 0.5°C warming (Vinnikov and Groisman, 1979). These conclusions have definite importance for forecasting. Figure 9.5 shows that the result of a more detailed calculation of changing mean annual climatic run-off in the USSR with a 0.5°C rise in mean global air temperature. This calculation by N. A. Lemeshko is based on an improved comprehensive method for determining evaporation from land. The map distinctly shows the vast areas of increased and decreased run-off that are almost unequivocally associated with regions of higher and lower precipitation for relatively small-scale warming.

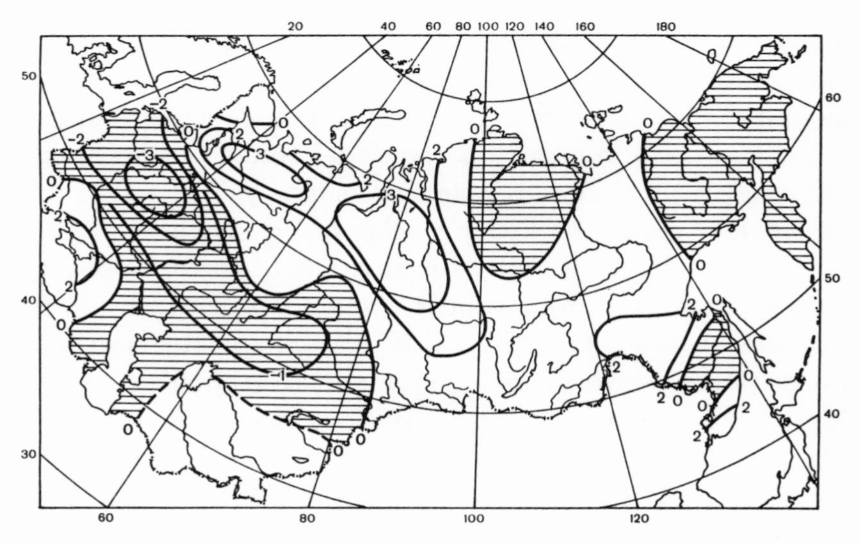

Figure 9.5. Variations in annual run-off (cm) with a 0.5°C warming.

The indicated type of variation in the annual river run-off is naturally assumed to be preserved also in case of a warming to 1°C compared with climate over the last 100 years, with temperature and total precipitation variations remaining mainly similar to those characteristic of a 0.5°C warming. However, during a greater global warming, the precipitation and annual run-off patterns will basically change (see Chapter 8). Annual run-off variations due to these conditions require further thorough calculations.

Possible variations in land hydrology due to anthropogenic global climatic warming detected by climatic theory models are discussed in Chapter 6. They indicate that global warming can be accompanied by a noticeable increase in annual run-off in high latitudes and a considerable change in moisture content of the active soil layer over a greater part of the continental surface. However, the accuracy of estimates made by climatic theory models is insufficient, while the available estimates cannot be used for forecasting because they are inconsistent.

Earlier, when hydrologists discussed anthropogenic variations in water resources, they ignored anthropogenic global climatic variations and restricted themselves to analysis of man's impact on catchment areas. Nevertheless, as climate warms up, the study of global aspects of anthropogenic changes in water resources becomes more important.

World Ocean Level. The problem of changing World Ocean level due to anthropogenic global warming has been widely discussed in recent years. There are some publications that cover various aspects of this problem (World Water Balance..., 1974; Grosval'd and Kotlyakov, 1978; Mercer, 1978; Gornitz, Lebedeff and Hansen, 1982; Hoffman, Keyes and Titus, 1983; Meier, 1984; Klige, 1985; Bolin et al., 1986, etc.).

According to the available estimates, 17–18,000 years ago, the World Ocean level was about 100 m lower than the present, however, in the last 5,000 years its variations have been relatively small (Fairbridge, 1961). Analysis of measurements at the network of gauging stations established that in the last 100 years, the World Ocean level has risen comparatively uniformly at an average rate of about 0.15 cm per year (World Water Balance..., 1974; Barnett, 1984; Klige, 1985). However, some studies assume this rate to be noticeably higher and for the last 50 years to equal 0.23 cm/year (Barnett, 1984).

Estimates obtained in the USSR during preparation of the monograph "World Water Balance and the Earth's Water Resources" (1974) and subsequently improved by Klige (1985) show that the current rise in World Ocean level (0.15 cm/yr) can be attributed to the lowering of the level of lakes (0.02 cm/yr) and ground water (0.04 cm/yr), the construction of reservoirs (–0.02 cm/yr) and melting ice cover (0.10 cm/yr).

This analysis was considered to be approximate by its authors since it ignores changing World Ocean level due to thermal expansion of ocean water during global warming. That factor can explain about one-third of the recorded World Ocean level elevation (Gornitz, Lebedeff, Hansen, 1982; Revelle, 1983).

It has recently been ascertained that from one-third to one-half of the ocean level elevation can be attributed to melting of small glaciers not related to Greenland and Antarctic ice sheets (Meier, 1984). The available estimates do not always agree about the contribution of these ice sheets to the current altered ocean level (Klige, 1985; Bolin et al., 1986).

What will happen to the World Ocean level in the future? Two factors will have an impact: thermal expansion of ocean water and melting of small glaciers. There are insufficient data, however, to provide a complete solution. Statistics usually help.

Klige (World Water Balance..., 1974, 1985, etc.) was the first to plot the graph relating altered World Ocean level to mean air temperature variations. The following linear equation corresponds to this graph

$$H(t) \approx aT(t - t_0) + b, \tag{9.1}$$

where H is the World Ocean level, cm; t is time; a, b and t_0

are empirical coefficients: $a = 15$ cm/K; $t_0 = 19$ years, $T(t)$ is mean global surface air temperature changes in the extratropical Northern Hemisphere (zone 17.5–87.5°N), which are viewed as characteristics of the Earth's global thermal conditions.

Equation 9.1 was obtained for 5–year mean values of $H(t)$ and $T(t)$. In the event that changes in $H(t)$ lag behind $T(t)$ by $t_0 = 19$ yrs, the correlation coefficient is fairly high, i.e., 0.94. These estimates were later repeated using independent data by Gornitz, Lebedeff and Hansen (1982), who obtained similar values of parameters: $a = 16$ cm/K and $t_0 = 18$ yrs.

It is an equation like 9.1 that was used in the report by experts of the World Meteorological Organization, the U.N. Environment Research Program and International Council of Scientific Unions (Bolin et al., 1986) to estimate the forthcoming changes in World Ocean level. As follows from this report, when mean global temperature rises by $3.5 \pm 2.0°$ C, the World Ocean level is apt to rise by about 80 cm. The use of an empirical expression like 9.1 for a greater parameter range than that for which it was obtained, requires additional substantiation.

It should be noted that in addition to slow evolution of the World Ocean level, because of a change in individual physical components of global water exchange, there are also possibilities of natural catastrophes. They primarily include mechanical destruction of the West Antarctic ice sheet at a definite stage of global warming (Mercer, 1978; Grosval'd and Kotlyakov, 1978), because of the particular nature of this area which rests on the sea floor and not on land as the East Antarctic sheet. This glacial destruction could take several centuries, although it could possibly be faster (Thomas et al., 1979). The 5-meter rise in the World Ocean level that would occur as a result would submerge many major cities and densely populated coastal areas in various parts of the Earth.

Although the statement of the aforementioned report suggests that an event like this will not occur before the end of the next century, careful study of this problem must continue.

9.4. Influence of Climatic Change on the Biosphere

Sea Ice. Strictly speaking, sea ice is a part of the climatic system, therefore evaluation of a decrease in its area during warming is an integral part of the general problem of predicting anthropogenic climatic change. However, since the possibility of destroying polar ice during global warming is a very important question in terms of the entire complex of natural conditions in high latitudes, it should be viewed as one of the aspects of the more general problem of future change in the biosphere.

Northern sea ice area and air temperature are directly and indirectly related to each other in high latitudes. Higher air temperature (in particular, in the warm season) reduces this area, thus decreasing the Earth's surface albedo and increasing the amount of absorbed shortwave radiation that causes an additional rise in temperature and further destruction of the ice cover. The opposite pattern occurs when surface air temperature initially drops (Budyko, 1968).

This shows that the area of sea polar ice can change considerably during comparatively small fluctuations in global climate. Empirical data confirm this well. Figure 9.6 depicts data by Zakharov and Strokina (1978), supplemented by Zakharov with up-to-date information on variations in ice area in the Arctic Ocean for July, August, and September over the period starting in the late 1930s. This graph demonstrates that from the 1940s to the mid–1960s, the ice area increased by more than 10% while beginning in the mid–1960s, it ceased to grow, and varied noticeably when it tended to shrink.

Data on the mean air temperature variations cited in Chapter 7 show that the ice area changes over the last sev-

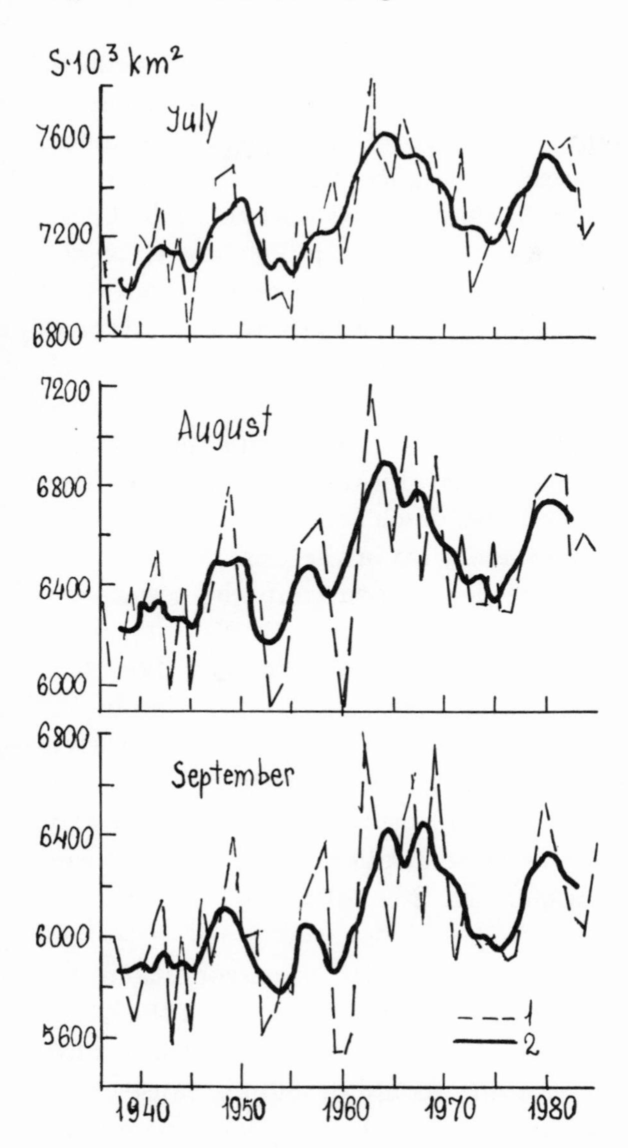

Figure 9.6. Variations in ice-surface area (S) in the Arctic Ocean in summer by monthly means (1) and 5–year means (2).

eral decades corresponded to decreasing and increasing mean time-smoothed global air temperature by a few tenths of a degree. Based on this comparison, it is natural to assume that Arctic ice will disappear completely as mean global temperature rises by a few degrees. This hypothesis can be verified by various methods. One approach to assessing the warming effects on ice cover is based on the numerical model of energy- and mass-exchange in ice, thus permitting a calculation of modern sea ice boundaries and their thickness. The findings agree well with observational data (Budyko, 1969, 1971). Calculations by this model demonstrate that sea ice conditions depend on solar radiation influx, air temperature in the warm season (the season of ice melting) and in the cold season when ice becomes thicker mainly due to freezing water at the bottom of ice. When air temperature varies comparatively little, ice thickness changes considerably in the warm period and much less in the cold season.

The time when permanent ice in the Central Arctic becomes annual ice can be determined from the results obtained by this model.

Empirical data on modern climatic changes show that during warming, the summer air temperature rise in the Arctic sea ice zone is approximately equal, on the average, to the increment for the entire Northern Hemisphere, whereas in winter it is 4 or 5 times greater. Consequently, a 2°C increase in the mean hemispheric temperature corresponds to a 2°C summer and 8 to 10°C winter in the sea ice zone.

As the calculations reveal, under these conditions the average summer ice thickness is reduced by more than 3 m. This is sufficient to convert permanent ice into annual, after which an additional warming should develop because of the lower Earth's surface albedo in high latitudes. Some studies concluded that iceless conditions in the Central Arctic can exist even without anthropogenic warming of the atmosphere (Budyko, 1971), therefore, it is probable that when this warming occurs, after the permanent ice melts, annual ice will either disappear or appear only in coastal regions of the Arctic Ocean during the coldest winters.

In addition to the semi-empirical method used here to estimate ice-melting conditions as warming develops, a purely empirical solution to the problem is possible.

We recall that permanent ice in the Arctic exists at air temperature in the warmest month of –1°to +2°C, and the bulk of the ice mass is in the zone with the temperature in the warmest month of 0°to 2°C. Climatic studies have established that during warming, the air temperature increase near sea ice edge is somewhat above the average temperature change over the entire ice cover zone and, as noted earlier, in summer, it is not below the average temperature rise for the Northern Hemisphere as a whole.

Consequently, if the mean hemispheric temperature increases by 2°C, permanent ice in the zone with the temperature of the warmest month of 0°to 2°C should melt first on the periphery where the temperature increase is the greatest, and then in the central regions. As ice melts, the air temperature in the regions, where it is somewhat below zero in the warmest month, apparently rises because of additional warming caused by lower Earth's surface albedo. This will cause complete ice melting. Considering that mean air temperature has already risen by several tenths of a degree compared with the end of the last century and that there is possible calculation error, one should think that a 2°to 3°C temperature increase above the pre-industrial level is sufficient to melt permanent sea ice in the Arctic. This higher temperature could be reached by the mid-twenty-first century, possibly, by the 2020s. A similar conclusion was drawn in some preceding studies (Budyko, 1972; etc.).

Partial or complete destruction or "disintegration" of polar ice will have a significant impact on the environment in high latitudes. There is no doubt that disappearance of sea ice will be followed by a drastic surface air temperature increase in the cold season, and this warming will envelope vast continental areas in Eurasia and North America adjacent to the seas that are currently freezing.

Noticeably milder climatic conditions in the current location of tundra and northern taiga will significantly alter vegetation in high latitudes. The major factor in this change will be migration of the more thermophilic plant species to the north, which can lead, in particular, to replacement of tundra by forest. As these boundaries change, the animal ranges will correspondingly shift northward.

More drastic changes will occur among the animals in polar seas, in particular, those permanently covered with ice. Many plants and animals will undoubtedly penetrate into these seas, that are currently inaccessible to them. Those species whose life is closely linked to polar ice will thus be threatened with extinction. Therefore the question of the future of animals and plants during global warming in high latitudes deserves particular attention.

Another pivotal point is early detection of the effect of high latitudinal warming on human economic activities. One of the obvious results of this warming will be improved navigational conditions in polar seas, with the possibility in a few decades of direct voyages in the northern Eurasian and Canadian ice-free range. Some consequences of the warming will be less favorable for countries with a cold climate, e.g., the problem of preserving new construction in regions of permafrost that is gradually being destroyed.

There is no doubt, however, that there will be greater opportunities to make economic use of the regions with cold climates as the Earth becomes warmer.

Natural Zones. Several studies have considered the question of changing location of geographical zones during global warming (Budyko et al., 1978; Budyko, 1980, etc.). Both paleogeographical data and relationships between the current boundaries of geographical zones and climatic conditions can be used to solve this question.

As mentioned in previous chapters, in the second or third quarter of the next century, the mean global surface air temperature will be close to the Pliocene climatic optimum. As warming develops, the climatic conditions will be closer to the warmer Miocene climates.

In the Miocene, the central regions of Western Europe were widely covered with forests, including evergreen plants, in particular, palms. In Northern Europe, up to Spitsbergen inclusively, there were rich coniferous-deciduous forests containing birch, beech, oak, pine, spruce and other species. Southeastern Europe, where the steppe zone is now situated, was initially occupied by beech-oak forests that included some evergreens later replaced by savanna. Northern Asia was covered with coniferous-deciduous forests.

In the Pliocene, the major geobotanical zones shifted to the south, although at that time they were in higher latitudes than now (Sinitsyn, 1965).

Similar conclusion can be drawn from the dependence of modern natural zonality on climatic conditions. Table 9.2 was compiled for this problem in the USSR. It shows data on thermal conditions in the warm season at the boundaries of geographical zones.

Based on Table 9.2 and previous data on future climatic changes, the conclusion can be drawn that in the first half of the twenty-first century, geographical zones should shift hundreds of kilometers farther north in some cases. This mainly concerns the geographical zones with sufficient and excessive moisture. Geographical zones in the regions of insufficient moisture will experience less change in their locations, since the higher potential evaporation due to warming in most cases will be accompanied by more precipitation. At the same time, in some arid regions, including deserts of Central Asia, there will be so much more precipitation that a stable vegetation cover similar to steppe or savannah will be maintained. Geographical zones in the twenty-first century will obviously differ sharply from the last centuries.

It should be emphasized that the components of the geographical environment are not able to vary simultaneously with comparatively rapid climatic changes. The Earth's surface water cycle will vary almost synchronously with climate. Vegetation cover, and especially soil, will adapt to new climate considerably more slowly.

Table 9.2. Thermal Conditions of the Warm Season

Thermal Conditions	Degree-Day Index*	Geographical Zones
Very cold	Annual temperature ¿ is not above 10°C	Arctic deserts
Cold	0–1000	Tundra and forest-tundra
Moderately warm	1000–2200	Coniferous forest, alpine meadows, mountain steppes and Siberian steppes
Warm	2200–4400	Mixed and broad-leaved forests, forest-steppe, steppe, northern desert
Very warm	> 4400	Subtropical vegetation, desert

*Defined as absolute temperature difference above 10°C (T − 10°) multiplied by number of days/yr where temperature > 10°C.

All the components of the geographical environment are thought to be able to only partially adapt within a few decades to new climatic conditions. At the same time, we should not underestimate the flexibility of many of these components for altered global climate. Grigor'yev (1956), in particular, indicated that during the comparatively short warming of the 1920s–1930s, the northern forest boundary in some regions of forest tundra shifted to higher latitudes.

It is obvious that displacement of geographical zones during rapid climatic alteration requires further studies.

Productivity of natural vegetation cover is an important feature of geographical zones. Since the effect of climatic and atmospheric chemical composition changes on photosynthesis is very important for agriculture, this topic is discussed in detail in Chapter 10. Here we only mention some studies that, based on the effect on productivity of natural vegetation cover with meteorological conditions and atmospheric CO_2 concentration, concluded that overall productivity of autotrophic plants could change in vast territories.

Altered plant productivity because of warming by the mid-twenty-first century was first estimated for all continents by N. A. Yefimova, who found the increment in total productivity could reach 50%. It was also noted at that time that this value would be the upper limit for productivity growth, probably unattainable, because the natural vegetation cover cannot adapt in such a short time to changing environmental conditions (Budyko, 1980). More comprehensive analysis by Lokshina (1986) showed that for the entire Northern Hemisphere the productivity of the continental vegetation cover in the Pliocene climatic optimum was 28% above the modern. This estimate may also be somewhat exaggerated for the actual conditions of the mid-twenty-first century, since the vegetation cover adapts comparatively slowly to the rapidly changing global climate.

The estimates by Yefimova and Lokshina are important for substantiating the conclusion that modern climatic change on the whole will have a favorable effect on the biosphere by increasing the energy source from autotrophic plants for the life of organisms.

9.5. Anthropogenic Effects on the Ozonosphere

Ozone actively absorbs ultraviolet (UV) radiation of the Sun, thereby influencing temperature distribution in the stratosphere, and ultimately the climate. In turn, climatic changes causing temperature and atmospheric chemical composition variations affect the ozonosphere. The atmospheric ozone concentration is 6×10^{-5}% (by mass); its total quantity reaches 3.3×10^9 T (1.16×10^8 T is in the troposphere).

Ozone started to form in the upper stratosphere due to the photodissociation reaction of molecular oxygen (UV solar radiation with wavelength less than 242 nm). Interaction of oxygen atoms with oxygen molecules in the presence of a third body M produces ozone.

The ozone is formed mostly by this mechanism at altitudes of 35 to 45 km. In the lower stratosphere and troposphere, the ozone-producing reactions are initiated by nitrous oxides that dissociate under the influence of UV- and visible radiation.

Ozone is mainly distributed at altitudes of 15 to 45 km with maximum at 15 to 25 km (the maximum level is lower in polar latitudes and higher in the tropics). The ozone content peaks in the Northern Hemisphere in spring, averaging 446 DU (Dobson units, 1 DU = 0.001 cm of ozone at standard pressure and temperature), whereas in autumn it is about one-half this value.

The average annual ozone content in the ozonosphere is 300 DU (i.e., about 3 mm under standard conditions), while at the equator it is even lower (up to 200 DU). There is great natural variability in atmospheric ozone because it is transported by uniform and turbulent fluxes (mainly below 25 km), and because various photochemical reactions (above the 25-km level) occur at different rates (Atmospheric ozone..., 1982). Anthropogenic modification of the ozone layer is real and dangerous, it is not easily detected against the background of great natural variability in ozone concentrations.

Atmospheric ozone can be destroyed as a result of the normal oxygen cycle (Chapman's cycle). This reaction occurs slowly; however, it can be accelerated catalytically by compounds of nitrogen, hydrogen and chlorine (nitrogen, hydrogen, and chlorine cycles, respectively). These reactions are the most likely at altitudes of 20 to 40 km.

Catalytic concentrations are thousands of times lower than ozone concentration, and the number of destructive cycles per catalytic molecule is 10^2 to 10^7. The contribution of various catalytic cycles depends on altitude: above 35 km, the chlorine cycle dominates and below, the nitrogen.

It is interesting that the ion cycle of ozone destruction exists either on account of ion-molecular reactions directly resulting in ozone destruction (ion cycle)

$$O_2^- + O_3 \rightarrow O_2 + O_3^-$$
$$qO_3^- + O_3 \rightarrow O_2^- + 2O_2,$$

or ion-molecular reactions producing catalyst-particles that destroy ozone.

Tal'roze and Larin (1982) considered, in particular, the formation of hydroxyl OH-radicals in the lower stratosphere (due to ion-molecular reactions of charged particles produced by galactic cosmic rays) and subsequent destruction of ozone by this radical in the known hydrogen cycle. This process is apparently the only one that leads to ozone destruction at night, and is very important when the high latitude atmosphere is intensely ionized by powerful solar flares.

Various natural and man-made substances coming from the troposphere have a significant influence on the ozone layer. For example, volcanic eruptions alone emit from 10,000 to 100,000 tons of chlorine into the stratosphere annually.

The ozone layer is primarily destroyed by chemical reactions with chlorine, nitrogen, and other compounds which come from operation of refrigerators and aerosol devices that emit chlorofluorocarbons (CFC); decomposition of mineral fertilizers that release nitrous oxide; operation of stratospheric jet aircraft that discharge nitric oxides and water vapor; nuclear bomb testing that forms enormous amounts of nitric oxides. Other man-made chlorine compounds also enter the stratosphere, e.g., methyltrichloromethane (CH_3 CC_3), carbon tetrachloride (CCl_4) and methyl chloride (CH_3Cl).

Chemically inert substances like CFCs and nitrous oxides come, as a rule, from the troposphere into the stratosphere, where under the influence of ultraviolet radiation, they release chlorine that destroys the stratospheric ozone layer.

It has been estimated that the residence time of F-11, F-12, F-113 in the atmosphere is within the range of 80–150 years and that of hydrogen containing CFC, such as CH_3Cl, CH_3CCl_3, is within the range of 5–20 years (see 5.3).

About 850,000 tons of CFC, mainly F-11 ($CFCl_3$) and F-12 (CF_2Cl_2), were produced in 1974, and almost the same annual amount was produced and released into the atmosphere globally: 650,000 tons of F-11 and F-12, 470,000 tons of CH_3CCl_3, and 140,000 tons of F-113 ($CF_2ClCFCl_2$) were released in 1985 (Stratosph. ozone, 1988).

The decrease of total ozone by several percent by 1986 due to CFC release is caused by F-11, 12 (70%); by F-113 (12%); by CCl_4 (8%), and by CH_3CCl_3 (5%). The remaining 5% is caused by Halous (bromine containing compounds) and other CFCs.

Model calculations (Chlorofluorocarbons..., 1979) demonstrate that if CFC production had been completely stopped by 1979, then in 5 or 15 years the ozone depletion would be 2%, if by 1983, the maximum depletion would be 2.5%. If future freon production continues at the 1975 level, the ozone depletion will be 11 to 16% (reaching half this value in approximately in 50 years). The ozone depletion will be the strongest at high altitudes.

It is important to determine total effects on ozone of increasing atmospheric concentrations of other halogen organic derivatives, as well as CO, NO_x, CO_2. Models indicate that annual production of 200 million tons of nitrogen fertilizers releasing N_2O increases the intensity of ozone destruction by nitric oxides in the stratosphere by 1.1–1.2 times, because the influx of tropospheric nitrous oxides into the stratosphere is a major source of stratospheric nitric oxides (Chlorofluorocarbons..., 1979). It was also calculated that higher atmospheric CO_2 concentration lowers stratospheric temperature, mainly because of the higher ozone content. Even now, the higher CO_2 concentration versus the pre-industrial level should have raised the ozone mass by 1 or 2%, and by 2030 by 6 or 7%.

It is important how the higher atmospheric CO_2 content considered here will affect the ozone concentration. In this case, we should take into account the influence on ozone of lower stratospheric temperature, that in addition to higher surface air temperature is an inevitable consequence of elevated CO_2 concentration.

Altered ozone concentration because of increment in CO_2 concentration from 275 to 600 ppm is discussed in the aforementioned study (Chlorofluorocarbons..., 1979). Calculations of the combined effect of CFCs and CO_2 (for fixed levels of CFCs and CO_2 concentration of 600 ppm) discovered a 40% decrease in ozone content in the upper atmosphere above 40 km and a 18°C temperature reduction at altitude 48 km.

An increase in CO_2 concentration to 600 ppm weakens the total ozone depletion caused by CFC emission from 10 to 7%. The effect on ozone from higher atmospheric CO_2 concentration in the upper stratosphere is thereby opposite to the influence of CFCs. The aforementioned study also indicates that annual production of 200 million tons of mineral fertilizers will cause a 1.4% ozone depletion (provided that the lifetime of N_2O in the atmosphere is 12 years) and 4.1% (if the lifetime is 40 years). The influence of all other anthropogenic halocarbons (except F-11, 12) will intensify the effect from CFCs by approximately one-third.

The effect of different combinations of anthropogenic factors on the ozone layer is covered in the Report of the 5th Session of the Coordination Committee on the Ozone Layer, 1981. It was assumed that CFC emissions would stay at the 1980 level; annual increment in N_2O concentration would be 0.2%; the number of subsonic aircraft flights would increase 10 times (from 1975 to 1990 with subsequent constant value). In this case, calculations show that the total ozone content in the Northern Hemisphere will change by less than 0.5% in the next several decades, however, stable ozone concentration will decrease by 40% at altitude of 40 km and increase by 25% at 10 km. This change in vertical ozone distribution naturally can have some effect on climate.

The anxiety of mankind about the future of the Earth ozone shield has resulted in the decision of the Vienna convention on the Protection of the Ozone Layer (1985) and the Montreal Protocol on substances that deplete the ozone layer (1987). According to this Protocol, its member states (among them are the United States and the USSR) agree not to produce annually and use the ozone-destroying substances (halocarbons) included in the Annex to the Protocol in amounts exceeding the 1986 production level by more than 10%. Beginning from July 1, 1993, the annual level of the production of these substances will be lowered to 80%, and, from July 1, 1998, to 50% of the 1986 production level, if no other decision is adopted before these dates (Montreal Protocol, 1987).

Model calculations show that the restrictions of the Montreal Protocol will have practically no influence on the rate of increase of CFC content in the atmosphere in the twentieth century and that they will slow this rate 1.5 times by 2050 compared with the case of preserving the contemporary rate of CFC release into the atmosphere. However, the calculations by two-dimensional models of zonally averaged troposphere and stratosphere simulating the seasonal and latitudinal changes of ozone content show that these restrictions do not allow the total ozone amount to decrease by more than 1% at all latitudes and seasons by the middle of the twenty-first century (Stratosph. ozone, 1988).

Data of combined ground and satellite measurements of atmospheric ozone concentration help to analyze the extent trends. The current accuracy of measuring mean annual ozone concentration with Dobson's instruments is estimated at 1.5%, however, the measurement system accuracy is actually lower (apparently about 4%). Satellites can measure both the total ozone content (using the method for measuring backscattering ultraviolet solar radiation or infrared radiation in the 9.6 μm band) and its vertical distribution.

The high natural variability of ozone makes it difficult to detect trends in changing ozone content. According to data of the ground measurement system, the total ozone content increased and decreased several times during the 1965–1980 period (within the range of 1% to several per cent). There are indications (however, unproved) of a possible ozone-solar cycle relationship.

After the Fuego eruption in 1974, a considerable (4%) ozone depletion was observed in the 32–48 km layer (the original values were only restored in 1980). According to ozone sounding data, in the 2 to 8 km tropospheric layer, the ozone concentration increased by about 20% from 1967 to 1980.

Thus, although anthropogenic factors will cause the total ozone content to change slightly in the next few years, a significant redistribution of it over altitude will possibly occur, which can produce climatic changes and other consequences. This indicates ozone sensitivity to the existing anthropogenic influence.

As noted in Chapter 5, the significant ozone depletion discovered recently in some regions of the Antarctic in spring (September-October) is especially alarming. Its content decreased from 320 DU in 1979 to 160 DU in 1985 ("ozone hole") (Ferman et al., 1985). Maximum depletion was recorded over the station Halley Bay, minimum values (260 DU) over the station Syowa. In the adjacent regions, ozone concentration was also depleted. An "ozone hole" was also observed in the Northern Hemisphere. This ozone depletion was followed by a drop in temperature at altitudes of 8 to 20 km to 196 K (this value is less than over the North Pole (205 K)), and the amount of ice aerosol over the South Pole in this case was an order greater than over the North Pole (Steele et al., 1983). Some authors (Molina et al., 1985) associate the appearance of the "ozone hole" with anthropogenic factors, others ascribe this effect to natural interannual variability with different cycles; several models have been already proposed to describe this phenomenon (Solar Cycle..., 1986).

Interpretation of these phenomena requires considerably expanded experimental and theoretical studies, improved measurements of ozone concentration from ground stations and satellites, pressure, wind velocity, trace gas content (study of the chemistry of the phenomenon is very important) and aerosol parameters.

Scientists in many countries are working hard on this problem, as well as within the framework of bilateral cooperation between specialists from the USSR, and Germany and the United States.

It should be emphasized that the upper atmosphere is very vulnerable. In addition to these effects related to the ozonosphere, we should mention zones with lowered electron concentration ("ionospheric holes") appearing in the upper atmosphere, namely, in the ionosphere. Both the appearance and enforcing of emissions atypical of natural luminescence of the upper atmosphere take place. This effect is caused by:

— accumulation in the upper atmosphere of different substances due to diffusion;

— accumulation in the upper atmosphere of different substances during the launching of powerful rockets;

— effect of electromagnetic radiation from transmitters.

We should especially focus on the emission of water, hydrogen-containing compounds (and other substances) into the upper atmosphere during the launching of rockets which can significantly alter the ionosphere. Anthropogenic change of the hydrogen content at high altitudes is so great now that the global balance of hydrogen is being disturbed in the upper atmosphere.

This indicates that anthropogenic impact is not only close to the "alarming" but even to the "critical" limit.

The change in the ozone layer will be very noticeable, if not catastrophic, during large-scale external effects on the atmosphere (the fall of large meteorites like the Tunguska one and in particular, in the event of possible nuclear war).

A nuclear explosion yields about 10^{32} nitrogen molecules per 1 megaton (MT) of explosion power, which interacts with ozone to destroy it; in this case the nitrogen content in the cloud is almost independent of the explosion power and is 10^{12} to 10^{13} molecules per cm^3. The influence of powerful nuclear explosions on the ozonosphere has been comprehensively studied with regard for various photochemical processes and atmospheric diffusion by Izrael', Petrov and Severov (1983). They used seven major reactions to estimate NO_x /O_x changes. The rates of photodissociation of molecular oxygen (O_2) and ozone (O_3) due to solar radiation were calculated as a function of the ozone concentration in the atmospheric column with a prescribed initial content of nitric oxides. The relative ozone concentration changes with time at different altitudes in a cloud formed after explosions of various power were also estimated. As seen from calculations for a vertical atmospheric column (with nitric oxide concentration of 10^{13} molecules per cm^3), three hours after a 1 MT explosion, about 75% of total ozone content remains; after a 35 MT explosion, about 25%; slow recovery follows over several days.

No more than 1% of the initial ozone content remains 3 to 10 hours after a powerful explosion at the height of the explosion cloud center; the total, comparatively high percentage of ozone content in the atmospheric column remains due to the ozone contained in the lower atmosphere (below the cloud) and in part, above the cloud.

Calculations for global conditions reveal the destruction of 30 to 60% of total hemispheric ozone for a long time depending on power of individual bursts, if the total power of the nuclear explosion is 10^4 MT. A later study (The Effects on..., 1985) has found ozone change to be 50 to 65% for the same explosion power.

It is clear that this drastic change in the ozonosphere will exert a pronounced effect on the Earth's climatic change, both on cooling during the period of "nuclear winter" and global warming, when the atmosphere becomes "transparent" (from aerosols) due to the influence of trace gases (Izrael', 1983).

As mentioned above, the destruction of the ozone layer results in an increasing flux of ultraviolet solar radiation, that is nonlinear and depends on radiation wavelength.

Ultraviolet radiation lies in a wide range of wavelengths, which is conventionally divided into three bands that are important for biological systems: UV-A: 400–315 nm, UV-B: 315–280 nm, and UV-C: < 280 nm.

UV-B radiation has a strong biological effect and at the same time is very sensitive to varying atmospheric ozone content. The most important components of living things are nucleic acids that absorb radiation in this very range. Even small variations in daily doses in the spectral range of active biological radiation can affect the life of organisms. Experiments have found that UV-B radiation exerts a harmful effect on productivity of such crops as wheat, rice, soybeans, barley, and potato. These plants are mostly vulnerable at the earlier stages of their growth.

UV-B radiation penetrates deep into water (to 20 m in pure water). The upper 20 m-layer is populated by a great number of marine organisms that are very sensitive to this radiation (roe, larva, and fry, shrimps, crabs and many species of plants). Even the existing levels of UV-B radiation suppress phytoplanktonic productivity. Increasing UV-B radiation causes changes in the exosystem composition.

UV-radiation significantly influences human health by promoting the synthesis of Vitamin D_3, causing sun burn, eye damage, allergic reactions and skin diseases, including carcinogenic. All of this, except for the formation of vitamin D_3, is harmful for man. A 1% increase in UV-B dosage increases non-melanoma skin cancer incidence by more than 2% (this is statistically well-verified). Increasing UV-radiation is found to cause malignant melanomas. Northern light-skinned individuals are particularly susceptible to these diseases. UV-radiation also alters immunological reactions in humans and animals.

CHAPTER 10

IMPACT OF EXPECTED CLIMATIC CHANGES IN AGRICULTURE

10.1. Agricultural Problems Due to Climatic Changes

Weather and climate exert a direct and significant effect on one of the most important spheres of social activity, agriculture, and especially, on farming. For example, unfavorable weather conditions in the United States lead to mean annual losses in agricultural production of about $10 billion, which is almost double the total loss in all other sectors of the economy (Thompson, 1976). Since year-to-year weather conditions in the crop regions of the USSR are more variable than in the United States, it can be concluded that similar estimates will be even higher for the USSR.

The impact of climatic changes and weather variations on food production is especially strong in many developing countries, where agriculture, even in favorable years, cannot meet the population's food demand (Cooper, 1982; Pulwarty and Cohen, 1984).

The need to study the multifactor effects of current climatic changes on agriculture was repeatedly emphasized in the reports of international and national organizations, materials of conferences and other publications. The World Climate Program and UNEP include, as one of major integral parts, the World Program of Studying Climatic Effects on Human Activity, in which agricultural climatology problems are primarily emphasized (Outline plan and basis..., 1979; UNEP Expert Group..., 1980; Joint WMO/ICSU/UNEP..., 1981; Report of the study..., 1984; Assessing the impact..., 1984; Bolin et al., 1986).

The problem of agroclimatic consequences of modern global climatic changes is the focal part to a large extent because of the global food shortage in the late 1970s. Analysis of the food production system that had been formed by that time, revealed many unfavorable trends in its development (The world food situation..., 1975; Food needs ..., 1977; Swaminathan, 1984). Many specialists consider that these trends were previously obscured by climatic conditions favorable, as a whole, for agriculture. This situation was aggravated in the 1970s by a series of consecutive crop failures in many agricultural regions of the world. The opinion that food shortage can easily be solved by intensification of farming was disputed.

FAO data indicate an improvement on stabilization in the food situation in some countries (for example, Asia) in the last 5 years. However, the global crop production system can hardly be considered successful. The critical food situation in Africa deserves special attention, while crop failures in the Sahel zone have already occurred for many years.

The estimated potential world agricultural production, assuming maximum efficiency, is fairly encouraging (Büringh, Ven Heemst and Staring, 1975; Jensen, 1978). At present, countries of Asia, Africa and South America have great possibilities for extending arable lands. However, development of new agricultural areas entails considerable expenses and introduction of modern agricultural technology. There is no hope of solving these problems in the near future because of the economics of the developing countries. The estimated additional land resources in the developed countries of Europe and North America are small. It should also be considered that more intensive farming in these regions will yield even smaller crop increments (Schneider, 1977). This problem will be more serious in Europe and North America because of greater environmental pollution caused by chemical products used in agriculture.

This merely outlines the difficulties encountered in trying to solve the current world food problem. It is very troubling because it coincides with expected rapid anthropogenic climatic changes. The following questions should consequently be answered:

— what are the agroclimatic consequences of current global climatic changes and anthropogenic atmospheric CO_2 concentration increment;

— how can possible harmful consequences be reduced and the total economic effect be optimized from climatic changes by transforming agricultural geography and introducing crops more tolerant to environmental stress.

These questions can be answered only if there are joint coordinated investigations by specialists in climatology, agrometeorology, plant physiology and economics (Chen, 1981; Managing climatic resources..., 1981; Ausubel, 1983). Since problems of agroclimatic consequences of global climatic changes are large-scale and of social importance, their solution should incorporate their international and national significance (Glantz, 1979; Wittwer, 1980; Meyer-Abich, 1980; Warrick and Riebsame, 1981; Mann, 1982, et al.).

An important link in any comprehensive interdisciplinary investigation of this problem is solution to the agrometeorological problem of assessing the effect of climate-induced environmental changes on vital activity of plants and crop yields. Such estimates should be used to substantiate the optimal planned agricultural development (Terjung et al., 1984; Parry and Carter, 1985).

The experts of the World Climatic Conference who quantitatively analyzed agroclimatic patterns in different regions focused their reports on the impact of modern climatic changes on agriculture (Swaminathan, 1979; McQuigg, 1979; Fukui, 1979; Mattei, 1979; Burgos, 1979). Many conclusions drawn by the experts deserve special attention, in particular, that recent trends of higher crop yields may be preserved in the future only if there are favorable climatic conditions. It is emphasized, however, that the considerable year-to-year harvest variability will have to be considered.

A number of studies by American specialists have evaluated the impact of climatic changes on crop productivity based on statistical forecasting models. U.S. wheat production in a number of states was computed for different combinations of altered temperature and precipitation by Thompson's regression model and its modifications (Thompson, 1975; Ramirez, Sakamoto and Jensen, 1975). It was shown there that climatic changes responsible for a slight temperature decrease during the growing season and more precipitation will be beneficial for higher crop yield in the central U.S. grain belt. It is also emphasized that cooler temperature will adversely affect spring wheat production in the northern United States, Canada and the USSR. It is most significant that the unfavorable consequences of climatic change in temperate agricultural regions should be attributed to undesirable higher year-to-year variability in crop yields rather than productivity level.

Extensive investigations of the influence of climatic changes in crop yields have been carried out in the United States within the framework of the CIAP program (CIAP Monograph 5, 1975; Jensen, 1976). Special mention should be made of the U.S. teamwork to analyze the agrometeorological consequences of global climatic changes under a special project in the late 1970s (Climate change..., 1978; Crop yields..., 1978; Crop yields, 1980; Bierly, 1980). The scenarios obtained for changing climatic conditions, in addition to crop productivity level, incorporate forecasts of its annual variability as well as some estimates of harvest growth due to economic factors. However, the conclusions drawn in this study are limited, since it ignored the direct physiological effects of productivity growth associated with higher atmospheric CO_2 concentration.

Mainly American specialists have especially shown more interest in the last 5 years in the problem of the impact of anthropogenic climatic change on agriculture. These problems are dealt with in numerous reports by groups of researchers and individual authors (Global impacts..., 1983; Changing climate, 1983; Blasing and Solomon, 1983; Waggoner, 1983, and others). This problem is the constant focus of the authors of the interdisciplinary magazine "Climatic Change" (Sakamoto et al., 1980; Rosenberg, 1981, 1982; Parry, 1985; Santer, 1985; Oram, 1985, and others).

Among the Soviet studies on this problem are the following works carried out in the All-Union Scientific Research Institute of Agricultural Meteorology (Sirotenko, Boyko, 1980; Sirotenko, Abashina, Pavlova, 1984; Pavlova, Sirotenko, 1985).

The State Hydrological Institute emphasizes the problems of agroclimatic consequences of modern climatic changes (Menzhulin, 1976, 1984; Menzhulin and Savvateyev, 1980, 1981; Koval' and Savvateyev, 1982; Nikolayev and Savvateyev, 1982; Menzhulin, Nikolayev and Savvateyev, 1983; Koval', Menzhulin and Savvateyev, 1983; Koval', Nikolayev and Savvateyev, 1985; Nikolayev, 1985 a,c; Nikolayev, Menzhulin and Savvateyev, 1985; Savvateyev, 1985). In sections 10.3 and 10.4 of this chapter, we will analyze in detail the methodology and basic results of the aforementioned works. These investigations were based on specific data regarding modern anthropogenic changes of global climate described in the previous chapters of this monograph. Updated forecasts of climatic condition indices allow us to define basic requirements for a scenario of their agroclimatic consequences in the following way.

1. Estimations of crop yield response to climatic changes should take into account physiological factors of CO_2 impact on photosynthesis and accompanying metabolic processes.
2. The most important characteristic feature of agroclimatic conditions, along with mean level of crop yields, is its year-to-year variability. Besides possible changes in

productivity, agroclimatic scenarios should incorporate estimated changes in variability indices as well.

3. Global anthropogenic warming will result in a longer warm season. Forecasts of this characteristic are needed to plan new territorial distribution of crops, to introduce new agricultural plants and to correct sowing periods.
4. Because various regions have different climatic conditions, computations of changes in crop yield indices should be based on detailed territorial scenarios of future climate. This will help to make forecasts more specific, which is most important for optimizing economic consequences of agroclimatic changes.

10.2. Evaluation of CO_2 Concentration on Aspects of Plant Physiology

Along with photosynthetically active radiation, carbon dioxide is an external photosynthetic substrate and a factor influencing plant morphogenesis; therefore, an anthropogenic increase in CO_2 content will directly affect their vital activity and productivity.

Results of recent experimental investigations allow us to estimate the direct effects of increased CO_2 concentration on plant productivity under different conditions.

Modern physiological concepts relate CO_2 assimilation by plants to differences in their photosynthetic metabolism. The main feature of C_3 plants, which includes the majority of widespread plant species (wheat, rice, potatoes, cotton, etc.) is that their photosynthetic rate increases significantly when CO_2 concentration is higher (Figure 10.1a). In C_4 plants (corn, sugarcane, sorghum), photosynthesis is slightly accelerated when the CO_2 concentrations exceed current levels. These kinetic effects are explained by the different chemical affinity for CO_2 by enzymes that catalyze the reaction of CO_2 fixation (Wong, 1980b; Laysk, 1981). According to modern data, the Michaelis constant is near 400 ppm for ribulosebisphosphate carboxylase (rubisco), while for phosphoenolpiruvate carboxylase (PHEP - carboxylase) in C_4 plants it is considerably smaller and is 50 ppm. Despite the fact that interaction between internal photosynthetic factors disguises kinetic effects of the initial carboxylation reaction to a certain extent, the low value of Michaelis' constant for PHEP-carboxylase is responsible for rapid plateau reaching of CO_2 -curves of photosynthesis rate plotted for C_4 plants. CO_2 exchange of mature corn leaves reaches its plateau when CO_2 concentration equals (or is even lower) than its present level. For wheat leaves, the plateau is reached when CO_2 concentrations are almost triple the modern levels (Downton, Björkman, Pike, 1980; Gifford, 1980).

When interpreting experimental kinetic dependences, it is often important to clarify whether the low plateau of photosynthetic rate for C_3 plants is due to the specific experimental conditions (Goudriaan and Ajtay, 1979). Insufficient

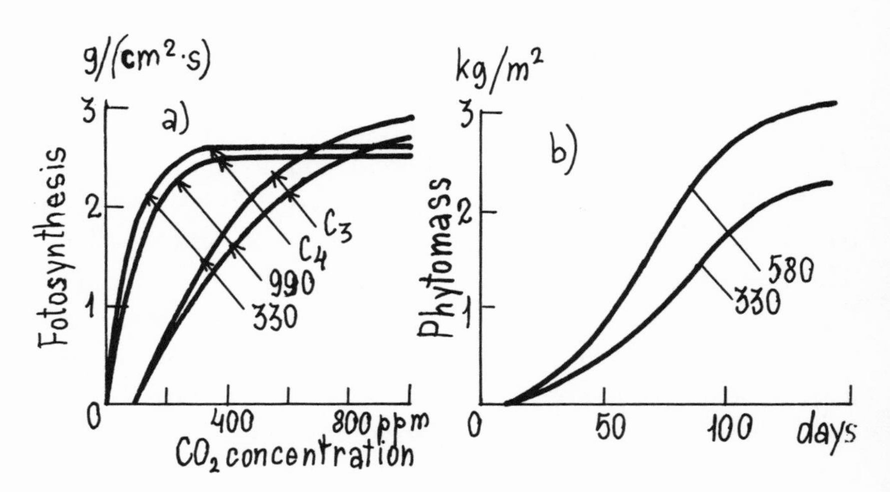

Figure 10.1. Experiments on growing plants with increased CO_2 concentration. a, CO_2 curves for C_3 (*Larrea divaricata*) and C_4 (*Tidestromia oblongifolia*) plants grown with normal (330 ppm) and triple (990 ppm) CO_2 content (Downton et al., 1980); b, dynamics of dry matter accumulation by wheat cultivated under normal and increased CO_2 conditions (Gifford, 1980).

ventilation in the case of high light depletes the CO_2 content in the air near leaves, that can restrict the assimilation rate (Laysk, 1982). Therefore, preference should be given to experiments where this effect is eliminated and the photosynthetic rate is related to CO_2 concentration in the intercellular air of the leaf tissue (Downton, Björkman, Pike, 1980; Gifford, 1980).

An important role in productivity of C_3 plants is played by photorespiration (Laysk, 1977). It is one of the possible pathways (glycolate pathway) for synthesis of glycine and serine in a plant cell. It starts from the oxygenation of the initial CO_2 -acceptor, RuBP in chloroplasts. Expenditure of RuBP leads to losses of organic matter that decrease total photosynthetic productivity. It was found that rates of carboxylation and oxygenation of RuBP depend significantly on the ratio of CO_2 and O_2 concentrations. Increased CO_2 content in the stroma of chloroplasts caused by higher atmospheric CO_2 level should, in the end, stimulate carboxylase activity of rubisco and decrease the portion of RuBP being oxidized. In this case, alternative biosynthetic pathways of glycolate and serine are activated, although their content in the biosynthetic products partially decreases (Keerberg and Viyl', 1982).

Investigations of the kinetics and substrate limitation of photosynthesis of plants grown with normal CO_2 concentration are undoubtedly important, however they cannot give a final answer to the question of productivity changes due to the lengthy exposure of plants to high CO_2 content. Brief experiments mainly concerning initial photosynthetic reaction cannot significantly affect the physiological processes responsible for subsequent, deeper biosynthetic reconstructions that are possible during the prolonged influence of higher CO_2 content on plants. A number of recent studies have allowed us to determine the morphogenetic impact of increased CO_2 content on growth, metabolism and productivity of plants.

One important problem that arises in analyzing the influence of higher CO_2 concentration on plants is transformation of the photosynthetic rate-CO_2 concentration curves. Figure 10.1 shows these CO_2-curves for two plants, Larrea (C_3) and Tidestromia (C_4), which were grown for a long time with CO_2 concentrations of 330 and 990 ppm (Downton et al., 1980). An indicative result of these experiments is that the CO_2-curves for plants of both types are slightly transformed as CO_2 concentration changes, while their metabolic features remain unchanged. Experiments also show that photosynthesis per unit area of leaves of plants grown in a greenhouse with triple CO_2 concentration is partially depressed. Analysis of characteristics of metabolic activity shows that this depression is mainly associated with lower rubisco content and activity in chloroplasts. Variations in other parameters of photosynthetic metabolism of such plants are insignificant.

The results of experiments on the influence of higher CO_2 content on plant respiration are ambiguous. For example, for wheat that has been grown for a long time with CO_2 concentration of 590 ppm, the intensity of respiration is approximately half that of plants raised under normal CO_2 conditions. Results of experiments with legumes were just the opposite (Hrubec, Robinson, Donaldson, 1984). Depressed respiration generally promotes a greater effect of high CO_2 concentration on assimilation.

Studies of plant transpiration and mineral nutrition in a CO_2-rich environment are of great interest. Such plants are characterized by smaller mean values of stomatal conductance of leaves. This is usually associated with stomatal closure (Raschke, 1975). The lengthening of diffusion pathways in a leaf, which should take place when leaves of test plants become thicker, also contributes to this effect. The lower exchange coefficient of chlorenchyma with the interleaf air can cause the noted partial photosynthetic depression.

Another rather important consequence of this effect is the decrease in evapotranspiration. Plants grown with a higher CO_2 content are generally characterized by more efficient utilization of water (Mauney, Fry and Guinn, 1978).

Table 10.1 presents values of the coefficient of transpiration efficiency (K_T) defined as the ratio between CO_2 assimilation and water losses through transpiration for corn (C_4) and cotton plants (C_3) raised under conditions of varying amounts of nitrogen and doubled and normal CO_2 content (Wong, 1980a, b). It is evident that coefficient K_T both for corn and cotton plants increases approximately two-fold when the CO_2 content is doubled and practically does not depend on the nitrogen supply.

Plants grown with varying water supply react differently to increased CO_2 concentration. It was noted that for experimental plants the relative improvement in assimilation is greater in the case of low moisture content, although the absolute increment in photosynthetic rate is higher when there is enough water. Sensitivity of wheat plants grown with increased CO_2 levels to moisture deficiency is diminished (Gifford, 1979a, b, 1980).

Experiments that revealed growth dynamics of plants are very important for an understanding of the CO_2-induced effects on plants. Results of practically all such experiments show that with optimal water supply, phytomass and leaf area of annual C_3-plants are much greater by the end of their intense growth period. It is interesting to note that the leaf area of C_4-plants, whose photosynthesis, as mentioned above, responds slightly to higher CO_2 concentration, also increases. Graphs in Figure 10.1b illustrate temporal variations in total phytomass growth in terms of dry matter weight of wheat, grown with natural and increased (up to 580 ppm) CO_2 content (Gifford, 1980). A 35% increase in phytomass by the 120th day after sprouting is mainly associated only with the increase in phytomass of the leaves.

Table 10.1. Changes in Transpiration Efficiency Ccoefficient (K_T) with Doubled CO_2 Concentration at Different Levels of Nitrogen Nutrition

Plants	NO_3^- Concentration in Nutrient Solution, m-mole/1	CO_2 Concentration, ppm	
		330	660
Corn (330 ppm)	24	7.0	13.4
	12	6.3	13.4
	4	6.4	13.5
	0.6	6.3	13.3
Corn (660 ppm)	24	6.8	13.1
	12	6.8	13.3
	4	6.6	13.3
	0.6	6.	14.0
Cotton plant (330 ppm)	24	3.3	5.9
	12	3.3	6.2
	4	3.4	6.0
	0.6	3.2	6.5
Cotton plant (660 ppm)	24	3.4	6.5
	12	2.9	6.5
	4	3.0	6.9
	0.6	3.0	6.2

Such an effect is observed both with sufficient and insufficient water supply.

Experiments showed that accelerated growth of C_3 plants cultivated with higher CO_2 content is observed during the active growing phase (Aoki and Yabuki, 1977). For example, bushing-out of wheat begins earlier in this case, is more intensive and more shoots are formed (Neales and Nickolls, 1978).

An important feature noted in some of the aforementioned experiments is prolonged phases of flowering and ripening of cereals. Stronger ears are formed if the leaf area is large and the ripening phase is long. A considerable increase in the seed and cotton mass is also peculiar to cotton plants, whose yield exceeded 80% of the control plant productivity if grown with doubled CO_2 concentration (Mauney, Fry and Guinn, 1978).

Valuable results were obtained in experiments to grow C_3-cereals. While doubled CO_2 concentration results in an average 26% increase in the crop yield and a 40% increase in dry matter weight for other C_3 plants (Kimball, 1983), the increment in the cereals yield is almost twice that of phytomass (Goudriaan and Ruiter, 1983). This effect can be especially important for the expected rise in atmospheric CO_2 concentration, considering the role of cereals in world agriculture.

A higher atmospheric CO_2 content is undoubtedly a powerful physiological and climatic factor in increasing crops productivity. However, some experiments have shown that it can have adverse effects on productivity, in particular, the senescence of some plant species (Chang, 1975; Omer and Horvath, 1983), or longer ripening phase of cereals. However, this harmful effect is likely to be compensated for by expected lengthening of the warm season during global warming.

The experimental data discussed above mainly concern plants growing in an environment with elevated CO_2 content and favorable conditions. Some studies have recently presented plant response to increased CO_2 content under the less favorable conditions such as lower soil moisture content, low light, extreme temperatures and shortage of minerals.

In the experiments on cultivating plants with different water supply, it was demonstrated that the relative increase in CO_2 assimilation is greater when there is a moisture deficit practically for all species. Sensitivity of wheat plants grown with a higher CO_2 level to a moisture deficit decreases (Gifford, 1980). Such effects are explained by the lower stomatal conductance of the leaves. Consequently, many authors have concluded that with other conditions equal, a CO_2-saturated atmosphere leads to higher mean crop productivity and can decrease its year-to-year variability in the regions subject to droughts.

Interesting results were obtained when plants were grown in an environment with increased CO_2 content and different light intensity, by itself a strict growth-limiting factor. It was found that for wheat the relative increase in phytomass with low light intensity can be larger than with higher illumination (Gifford, 1979a, b; MacDowell, 1973). This can probably be explained by suppressed respiration and lower light compensation level. For plants that do not respond to increased CO_2 by intensifying respiration, the relative increase in gas fluxes does not depend on the degree of illumination (Sionit, Hellmers and Strain, 1982).

The results of the majority of experiments on the influence of temperature changes on photosynthesis and related metabolic processes indicate a positive impact of higher temperature on photosynthesis, although this effect can be neutralized in ontogenesis (Berry and Raison, 1981). Moreover, temperature is a very important factor that regulates the intensity of metabolic run-off of assimilates, and a considerable increase in it can adversely affect seed formation. Experiments have revealed that in a number of cases, high CO_2 concentrations lower the critical temperature at which plant development is normal (Sionit, Strain, Backford, 1981).

Some studies have investigated the response of plants to increased CO_2 when there are insufficient or excessive minerals. Results of such experiments depend a lot on the plant species and the nutrient under consideration. Whereas under normal CO_2 conditions, nitrogen deficiency limits the growth of all plant species, except for leguminous C_3 nitrogen-fixing plants, doubling CO_2 causes an increment in plant dry matter with even lower nitrogen supply (Sionit et al., 1981; Wong, 1980a, b). Shortage of such important mineral components as phosphates and potassium adversely affects plant growth when the CO_2 level is raised (Goudriaan and Ruiter, 1983). It is a matter of general experience that sodium excess often leads to lower crop yields and even to their poisoning. It has been demonstrated that tolerance of some plant species for soil salinization by sodium compounds with a strong rise in CO_2 concentration (up to 3500 ppm) increases markedly (Schwarz and Gale, 1984).

Overall, these experiments allow us to conclude that in the majority of cases, increased atmospheric CO_2 level has a favorable effect on plant growth and productivity. Investigations yielded useful data for models of crop productivity for higher atmospheric CO_2 content.

10.3. Modeling of Changes in Crop Productivity

In addition to developing future climatic scenarios and studying plant physiological response to changing environmental conditions, it is necessary to develop methods for agroclimatic forecasting. This section will consider this problem in terms of altered crop productivity level.

The problem of estimating the agroclimatic consequences of modern climatic changes as it is defined above could not even be formulated earlier, since sufficiently substantiated scenarios of future climate were only developed recently.

A forecasting methodology has been developed for estimating crop productivity under meteorological conditions in the vegetation period of a year. Major advances in solving this problem were made using empirical statistical methods (Chirkov, 1969; Kulik, 1971; Ulanova, 1975; Moiseychik, 1975, and others). Modern empirical statistical models of this type are "adjusted" to definite crops and climatic conditions. They are widely used for agrometeorological forecasting.

It should be mentioned that it is difficult to apply empirical statistical methods directly to estimating agroclimatic consequences of modern anthropogenic climatic changes. There are three reasons for this fact. First, the scenario of future climate is not (and cannot be) so detailed that it could be used directly in empirical statistical models. Secondly, possible alterations of local meteorological conditions caused by global climatic changes can disrupt empirical relationships between crop productivity and predictors. And thirdly, empirical statistical models ignore the physiological effect of higher CO_2 content on productivity.

In the last two decades, considerable progress has been made in one of the branches of agrometeorology that is based on physical concepts (Budyko, 1949, 1956, 1964; Monsi and Saeki, 1953; Ross, 1964; Wit de, 1965). Multiparametric dynamic models and simulation schemes of plant productivity have been developed, using principles of physical agrometeorology (Wit de et al., 1971; Sirotenko, 1981; Modeling productivity of agroecosystems, 1982). Like empirical statistical forecasts, these models are primarily constructed to predict crop productivity for a specific year.

Detailed dynamic models are quite exacting for the initial hydrometeorological and physiological data. They are difficult to apply directly for estimating the influence of global climatic changes on crop yields, however, when only generalized estimates are incorporated in climatic scenarios. It should also be remembered that many correlations used in the detailed simulation models are empirical. Their stability during environmental variations due to climatic changes has not been investigated completely. Apart from this, when such models are used to solve problems we are interested in, we have to resort to formal temporal interpolation of hydrometeorological parameters, that is not provided for by available scenarios of climatic changes.

Specialists from the State Hydrological Institute have developed and used a model in which input parameters are compatible with hydrometeorological characteristics used in climatic scenarios. This specialized model is based on parametric description of energy- and mass-transfer in the agrometeorological system.

The model is a two-block algorithm that implements the following sequence of computations:

1. Solving the inverse problem of vegetation hydrothermal regime. Values of temperature, air humidity, solar and heat radiation fluxes during the period of plant growth are used to compute the relationship between leaf-area index and stomatal conductance for vegetation capable of transpiring given amounts of water. In this case, total evapotranspiration is linked by an empirical correlation to radiation, air humidity deficit, and precipitation; data on their changes are contained in climatic scenarios.
2. Calculation of total photosynthesis rate of crops with different leaf-area indices and stomatal conductance, but with the same transpiration. Carbon dioxide concentration, parameters of photosynthetic activity, phytomass temperature (computed in the first block) and the flux of photosynthetically active radiation serve as input parameters. The productivity index was evaluated from the intensity of the total biosynthetic rate depending on the rate of photosynthesis and parametrically related to the mineral nutrition level.

Unequivocal analysis of the combined changes in stomatal conductance and leaf area index requires an additional condition that characterizes plant adaptation. Three model variants of this reaction were investigated:

— plants respond to changing environmental conditions only by altering total leaf area;

— these variations only change the coefficient of stomatal conductance;

— in the process of adaptation, leaf area of plants and stomatal conductance change to ensure maximum photosynthesis rate.

We emphasize that these possible physiological responses of plants reflect their general adaptability. Consequently, phytometric parameters calculated in the model should be interpreted as effective values of stomatal conductance and leaf area in the period of maximum growth.

Calculations showed that when the third model adaptation mechanism of vegetation is applied to varying external conditions general agreement between the computed results and empirical data on changes in leaf area per unit of phytomass (see Figure 10.2c) and stomatal conductance on CO_2 concentration can be ensured.

This model was put through thorough numerical tests to compute sensitivity of productivity index to variations in

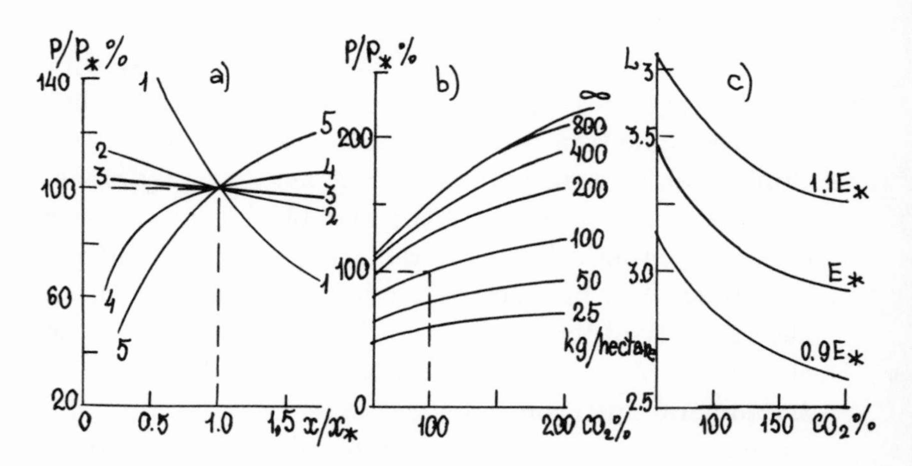

Figure 10.2. Sensitivity of model productivity index. a: relative values of the productivity index versus physiological constants: 1, coefficient of PAR attenuation; 2, photorespiration parameter; 3, respiration coefficient; 4, coefficient of mesophyll conductance; 5, empirical quantum photosynthetic yield. b: CO_2-curves of relative productivity index at various levels of mineral supply. c: dependence of relative plant leaf-surface area on CO_2 concentration with different level of moisture supply (E_*, evaporation norm).

parameters: quantum yield of photosynthesis, leaf mesophyll conductance constant, respiration parameter, photorespiration coefficient, extinction coefficient for photosynthetically active radiation (PhAR) (see Figure 10.2a). In the figure, these parameters and productivity index are represented as relative values. Series of such experiments revealed those parameters that should be preselected most precisely by the model.

Model estimates were made of productivity indices for wheat crop with preselected vegetation dynamics of the physiological parameters and hydrometeorological factors, as well as of the rate of phytomass growth within short time-intervals of the growing season. Correlation coefficients between productivity indices and long-term data on spring and winter wheat crop yields at experimental stations were determined. Productivity indices computed by the model and based on mean values of hydrometeorological parameters in the period of maximum seasonal evapotranspiration level were analyzed.

Calculations showed that use of indices computed with regard for phytomass growth dynamics as the productivity indicators does not improve their correlation with long-term mean crop yields. Correlation coefficients determined from monthly mean hydrometeorological parameters equalled 0.83 for winter wheat and 0.79 for spring wheat (Figure 10.3).

Figure 10.2 shows a set of integral CO_2-curves illustrating productivity index for the wheat crop with different mineral levels. Sensitivity of crop productivity to increased CO_2 concentration rises noticeably as more fertilizer is applied.

Initial calculations of the impact of modern climatic changes on productivity included analysis of the geographical distribution of yield. Figure 10.4 shows the climatic productivity index for wheat with unlimited fertilizer supply. These patterns agree well with the known empirical data on modern agroclimatic resources of different territo-

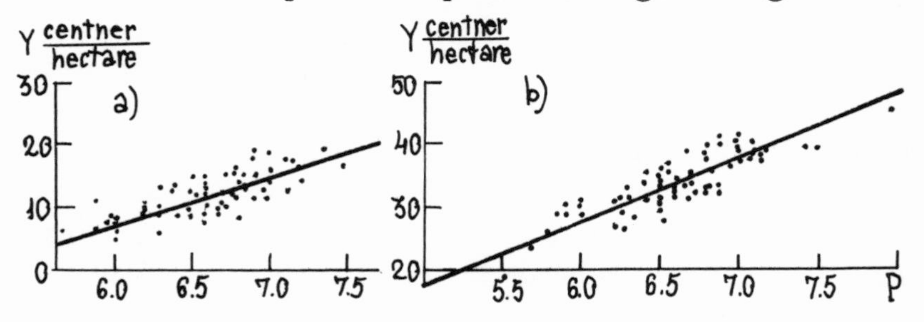

Figure 10.3. Correlation graphs for the dependence of productivity index P on crop yield Y of spring (a) and winter (b) wheat cultivated at experimental stations.

ries. This confirms that the model is plausible for processes governing climatic productivity level.

10.4. Empirical Analysis of Trends and Variability of Crop Yields

Possible agroclimatic consequences of modern climatic changes are not revealed only through changes in potential crop yield. Climatic changes can also significantly affect the characteristics of temporal variability, in particular, year-to-year variations in crop yields.

Changes in environmental and climatic factors that are beneficial for higher productivity are not always accompanied by favorable changes in other agroclimatic indices. Such effects are due to the multifactor impact of environmental factors on vegetation. For example, a higher atmospheric CO_2 content that is conducive to greater productivity can impair moisture conditions in some regions, that in itself will adversely affect yields. But even if the beneficial impact of increased CO_2 on crop yield outweighs such negative consequences, it is possible that the increase in temporal variability of precipitation will cause considerably greater changes in crop yield. In addition, changes in year-to-year variability of crop yield have to be computed for more complete economic assessment of the consequences of modern climatic changes.

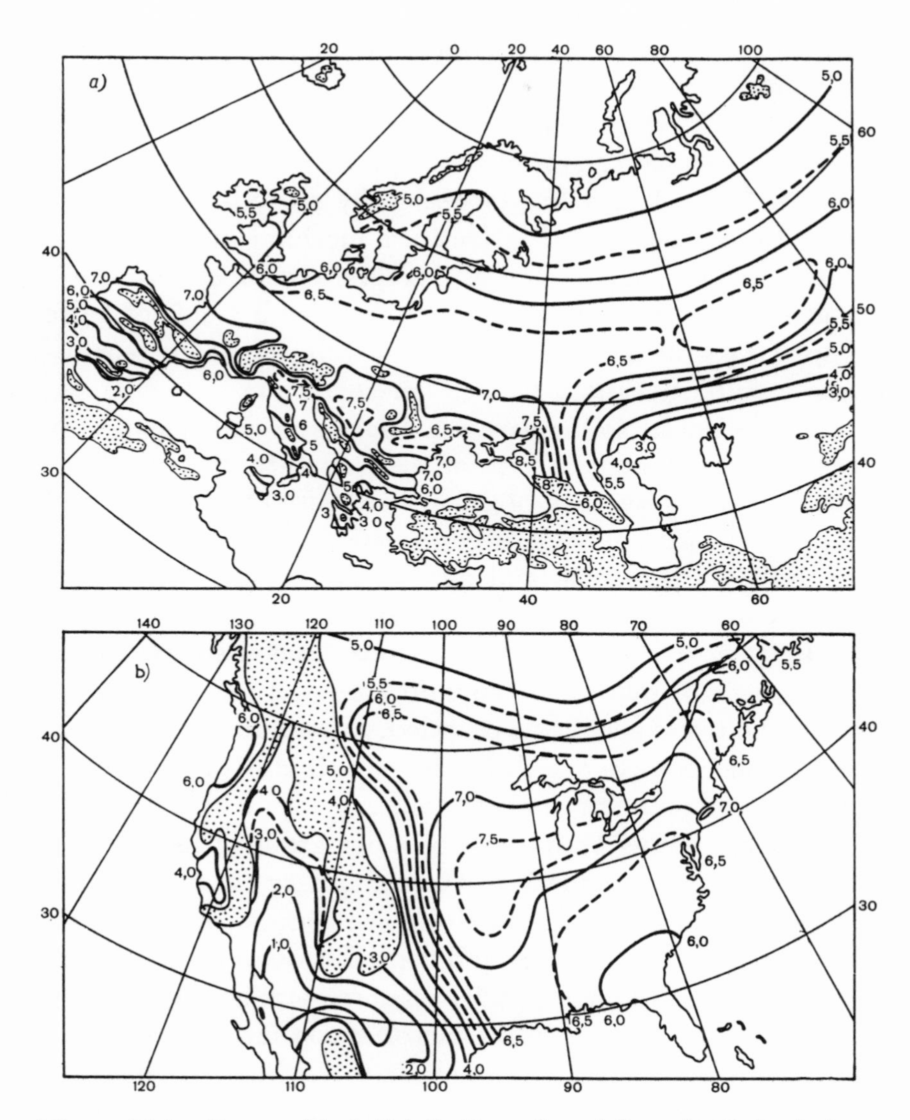

Figure 10.4. Geographical distribution of model productivity index for wheat crop in Eurasia (a) and North America (b). Figures at isolines indicate CO_2 assimilation (mg/(m^2 × sec) during the phase of ear formation.

Analysis of available crop yield data shows that during the last 50 years, a considerable increase in basic crop yields, cereals included, was observed in developed and some developing countries. The "green revolution" was associated with a rapid increase in the amount of fertilizers and energy used in agriculture. For example, production of mineral fertilizers in the USSR has increased by more than 50 times since 1945. The level of agricultural mechanization in terms of the number and power of tractors, combines and other machines has also risen. However, although yields of basic crops have grown considerably, they still show high year-to-year variability caused by weather conditions.

The effect of soil cultivation factors on crop yields to a large extent impedes study of their climate- and weather-related variability. The problem of investigating trends and variability of crop yields becomes complicated, incorporating both agroclimatic and agrotechnological components, since it deals with data on crop yields.

In order to obtain a homogeneous series used to analyze the impact of weather factors on crop yield variability, an economic trend should be revealed. We applied a technique which incorporates the dynamics of soil cultivation factors of crop yields.

It was assumed that the impact of a group of such factors on crop productivity and, in the end, on its yield can be described on the basis of the well-known concept of limitation. For example, the quantity and composition of fertilizers figure in the formation of conditions of mineral supply for crop productivity. Increase in the factor of specific tractor power (number per unit of area) determines the quality of soil cultivation and optimal sowing. Use of more harvesters is an indication of better and faster harvesting to prevent losses and produce higher yield. Analysis of the impact of other soil cultivation factors on crop productivity and economic yield also shows that this effect can be interpreted by the limitation concept.

Suppose we are interested in the yield of a certain crop that is composed of the yields of its individual varieties. We determine the trend value of crop yield $Y_t^{(k)}$ of each variety k, using the following expression:

$$Y_t^{(k)} = a_0^{(k)} + \frac{1}{a_1^{(k)} + \sum_i \frac{a_i^{(k)}}{\eta_i}},$$

where $a_0^{(k)}$, $a_1^{(k)}$, $a_i^{(k)}$ are empirical constants; η_i is specific values of limiting factors.

After introducing the term $P^{(k)}$ that expresses the part of the total area of crop cultivation occupied by variety k, and assuming its maximum economic yield max $Y^{(k)}$ is reached with $\eta_i \to \infty$, we can write the following formula for the harvest of the given crop:

$$Y_t = \sum_k P^{(k)}(Y_{\max}^{(k)} - \frac{1}{a_0^{(k)}}) + \sum_k \frac{P^{(k)}}{a_1^{(k)} + \sum_i \frac{a_i^{(k)}}{\eta_i}}.$$

It describes the impact on crop yield not only of soil cultivation factors, but also of changes in mean yield of different varieties of crops.

We will consider trends of wheat yield from 1945 to 1980, computed by this technique for the United States and USSR. Since we lack sequential data on changes in crop varieties for this period, coefficients a_i in the last formula were considered to be equal for all varieties of wheat.

The following factors determined the trend in wheat crop yields in the USSR:

— mineral fertilizer consumption,
— normalized number of grain combines,
— number of tractors in terms of standard power,
— electricity expenditure in agriculture,
— doses of pesticides.

The computations for the United States add two more factors;

— profit from wheat sales,

— conventional number of transportation vehicles in agriculture.

Figure 10.5 shows empirical data on wheat yields in the USSR and United States, as well as calculated economic trends.

The most significant result of the computations based on all five soil cultivation factors is that during the studied period fertilizers and the number of harvesters were the most important for increasing wheat yield both in the USSR and the United States. It has been demonstrated that in contrast to other methods of approximating crop yield trends by small-parametric functions of time, the approach described above also incorporates lower crop yield caused by the specific economic situation.

The use of the complete set of economic factors to compute economic trends of crop yields is more accurate than the computations that use only one of the factors or small-parametric trend approximation. However, the whole set of soil cultivation data is not always available. For this reason, it becomes important to compute the trend from limited information. This possibility was determined by computations based on only one of the above factors, as well as various combinations of two, three and four factors. The focus was substantiation for using such an index as mineral fertilizers as the only factor, since information about their production and doses is the most accessible and is presented in homogeneous statistics.

Figures 10.6 and 10.7 show computed wheat yields in the USSR and United States versus each of aforementioned factors. Broken lines indicate the computed trend that incorporates all factors, but is represented as a function of each of them. It can be seen that the position of empirical points relative to computed curves generally confirms the hypothesis that soil cultivation indicators are limiting factors for crop yields.

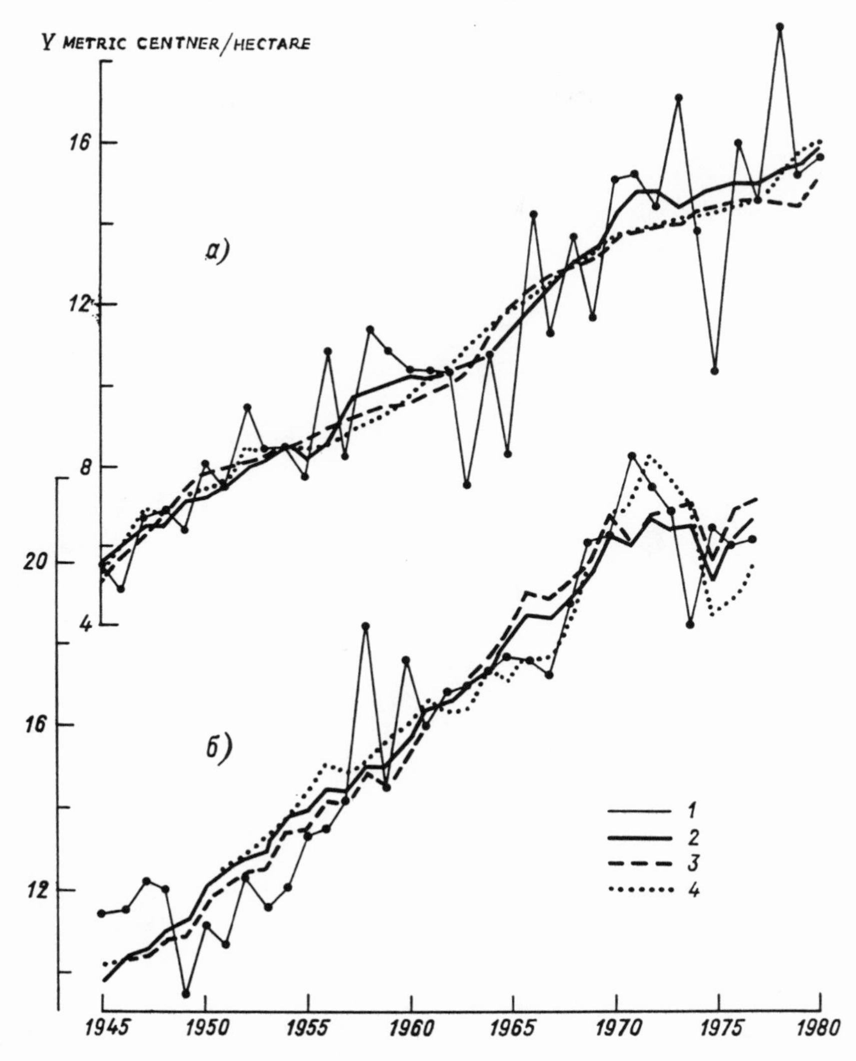

Figure 10.5. Facts (1) and economic trends (2–4) of wheat yields Y in the USSR (a) and United States (b). 2, trends calculated by incorporating all soil engineering factors; 3 and 4, by using data on the dynamics of fertilizer consumption or harvesters.

Computations for the USSR revealed that the response of wheat yields to such altered factors as the consumption of mineral fertilizers and electricity has considerably decreased in recent years. This is not true of such a factor

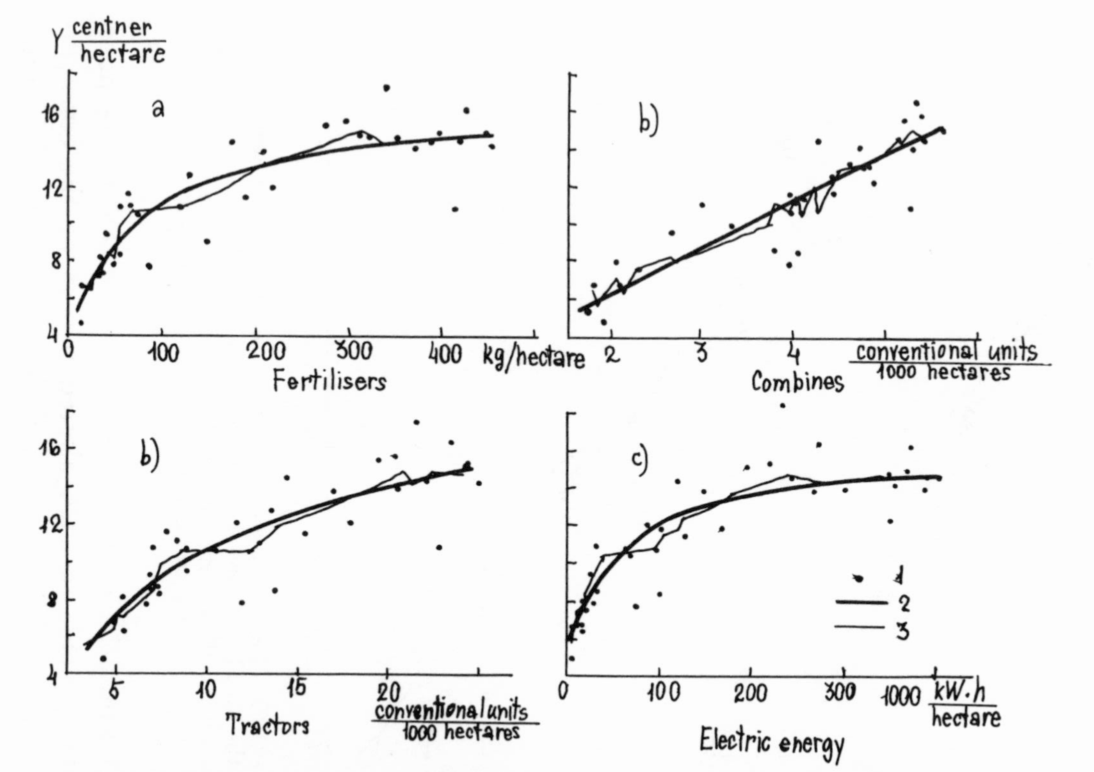

Figure 10.6. Impact of soil-engineering factors on wheat yields Y in the USSR. 1, facts; 2, calculated dependences on each factor; 3, parametrically represented trends based on all factors.

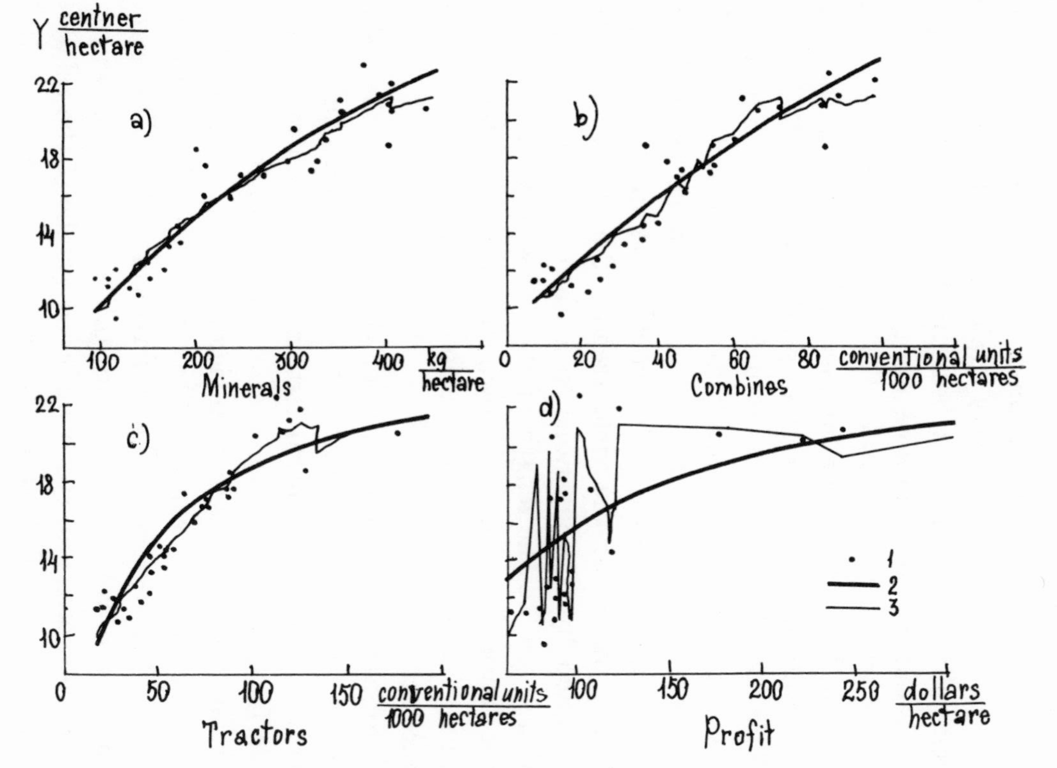

Figure 10.7. The impact of soil-engineering factors on wheat yields in the United States. Conventional designations are the same as in Figure 10.6.

as harvesters. Consequently, higher grain production in the USSR should apparently be related to greater usage of harvesters. Analysis of the impact of economic factors on U.S. wheat production shows that its response to fertilizer and electricity consumption is still significant.

Table 10.2 contains correlation coefficients between trends and empirical data on wheat yields and standard deviations in multifactor computations of trends. In order to resolve the problem of major influential factors, it is especially important that both in the USSR and United States, mineral fertilizer consumption was the most significant contributor to deviations in empirical data from the trends. In both cases, the number of harvesters is the next most significant factor for grain production.

The use of this technique in forecasts is an important problem encountered in studying the impact of economic factors on crop yields. This is possible when the series is long enough to ensure stable coefficients in the approximation formula for the yield trend. This possibility was determined by computations to analyze the sensitivity of coefficients a_i to the length of the series of yield and economic factors used. It was shown that the use of data series spanning 32 years and more ensures good reliability of these coefficients.

Information on soil cultivation factors prior to the 1940s available in statistical information publications primarily concerns the use of mineral fertilizers. Data on other economic factors are incomplete and inhomogeneous. Trends and relative deviations in crop yields $(Y - Y_t)/Y_t$ since 1920 for rye, spring and winter wheat in the USSR were computed on the basis of mineral fertilizer data. The same characteristics were computed for U.S. crop yields of wheat, corn, barley, oats and rye. Combined graphs of the relative yield deviations of these crops are shown in Figures 10.8 and 10.9.

Table 10.2. Impact of Technological Factors on Wheat Yields

	Correlation Coefficient of the of the Trend and Actual Yield		Standard Deviation of Actual Data from Trend, MT/ha	
Factors	USSR	United States	USSR	United States
All economic factors	0.90	0.95	0.145	0.130
Mineral fertilizers	0.88	0.94	0.155	0.132
Grain combines	0.86	0.93	0.160	0.134
Tractors	0.82	0.92	0.166	0.156
Electricity	0.78	—	0.170	—
Pesticides	0.76	—	0.184	—
Motor Trucks	—	0.86	—	0.194
Profits of previous year wheat harvest	—	0.59	—	0.309
Predictable profits of current year wheat harvest	—	0.48	—	0.394

Analysis of these computed series revealed patterns of temporal variations in crop yield. The most important feature of this dynamics is that negative anomalies of relative deviations become more frequent in the 1930s, which

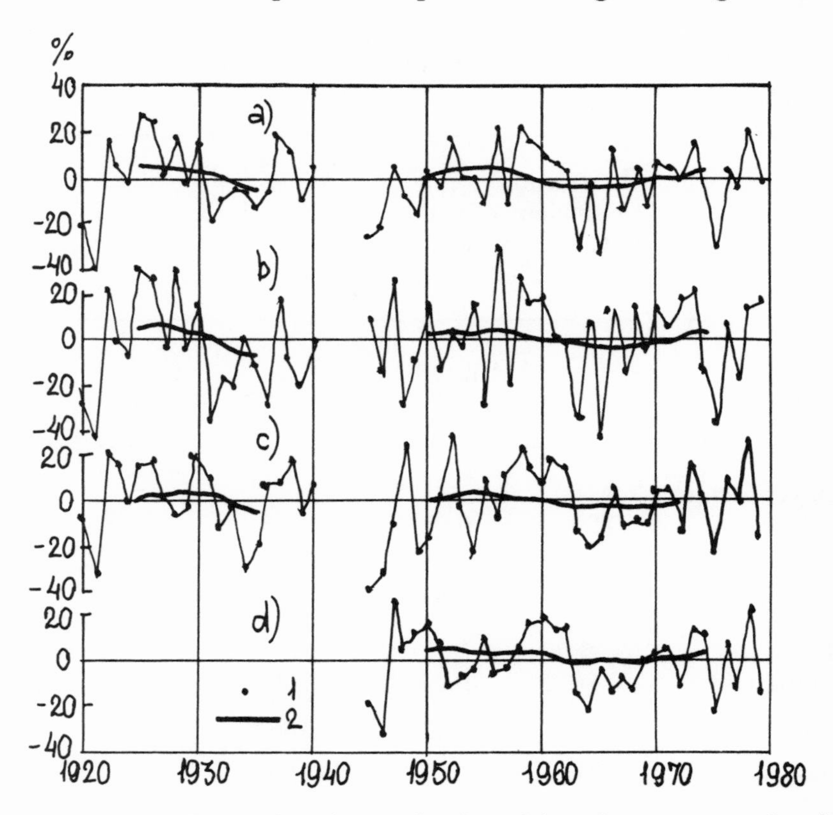

Figure 10.8. Dynamics of annual values (1) and 11-year running (2) mean relative deviations (%) of cereal yields in the USSR. a, winter wheat; b, spring wheat; c, wheat; d, rye.

affected practically all studied crops. For this reason, we can state that global warming in the 1930s entailed similar changes in atmospheric precipitation in the two largest grain-producing regions of the Northern Hemisphere. This is also confirmed by other authors (Vitel's and Drozdov, 1975; Rauner and Lazovskaya, 1978; Rauner, 1981).

Positive values of mean relative deviations in yields of wheat, corn and oats in the United States occur in the period from 1939 to 1947, followed by a phase of negative mean deviations (1948 through the late 1950s). The dynamics $(Y - Y_t)/Y_t$ for barley and oats in the United States somewhat resembles that for the first three crops,

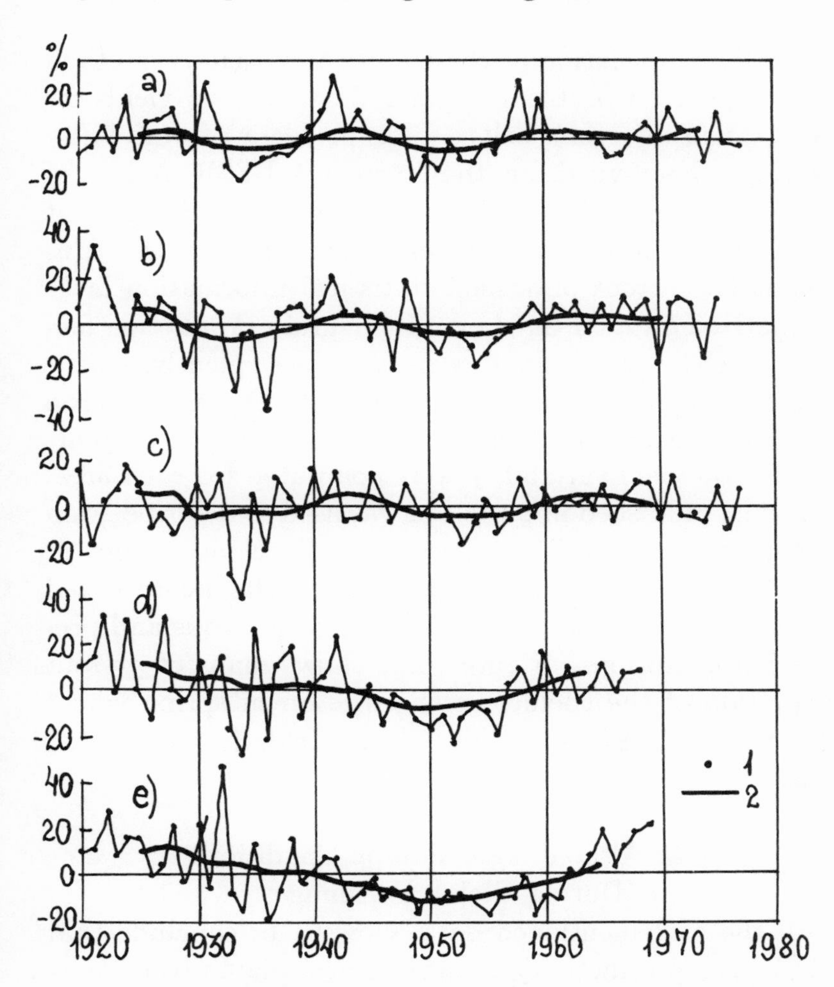

Figure 10.9. Dynamics of annual values (1) and 11-year running means (2) relative deviations (%) of cereal yields in the United States. a, wheat; b, corn; c, oats; d, barley; e, rye.

but has longer declines; negative values of relative deviations for barley occur during the 1940s–1960s, for rye in 1939–1963. This difference in dynamics of relative deviations in yields of these crops occurred because climatic changes in atmospheric precipitation in different regions of North America were not completely synchronous.

It is interesting to define the patterns of current territorial distribution of characteristics of year-to-year crop yield variability. Application of the empirical-statistical method to solving such a problem requires data on crop yields for small regions of grain belts. The problem of variability in spring wheat yields in the European USSR and total wheat yields in the United States was chosen because it is possible to obtain long series of statistical data. Unfortunately, current published statistical information lacks systematized data on the dynamics of grain production factors in each of these small regions. Consequently, small-parametric approximations were used to reveal the crop yield trends. Standard deviations characterizing mean year-to-year crop yield variability were computed for each area. Additional published information was used to describe grain belt regions in the USSR (Pasov, 1973a, b; Konstantinov et al., 1981). Analysis of the computation results incorporated data on the territorial distribution of wheat crops and precipitation conditions. Figure 10.10 shows maps of present distribution of the coefficient of variation in spring wheat yields in the USSR and total wheat yields in the United States.

10.5. Altered Agroclimatic Conditions During Global Warming

All the aforementioned data allow us to conclude that while future productivity of agricultural plants (especially the C_3 species) should increase under the impact of higher CO_2 content it will also vary because of global warming. In this case, for regions situated in the middle latitudes, the interrelationship between temperature changes in the lower air layer and precipitation will be very important.

It is easy to understand that in areas of excessive precipitation, the higher temperature will favorably affect agriculture, mainly due to a longer warm season. This impact will be most important for regions with a colder climate. In mid-latitude regions with warm climates, the expected

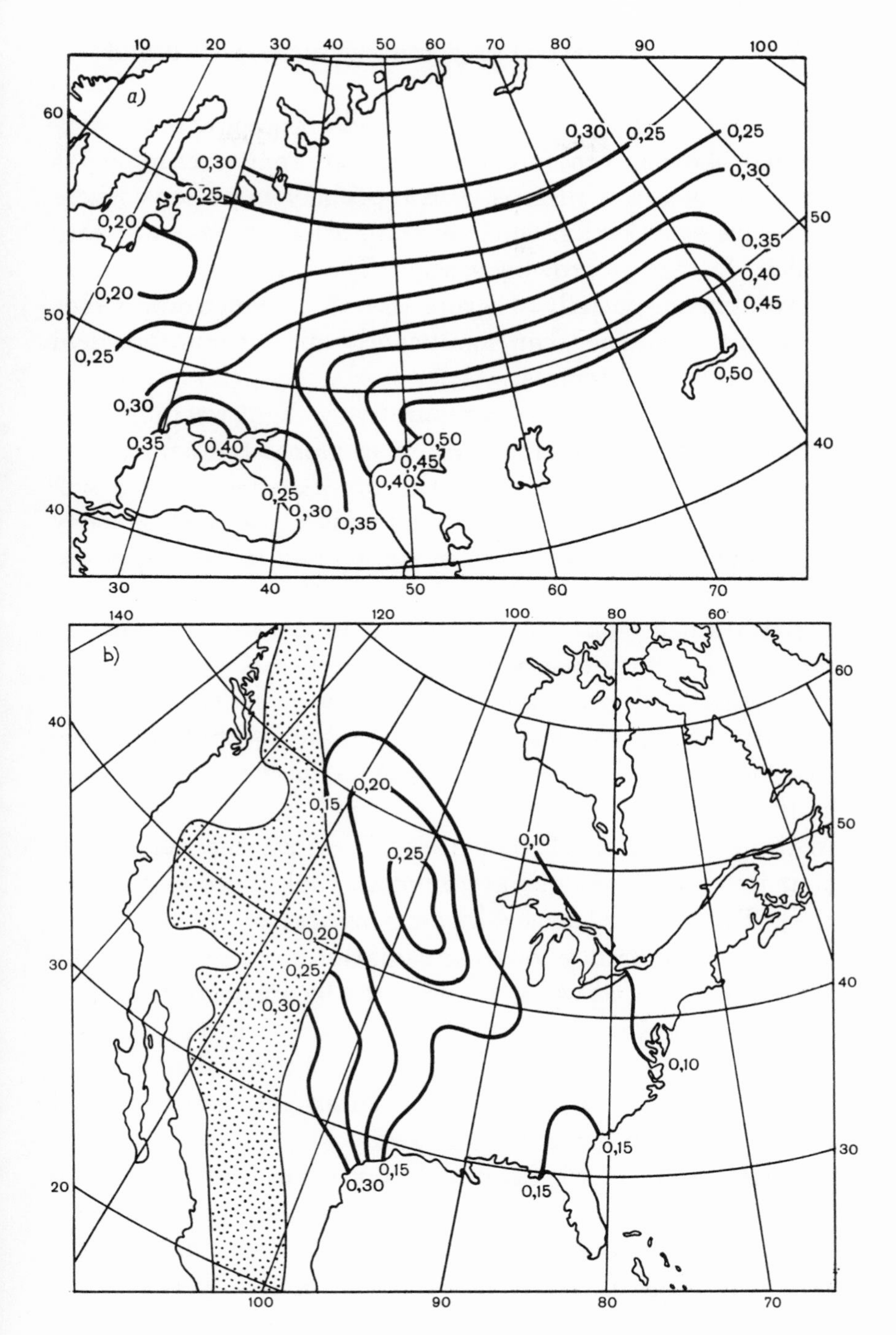

Figure 10.10. Geographical distribution of the coefficient of year-to-year variation in spring wheat yields (USSR) (a) and wheat as a whole (North America) (b).

crop yield change will mainly depend on precipitation variations. Lower total precipitation can reduce crop yields even in fairly wet regions, and especially in regions where excessive evaporation versus precipitation causes soil water deficit during the growing season. The change in frequency of droughts, especially of severe ones that reduce crop yields over vast territories, can significantly affect the mean yield 5- and 10-year intervals.

Chapter 9 shows that comparatively slight warming that corresponds to an increase in mean global temperature of about 1.5°C compared with the pre-industrial epoch will lower the amount of precipitation and increase the frequency of droughts over large areas in steppe and forest-steppe regions in the middle latitudes. Since these are important agricultural regions, the impact of moderate warming on crop productivity deserves a lot of attention.

Analysis of patterns of agroclimatic consequences of global warming can use data on crop yield variation in the two largest grain-producing countries in the mid-latitudes of the Northern Hemisphere, the USSR and United States in the period from 1920 to 1985. Previous chapters noted that two significant global warmings took place in this period. The first was caused by a natural factor (higher atmospheric transparency for solar radiation) and peaked in the 1930s. The second was mainly caused by anthropogenic factors (increased concentration of CO_2 and trace gases) and occurred in the mid-1970s. In both cases, the rise in mean global temperature versus the pre-industrial epoch reached 0.5°C, and even more in some years.

Data on mean wheat production in the USSR in Figure 10.11 are important for an overall assessment of the impact of global warmings on crop yields. It is clear that year-to-year variations in crop yields are determined by two prime factors, agrotechnology level, and weather and climatic conditions. The simplest method of distinguishing the impact of these factors can be based on data on maximum crop yields, from which the trend has been calculated (curve 2 in Figure 10.11). It is natural that even in the years of maximum productivity, weather conditions were not beneficial for agriculture over the entire USSR grain-growing area. There are grounds, however, for considering crop yields in main cereal belt regions (wheat in this case) to be limited to a large extent by technological factors rather than meteorological. This conclusion is confirmed by computations from formulas for estimating production response to technological factors and, in particular, greater use of mineral fertilizers (see section 10.4). These calculations have led us to conclude that crop yield increased naturally under favorable weather conditions in recent decades.

Returning to analysis of productivity data represented in this figure, we note that crop yields decreased by 25–30% in the 1930s for six consecutive years. Such a drop was primarily caused by meteorological factors, since there was low precipitation in the main USSR agricultural regions in that period. We note also that this climatic change was global; a series of droughts was observed in the United States in the same years, which reduced mean wheat production by approximately the same amount. In the post-war years, atmospheric precipitation conditions influencing crop yields were comparatively favorable for a rather long time. Some reduction in wheat production in the USSR in the early and mid-1960s was apparently caused not by climatic factors, but by the development of vast new areas for growing cereals in the east (virgin lands), where wheat productivity was noticeably lower than in regions with more favorable climatic conditions. Although this difference in productivity was not eliminated, the continuing higher level of agrotechnology soon restored the higher wheat yields for the country as a whole.

The second significant deterioration in atmospheric precipitation conditions in the main agricultural regions took

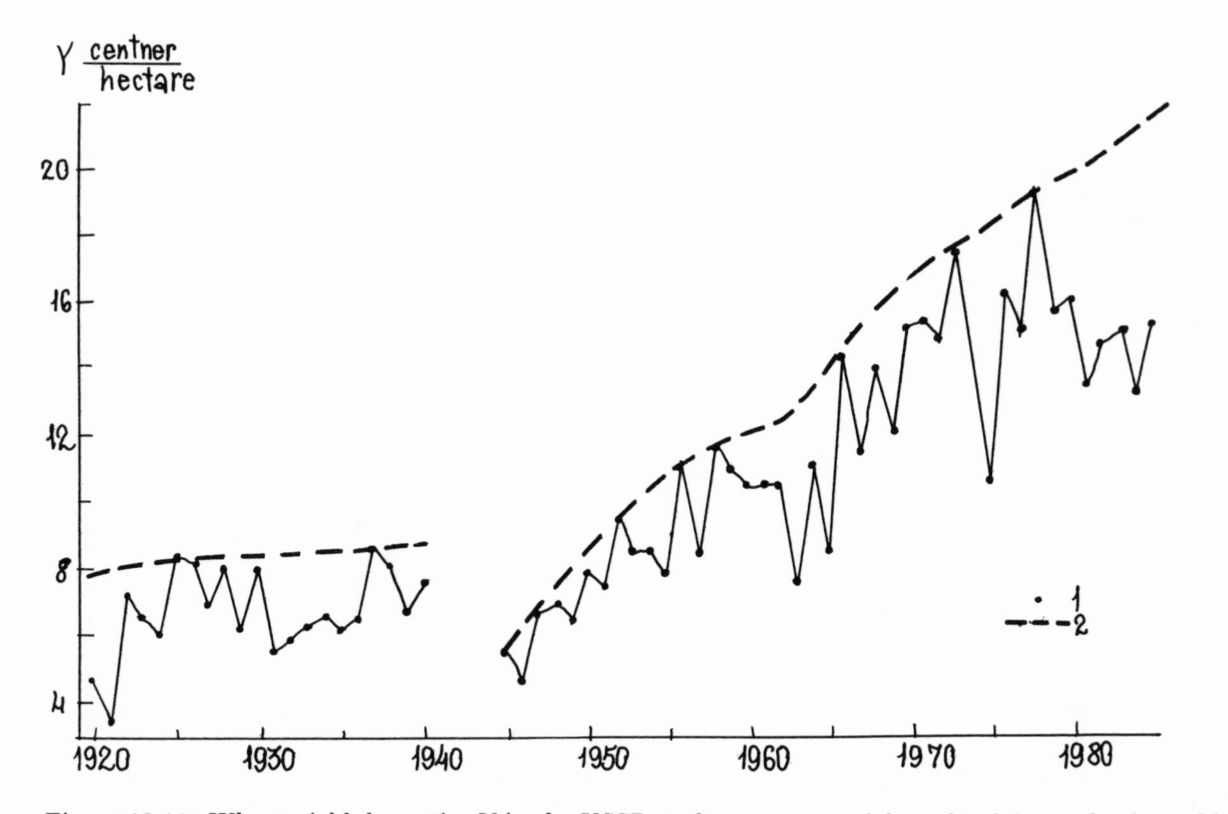

Figure 10.11. Wheat yield dynamics Y in the USSR. 1, facts; 2, potential productivity under favorable weather conditions.

place in 1975–1985. From data presented in the previous chapters, it is apparent that the mean global temperature at the Earth's surface had risen by approximately 0.5°C by the early 1980s. Such warming was comparable to that in the 1930s, which, as it has been shown, caused considerable agrometeorological changes. It can be safely said that agrometeorological conditions in the agricultural regions of the USSR changed quickly as warming occurred. In this case, the changes in its different regions were synchronous, to a large extent. Table 10.3 contains mean values of aridity index S for the periods 1964–1974 and 1975–1985, computed from data on temperature and precipitation in May–July (Ped', 1975; Long-term series..., 1985).*

Table 10.3 shows that in seven out of nine main agricultural regions, precipitation was noticeably lower in the last decade than that in the previous 10 years. In two regions, Southwest and Central Chernozem, precipitation did not decline. All of this allows us to conclude that agroclimatological conditions for the country as a whole deteriorate considerably as global warming develops. The same conclusion can be drawn from data presented in Figure 10.11, that show wheat yield dropping in the period 1975–1985 to 75% of its values under favorable climatic conditions, while it was 86% in the previous decade.

Data for the United States indicate a slight decrease in cereals production for the indicated decade compared with its potential under favorable conditions. In this case, however, the variation in crop yields was less, possibly because the U.S. agricultural regions are less arid than the USSR.

---

*Index S introduced by D. A. Ped' is defined as $S = \Delta T/\sigma_T - \Delta R/\sigma_R$, where ΔT, σ_T and ΔR, σ_R are anomalies and standard deviations in temperature and precipitation. The index S is used in agrometeorology to characterize aridity conditions in the growing season. Index S increases as temperature rises and precipitation decreases.

Table 10.3. Averaged Aridity Indices S.

Economic Region of the USSR	1964–1974	1975–1985
South (Ukraine)	0.03	0.23
Southwest (Ukraine)	0.12	0.05
Central Chernozem	0.19	0.15
North Caucasus	0.23	0.60
West Siberia	- 0.02	0.38
Donets-Dnieper region	- 0.01	0.42
Urals	- 0.37	0.44
Volga region	- 0.03	0.01
Kazakhstan	0.14	0.39

Passing on to the problem of expected changes in precipitation conditions over the USSR in the near future, we focus on the fact that the patterns of precipitation anomalies (see Figures 9.5 and 9.6) corresponding to a global warming of 0.5°C (which has already been reached) and of 1.2°C (possible in 10–15 years) do not differ considerably. This means that the trend toward deteriorated precipitation in the USSR cereal production regions will be maintained in the next decade.

In addition, we should mention two more factors that will aggravate the poor precipitation conditions. The first of them is the greater likelihood of positive temperature anomalies during global warming (see Chapter 9). The second is lower soil moisture content related to higher evaporativity caused by rise in air temperature.

The destructive effect of these factors on crop yields will be accompanied by the beneficial impact of high CO_2 level that enhances physiological productivity of the majority of crops. This influence will be the greatest in years with more or less favorable precipitation conditions.

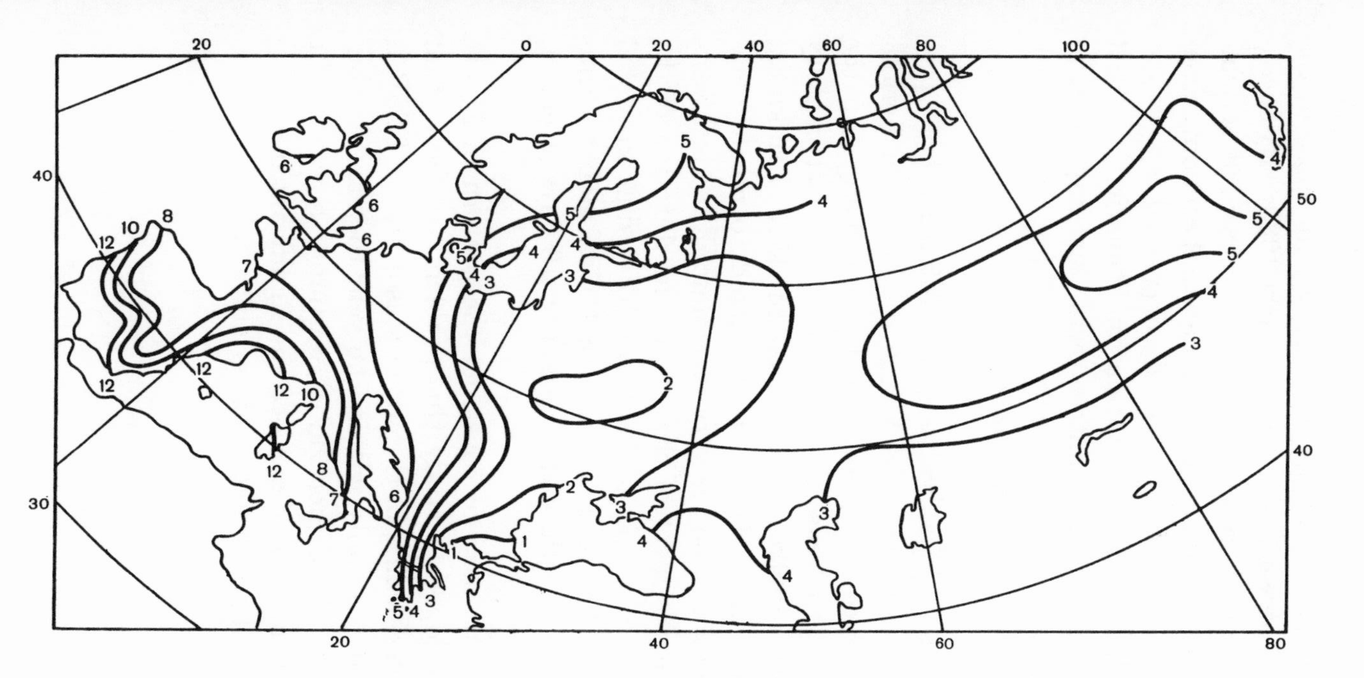

Figure 10.12. Increase in duration of the warm season with temperature above 10°C with a 0.5°C global warming. Figures at isolines are the number of days of longer warm period compared with the modern.

In summary, the conclusion can be drawn that precipitation conditions will not improve before the close of the century compared with those in the early 1980s. By incorporating this conclusion into long-term forecasts of agricultural production, the damage to the national economy from climatic changes can be substantially decreased.

It is very likely that precipitation conditions in the USSR (as well as in many other countries) will start improving in the beginning of the next century (see Chapters 8 and 9). This climatic change will have a noticeable impact on crop yield growth.

Anthropogenic global warming will influence both precipitation conditions and the length of the warm period. The results of recent studies indicate that changes in the length of the warm season took place in the last century (Uchijima, 1976; Pielke, Styles, Biondini, 1979; Brinkman, 1980; Newman, 1980, 1982). It is related to mean hemispheric temperature, and is very pronounced in southwestern Europe. Assessment of changes in the duration of the season year with temperature above 10°C due to global warming led to the conclusion that a 1°C of global warming increases the length of this period by 10 days (Kellogg, 1978).

The technique used to construct the pattern in Figure 10.12 was developed to study this problem. It shows that the current lengthening of the warm season in the USSR is not major and is mainly within the year-to-year variability interval. With further global warming, however, the lengthening of this period can become significant in adjusting climatic zoning in practice of agriculture.

CHAPTER 11
RELIABILITY OF DATA ON EXPECTED CLIMATIC CHANGES

11.1. Verification of the Calculated CO_2 Climatic Effect: CO_2 Impact on Temperature Conditions in the Geological Past

Chapters 9 and 10 deal with important practical problems whose solution requires reliable data on future climatic changes. Consequently, it is very important to have accurate estimates of future climatic conditions.

As an example of estimating reliability of data on climatic changes induced by varying greenhouse effect in the atmosphere, we consider the problem of calculating fluctuations in mean surface air temperature in the geological past by methods of physical climatology. This problem is similar in some respects to the problem of forecasting climatic change due to modern anthropogenic factors.

Quantitative data on changing atmospheric CO_2 concentration in different Phanerozoic epochs (i.e., for the last 570 million years) have been obtained in a series of studies published since 1977 (Budyko, 1977a; Budyko and Ronov, 1979; Budyko, Ronov and Yanshin, 1985, etc.).

Data on changes in CO_2 concentration given in the last study are presented in Figure 11.1 and show that throughout the Phanerozoic the atmospheric CO_2 concentration noticeably exceeded its current value and had a few peaks. They all correspond to the epochs of increased volcanic activity, when more volcanogenic rocks were formed per unit of time.

It should be mentioned that since the second half of the Cretaceous period, the CO_2 concentration has been declining naturally, and in our era reached the lowest level for the entire Phanerozoic. Figure 11.2 shows calculations of CO_2 concentration and variations in volcanic intensity at the end of Phanerozoic indicating that the reduction in CO_2 con-

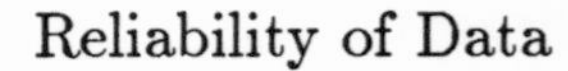

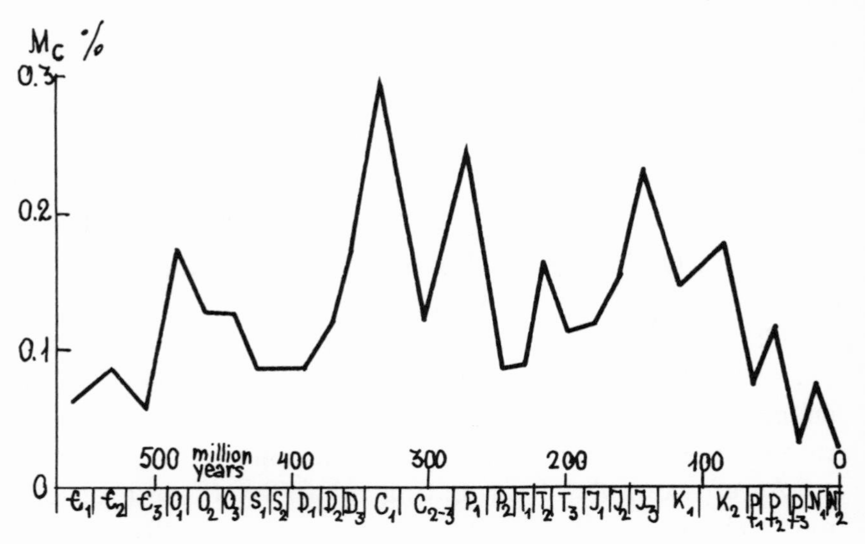

Figure 11.1. Variations in the Phanerozoic CO_2 concentration.

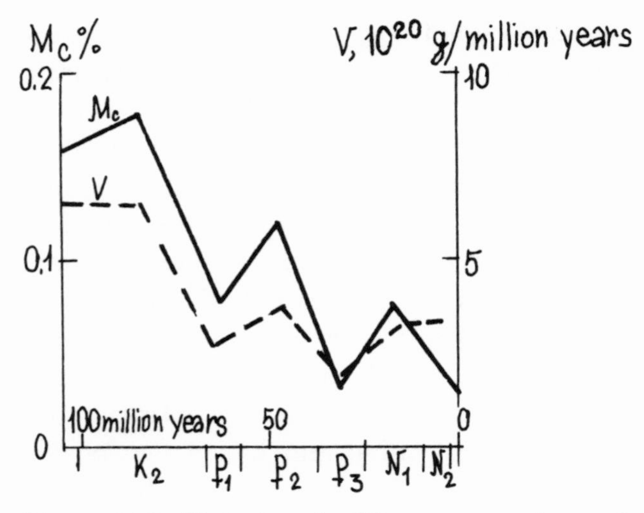

Figure 11.2. Variations in CO_2 concentration and volcanic intensity at the end of the Phanerozoic expressed by mass of volcanogenic rocks formed over a million years.

centration that started in the Late Cretaceous was not continuous.

Even with limited time resolution it is apparent from the available data that after the peak CO_2 concentration in the Late Cretaceous, there were two noticeable peaks in the Eocene and Miocene. However, the level of concentration during these peaks decreased gradually. At the same time, CO_2 concentration also dropped in subsequent depressions, beginning with the Early Cretaceous and including the Paleocene and Oligocene. We should focus on the fact that the atmospheric CO_2 concentration variations were similar to those in volcanic activity (Figure 11.2) (data for the Pliocene are partially excluded). It thus seems obvious that the trend of decreasing CO_2 concentration dominating throughout the Phanerozoic can be attributed to lower volcanic activity.

Mean surface air temperature dependence in the geological past on atmospheric CO_2 concentration is considered in Chapter 1, where it is shown that CO_2 concentration variations were the main cause of climatic change in the Phanerozoic. In this chapter, the data obtained by empirical methods have been used to estimate climatic sensitivity to atmospheric CO_2 concentration variations. These data can also be used to solve the inverse problem. Assuming that the relationship between mean global surface air temperature and CO_2 concentration has been determined in the studies on climatic theory, we can verify how accurately this relationship allows us to describe those temperature changes that occurred in the geological past.

As indicated in previous chapters, the dependence of surface air temperature on CO_2 concentration is logarithmic, and the mean value of parameter ΔT_c (the temperature increase as CO_2 concentration doubles), found in climatic theory calculations is close to 3°C.

More reliable information on geological past climates is available for the last part of the Phanerozoic, i.e., mainly for the Cenozoic. In addition, the data on the quantity of carbonate rocks in sediments used to calculate CO_2 concentration for this period of time are more reliable. We should also take into account that the distribution of continents and oceans in the Cenozoic (especially, in the Neogene) was closer to the modern as compared with earlier epochs. Consequently, data on the Cenozoic, and partially for the second half of the Mesozoic are more important for verifying the results of CO_2 calculation.

Sinitsyn (1965, 1966, 1967, 1970, 1976) was the first to create detailed maps of climatic changes in the Phanerozoic on the basis of generalized data on lithogenesis, floras and faunas of the geological past. He constructed a series of paleoclimatic maps for different Phanerozoic periods and their parts. These maps covered the territories of Europe, extratropical Asia and North Africa. Although a lot of new information has recently been obtained on Phanerozoic climate, later studies usually refer to limited regions and do not contain paleoclimatic charts that cover the Earth or most of it. We therefore have to use Sinitsyn's maps often to calculate mean global temperature during the Phanerozoic.

These maps are the basis for calculating the difference between mean annual air temperatures for various geological epochs and at present for the 30–80° latitudinal zone (ΔT_{30-80}). Then by extrapolating data on mean latitudinal temperature distribution obtained from these maps and by using the results of climatic theory studies we can determine mean air temperature differences for the Northern Hemisphere (ΔT_{0-90}). The results of this calculation for the end of the Mesozoic era and some Tertiary epochs are presented in Table 11.1. It also includes data on mean global air temperature ΔT obtained theoretically for the same time intervals.

Calculation of the values of ΔT, in addition to the influence of variations in CO_2 atmospheric concentration, incorporated the mean temperature effects of the Earth's albedo fluctuations and the Sun's luminosity. Assessment of the effects of these two factors that are less important than CO_2 is given in Chapter 1 for the last 100 million

Table 11.1. The Difference Between Past and Present Mean Air Temperatures (in Degrees Celsius)

Time interval	ΔT_{30--80}	ΔT_{0--90}	ΔT
Cretaceous	17.4	11.0	9.8
Paleocene-Eocene	15.2	8.2	7.9
Miocene	11.8	6.0	6.4
Pliocene	9.2	4.8	3.4

years (see Table 1.3).

As seen from Table 11.1, the average discrepancy in air temperature differences for those four epochs obtained by calculation and from Sinitsyn's maps does not exceed 1°C, i.e., comprises only a small portion of the range of indicated temperature variations. This proves that the data on atmospheric CO_2 concentration in the geological past used in this study are reliable. Similar calculations were made in a number of earlier publications (Budyko, 1980, 1984, etc.).

The results of calculated paleotemperatures can also be compared with empirical data on temperature changes in individual geographical regions. Figure 11.3 presents temperature differences in various epochs of the Cenozoic and at present for three regions: Western United States (curve 1), mid-latitudinal oceans of the Southern Hemisphere (curve 2), and southern part of the North Sea (curve 3). Curve 1 is based on the results of paleoclimatic studies from palynological data (Axelrod and Baily, 1969), curve 2 (Shackleton and Kennett, 1973) and curve 3 from isotope content of marine organisms (Buchardt, 1978). Data on changes in the Cenozoic CO_2 concentration are also presented in the figure for comparison. There is good agreement between independent data on temperature changes and CO_2 concentration variations. Thus, in particular, all the curves reflecting temperature changes show warming at the end of the Pliocene-

Eocene as well as drastic cooling in the Oligocene, temperature maximum in the Miocene and cooling in the Pliocene versus the Miocene. Some discrepancies in the position of these temperature peaks and depressions obtained in different studies are explained by the noncoincidence of the temporal scales used in each study (in these scales the age of the Tertiary epoch boundaries somewhat differs from our calculations).

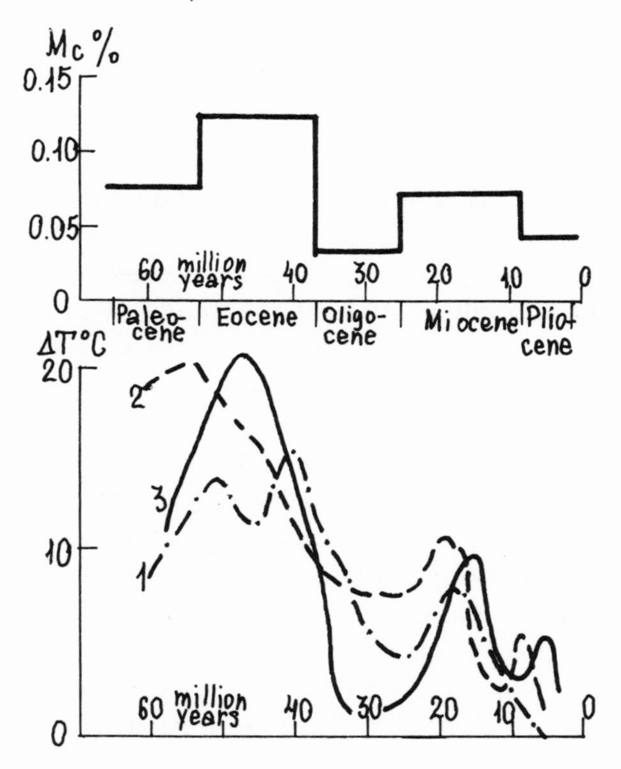

Figure 11.3. Variations in CO_2 concentration (M_C) and mean air temperature (ΔT) in the Cenozoic.

Data on Fig 11.3 show some features of the Tertiary climatic change and of temperature variations occurring during time intervals shorter than the geological epochs, therefore it is impossible to compare these variations with CO_2 concentration fluctuations.

The available data on changing CO_2 mass throughout the Phanerozoic were compared with those on past climatic conditions in a number of other studies (Tenyakov and Yasamanov, 1981; Frakes, 1984, etc.), and there was satisfactory agreement between the data on climatic changes and on CO_2 concentration fluctuations. Similar problems were discussed in two recent papers (Manabe and Bryan, 1985; Manabe and Broccoli, 1985).

The first was based on the general atmospheric and oceanic circulation model to estimate climatic effects of various carbon dioxide concentrations, an eight-fold increase compared with the current level. This allowed verification of the hypothesis that the Cenozoic climatic change towards cooling could be attributed to a considerable decrease in atmospheric CO_2 concentration. Since most of parameters describing changes in the physical state of atmosphere and ocean with lower CO_2 concentration were close to the empirical results of studying climatic evolution in the Cenozoic, the authors concluded that their analysis mainly corroborates this hypothesis.

The second study used two versions of the general atmospheric circulation model to incorporate heat exchange in the upper oceanic layers and verify the assumption that the major factor in developing the last Quaternary glaciation (18,000 yrs. ago) was a drop in CO_2 concentration to 200 ppm, i.e., approximately one-third of its modern level.

This assumption was confirmed by comparison of the calculations with the available paleoclimatic evidence, and important conclusions were drawn simultaneously about the influence of cloud cover changes on the temperature – CO_2 concentration relationship. They are discussed below.

In summary, we can conclude that the most effective modern climatic theories allow us to reliably estimate the major features of climatic changes with increasing or decreasing atmospheric CO_2 concentration. This conclusion, however, refers to the intercomparison of stationary states of the climatic system with different chemical composition of the atmosphere. The forecast of anthropogenic climatic change differs somewhat from this statement of the problem, since it must predict atmospheric variations in chemical composition and incorporate climatic effects of climatic thermal inertia (the latter is discussed in detail in section 6.2).

The efficacy of calculating climatic change with gradual increment in atmospheric CO_2 concentration based on instrumental observations of mean temperature variations throughout the period of existence of the world meteorological station network can be verified.

Detecting anthropogenic climatic change. This question can be clarified by data on detection of anthropogenic climatic change in the twentieth century. The results of a series of works on this problem presented in Chapters 1 and 7 should be supplemented by data from the recent study by Budyko, Byutner and Vinnikov (1986) that made a more definite assessment of the accuracy of the calculated anthropogenic change in mean surface air temperature.

The theoretical altered mean temperature due to higher CO_2 concentration can be estimated by solving three problems: determining the increase in CO_2 concentration for the last hundred years, evaluating the sensitivity of mean temperature to changing CO_2 concentration under steady-state conditions, and incorporating the effect of thermal inertia of the climatic system on anthropogenic climatic warming. Correspondingly, the theoretical estimate of increased mean surface air temperature ΔT_T due to anthropogenic CO_2 increment in the atmosphere over the last hundred years can be expressed as:

$$\Delta T_T = \Delta T_c[\ln(C_2/C_1)/\ln 2]\mu \qquad (11.1)$$

where ΔT_c is the change in mean temperature with doubled atmospheric CO_2 concentration obtained in climatic theory studies; C_1 and C_2 are the initial and final CO_2 concentrations for the periods 1881 and 1984, respectively; μ is the coefficient that describes the thermal inertia of the climatic system.

Each of these three terms in the right side of the equation is known with limited accuracy. According to the data of several collective reports, the most probable range of values for first term ΔT_c is 3.0 $\pm$1.5°C. We will consider that this range describes the 95%-confidence range for theoretical estimation of this parameter. Then the mean value of ΔT_c-estimates is equal to 3.0°C and the root-mean-square error, based on a hypothetical standard distribution of the estimates, equals 0.75°C.

There are grounds to believe that the given mean value is much more accurate. This conclusion can be drawn if we exclude those estimates that were obtained by the models that use insufficiently realistic parameters. Such a study found that the root-mean-square error of the model estimate of the ΔT_c-parameter does not exceed 0.5–0.6°C, and consequently after choosing the value 0.75°C, we noticeably exaggerate the measure of uncertainty in the existing theoretical estimate of the ΔT_c parameter.

The second term in formula 11.1 contains an imprecise value of atmospheric CO_2 concentration pertaining to 1881. It is known that joint usage of several independent methods for estimating atmospheric CO_2 concentration for the pre-industrial period considerably decreased uncertainty regarding the atmospheric CO_2 concentration at the end of the nineteenth century. In the early 1880s, this value did not exceed the range $C_1 = 285 \pm 5$ ppm. With 95% probability for this range, we find that term $(\ln \frac{C_2}{C_1} / \ln 2)$ has an average value of 0.267 with the root-mean-square error of 0.0125. The value of C_2 equal to 343 ppm in 1984 is assumed to be accurate.

The third term in the right part of the expression μ describes the attenuation in the climatic system reaction to increasing CO_2 concentration from C_1 to C_2 on account of climatic system inertia. As indicated in Chapter 6, according to the simplest models, whose major parameters can be obtained empirically, the value of this coefficient is about 0.7. By somewhat exaggerating the uncertainty of our knowledge about this problem, we can adopt the value of 0.1 as the root-mean-square error of estimating μ. This means that the true value of μ ranges from 0.5 to 0.9 with about 95% probability, which represents practically all the variety of the results obtained by models that realistically reflect the properties of the global climatic system.

Using the selected parameter values, we find from 11.1 the mean value of the theoretical estimate for the expected change in mean surface temperature ΔT_T for the 1881–1984 period to be 0.56°C.

If we take into account the independence of errors in determining all three terms in 11.1, the root-mean-square error of this estimate ΔT_T is $\sigma_T = 0.17$°C. The distribution of the ΔT_T value can be assumed to be very close to normal.

We will now discuss the empirical estimates of mean surface air temperature variations ΔT_c for the same 1881–1984 period. Assume that during this time there is a linear trend in mathematical expectation of mean temperature, possibly caused by anthropogenic rise in atmospheric CO_2 concentration. Random variations due to natural causes took place against the background of this trend.

For the available series of mean annual surface air temperature in the Northern Hemisphere it is possible to obtain an estimate of the linear trend parameter β for the 1881–1984 period equal to 0.47°C/100 yrs and an estimate of the root-mean-square error of this parameter caused by random errors and natural variability in climate equal to $\sigma_\beta = 0.11$°C/100 yrs.

Thus, we can conclude that mean global air temperature changed during 1881–1984 by the magnitude of ΔT_e equal to 0.49°C at $\sigma_e = 0.12$°C. Estimation errors in this case are distributed according to normal law.

Both methods have thus revealed that a global warming occurred in this period, with over 99% reliability. The extremely close agreement of the most probable values of temperature variations obtained by two different methods is particularly important.

The difference between these values 0.56–0.49 = 0.07°C is several times less than the probable error of determining each of them. Actually, they are indistinguishable. It can therefore be concluded that global climatic warming has been caused by anthropogenic factors. The preceding assumption that the error in determining ΔT_T has been somewhat exaggerated in this calculation is also confirmed.

An important difference between this calculation of anthropogenic climatic change and previous ones is that fewer assumptions were used to interpret the empirical data, in particular, the influence of atmospheric transparency fluctuations on the total change in mean air temperature was ignored. On the one hand, this approach somewhat decreases the accuracy of theoretical and empirical estimates of changing mean surface air temperature, but on the other hand, it makes the detection of anthropogenic climatic warming much more general.

The results of this analysis confirm the possibility of very accurate prediction of an anthropogenic temperature trend in the Northern Hemisphere averaged over a hundred-year interval by using empirical data that partially describe the increment in CO_2 concentration for this period. It is more difficult, but at the same time more important to forecast the altered anthropogenic trend for shorter time intervals at the beginning of the prescribed time period. This rules out the possibility of using empirical information about the altered atmospheric chemical composition at this time.

Checking the projected anthropogenic climatic change. The first realistic forecast of anthropogenic mean global temperature change for 1970–2070, mentioned in Chapter 1, was made in 1971 and published in early 1972. We can now verify this prediction for about 15 years. The results

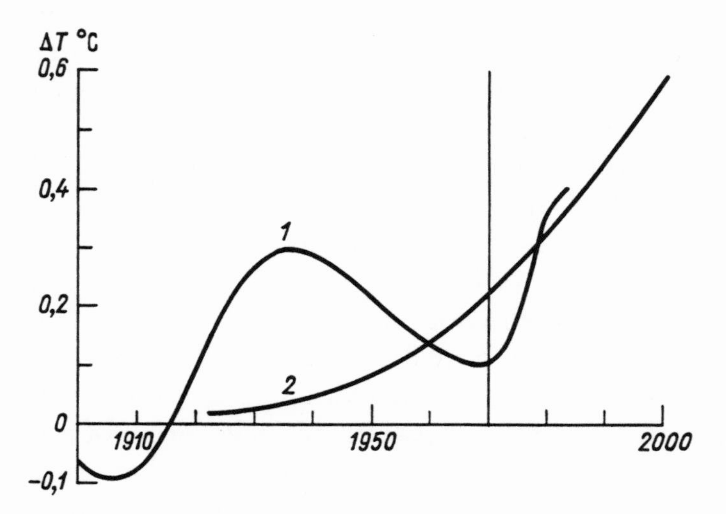

Figure 11.4. Variations in air mean temperature in the twentieth century. 1, smoothed temperature changes, found from observational data; 2, forecast of anthropogenic temperature changes.

of this analysis are presented in Figures 11.4 and 11.5.

Figure 11.4 shows that the predicted global warming of the 1980s, comparable to the 1930s warming, the largest in the last 100 years, has been fully confirmed. Moreover, the mean smoothed values of temperature anomalies in the 1980s essentially coincided with the prediction. It would seem from this agreement that future changes in mean global temperature can be predicted 10 to 20 years in advance. However, this conclusion would not be quite correct. To obtain a realistic figure for the above forecast, one should consider Figure 11.5, which depicts temperature changes smoothed over 5-year time intervals (curve 2 coincides with a portion of curve 1 in Figure 11.4) and its anomalies for each year (curve 1). We see that in individual years throughout the last decade, positive temperature anomalies were very high. In 1981 in particular the mean temperature was the highest over the entire monitoring period and probably the highest for the last several centuries.

It is also obvious that the year-to-year variability in annual temperature anomalies caused mainly by unstable at-

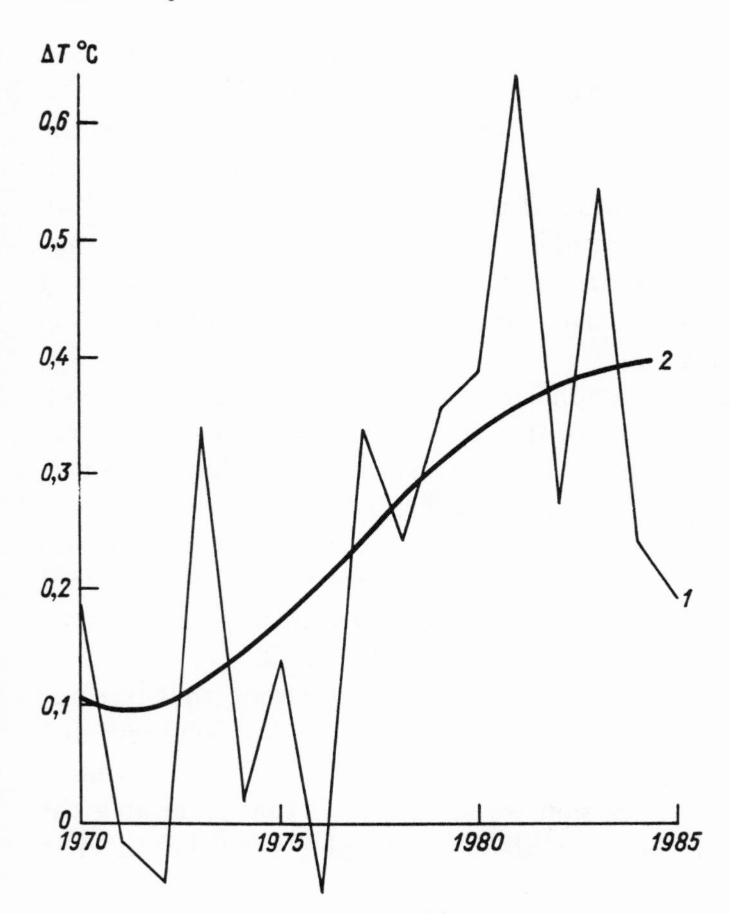

Figure 11.5. Variations in mean air temperature at the end of the twentieth century. 1, temperature anomalies for individual years; 2, smoothed temperature changes.

mospheric circulation was enough to noticeably limit the accurate determination of mean temperature anomalies smoothed over the 5-year time intervals. In other words, it is easy to find that random factors can make the curve 2 data differ from the anomalies produced by anthropogenic climate change, by an average of 0.1°C. A less than 0.1°C discrepancy between the forecasted and empirical anomalies is consequently an undoubtedly random event.

Updated calculations of anthropogenic climatic change for this time interval do not fully agree well with the first observations of mean temperature as compared with the first forecast. The calculation results in Chapter 9 show that the mid-1980s temperature should be somewhat higher than the observation results. This discrepancy seems to be unavoidable in the years when anthropogenic warming only slightly exceeds natural variability of mean temperature due to atmospheric transparency fluctuations. As global warming develops, the relative importance of this source of errors will diminish.

It should be emphasized that good agreement between the forecasted scale of the 1980s warming and observational data is important proof that the forecast is reliable. Consequently, we should recall, as indicated in Chapter 1, that the first prediction of anthropogenic change in mean temperature was published when almost everybody believed that the global cooling following the 1930s warming would continue. Under these conditions, the probability of chance coincidence between the preforecasted warming that in some years exceeded all the warmings of the last century, and the observation results was obviously insignificant. In mentioning this conclusion we should note that it refutes the assumptions that it is difficult to reliably foresee changes in atmospheric chemical composition affecting climate and to obtain accurate estimates of the relationship between mean air temperature and CO_2 concentration, etc.

11.2. Accurate Forecasts of Anthropogenic Climatic Change

The major requirements for early and accurate forecasts of anthropogenic climatic change were formulated by the early 1970s. They were based on the fact that national economies depend significantly on modern climate, and a noticeable climatic change will require enormous investments to adapt the economy to new conditions. The same length of time is needed for information on climatic changes, probable operation of projected industrial and agricultural structures, and adaption of the national economy to altered climate. This is apparently no less than several decades.

Taking into account that the prediction of future climatic conditions not only incorporates physical and chemical processes in the climatic system, but also the rates of economic development, we can conclude that climatic forecasts are conventional, since unfavorable forecasts could generate suggested changes in economic development plans. But since it is difficult to make a quantitative prediction of economic development for many years ahead, this forecast is not very accurate. Consequently, the use of schematic methods for estimating future climatic conditions is justified, however they should correctly assess the sign and magnitude of possible climatic changes (Budyko, 1974).

Subsequent studies somewhat altered the previous evaluation of permissible error in predicted climatic change. Although this error is inevitable for the distant future, more accurate information about expected climatic conditions can apparently be obtained for the next 15–20 years.

This requirement can be met because of the following simple considerations.

Section 11.1 data show that variations in mean surface air temperature due to certain changes in atmospheric chemical composition can be calculated with sufficient forecasting accuracy. Of particular importance are data proving the possibility of predicting anthropogenic changes in mean temperature for a 15-year period of time without using empirical data on variations in the atmospheric chemical composition. The possibility of reliable prediction of expected anthropogenic changes in the mean temperature will subsequently increase. If this change even now exceeds the natural variability of the Northern Hemispheric mean temperature, in 10 or 20 years it will be noticeably above the natural variability. After this, predictions of expected changes in the mean temperature will be made both by physical calculations and simple extrapolation of the trend that has been observed over the last decades. It should be mentioned that contrary to explicitly erroneous attempts to predict climatic change by extrapolating mean temperature trends in the epoch when these trends largely depended on natural rather than anthropogenic factors, extrapolation of anthropogenic trends is quite permissible for short time intervals (in any case up to 15 or 30 years). This is possible, in particular, because the majority of studies on future anthropogenic increase in mean air temperature have concluded that in the next decades this increase will have an almost linear time dependence.

The possibility of using this empirical method to estimate future changes in mean air temperature is of basic importance. As noted in previous chapters, anthropogenic warming is caused by two major factors: higher atmospheric CO_2 content and more trace gases. The accuracy of estimating the influence of each factor on climate varies significantly. It is obvious that in the next decades major variations are not very likely in the extant structure of power that is based mainly on fossil fuel. Consequently, as indicated in Chapter 2, the predicted fossil carbon fuel consumption up to 2010 is fairly reliable. Numerical experiments also show that changing these predictions in reasonable limits has little impact on the estimated expected increase in mean global temperature for the next 25 years. The methods used in global warming calculations to incorporate the relationship between mean temperature and CO_2 concentration have a reliable empirical basis stated in some chapters of this book. The estimated impact on warming of higher atmospheric trace gas content is less substantiated. The greatest errors in these estimates can be caused by insufficient accuracy in determining expected rates of the increase in trace gas content. Moreover, unlike CO_2 , the increment in trace gas concentration can be retarded by current technology, although the possibility of solving this problem on an international basis is still obscure.

After global warming has become stronger, it will become clear whether the estimated climatic effect of trace gases is reliable. Nevertheless, there are some data even now to partially solve this problem. According to estimates of current global warming presented here, the mean temperature should have increased by 0.25°C during the 1970–1985 period due to higher CO_2 concentration, and by 0.45°C due to higher content of CO_2 and trace gases. Empirical data in Figures 11.4 and 11.5 are the average of these two estimates. This conclusion on the one hand confirms the impact of increased amounts of trace gases on modern global warming, and on the other hand, assumes that these estimates of expected warming are possibly somewhat exaggerated. It should be explained that this exaggeration (if it actually takes place) does not alter the prediction results very much. According to the prediction made here, in early 2000 the mean air temperature will rise by 1.4°C versus the pre-industrial level. Assuming that this prediction doubled the effect of trace gases on mean temperature, we find that warming by the year 2000 in this case will be 1.2°C, i.e., it will be only 14% less than the predicted value. This example demonstrates the low sensitivity of global warming estimates even to large errors in the assessments of important climatic-changing factors.

It thus seems possible to estimate the increasing mean global temperature for the next 20 or 25 years with error considerably less than the value of expected temperature change. The probable error of such a calculation will certainly increase for the more distant future. Nevertheless, there are grounds to believe that before 2030–2050 not only the estimates of increasing mean temperature will have the correct sign and order of magnitude (which is almost trivial), but the errors in these will not exceed 50% of the expected temperature change. This conclusion can be drawn from analysis of errors when taking into account the major factors that influence the temperature change similar to the above calculations on the reliable detection of anthropogenic climatic change that has already occurred.

Passing to the question of accurate assessments for future regional climatic changes, it should be noted that this accuracy can be assessed more easily for surface air temperature whose regional changes, as indicated in section 9.1, can be defined by several independent methods. Considering the discrepancy between independent estimates, we find that errors in calculating their average values in most cases are noticeably smaller than corresponding temperature changes. It is more difficult to find errors in determining precipitation changes.

It should be mentioned that the precipitation change detected by paleoclimatic methods really often describes the changes not in precipitation but in soil moisture that directly influences vegetation, lithogenesis, etc. This feature of the method for estimating total precipitation in the geological past can result in systematic errors in determining their changes. In particular, the decrease in precipitation during global warming can be exaggerated for this reason, while increase in precipitation can be underestimated due to the influence of increased evaporation on soil moisture.

It would be very desirable if empirical data on precipitation could be verified by comparing them with other independent information. Consequently, we should focus on the good agreement between precipitation variation maps in Figures 9.4a and b obtained by different methods (these maps pertaining to different levels of global warming and could be differentiated, however, from the point of view of climatic theory their similarity is quite possible and is hardly accidental). The possibility of comparing empirical estimates for precipitation changes with similar estimates obtained from climatic models is very important.

Changes in total precipitation on the continents during smaller global warming (0.5–1.5°C) obtained by different empirical methods are similar to each other and, to a certain extent, agree with the model calculations of precipitation variations. At the same time the available paleoclimatic data on changing precipitation during greater warmings have not yet been confirmed by other independent data. It can be assumed that, since the estimates of regional temperature changes obtained from paleoclimatic analogues were quite reliable, data on regional distribution of altered precipitation obtained by similar method can be expected to be quite realistic.

11.3. Discussion of Future Climatic Changes

In the early 1970s it was commonly thought that anthropogenic climatic change can only be predicted by using an exact climatic theory. Since there was no such theory at that time, and it was clear that one would not appear soon, it was sometimes concluded that reliable information on future climatic conditions could not be obtained. This viewpoint was particularly advanced in the proceedings of the 1974 conference on the physical basis of climate (Physical Basis of Climate..., 1975). Such an opinion has raised some objections. In particular, it was noted that "... most fields of technology and other applied sciences make extensive use of approximate physical theories, since exact theories of corresponding processes are either not available or their use is unnecessary. The example of hydromechanics which is better known to specialists in the field of atmospheric sciences is enough to illustrate this. In spite of the outstanding achievements in developing the theory of streamlining different bodies by liquids and gases, designing of ships and aircraft makes extensive use of approximate theories that yield fairly reliable results. Experimental studies that occupy a significant place in all the branches of applied sciences are very important in substantiating these approximate theories, determining the included parameters and verifying the findings.

"We believe that it is natural to apply a similar approach to developing the problems of climatic theory, since in this field we have a lot of empirical information, that is quickly supplemented as the experimental technique of studying modern and past climates develops.

"An example of the comprehensive approach to studying the problem of physical climatology is, in particular, investigations of thermal balance of the Earth's surface that began in the 1940s. Not only the values of the heat balance components, but their sign and magnitude were considered unknown before these studies began. Since the accuracy of the available methods for theoretical and experimental determination of thermal balance components was limited, the first studies proposed using several independent methods to determine the heat balance terms, and the results were considered reliable if they agreed. This was a very effective method of studying the thermal balance of the Earth's surface. Although subsequent studies constructed numerous new charts, all these data on the balance components usually differed from the initial values by no more than 10 or 20%. The results of these first studies on thermal balance were thus fairly reliable.

"We should recall that the mass on the thermal balance components of the Earth's surface constructed later using detailed climatic theory (Holloway and Manabe, 1971) were also close to the earlier maps based on approximate methods.

"Whereas study of thermal balance was mainly of cognitive importance and the more or less rapid influx of data on thermal balance components did not exert a pronounced effect on man's economic activity, now a delay in obtaining reliable information on future climate when it changes so fast can have severe economic consequences. It is consequently necessary to use more fully modern scientific methods to clarify the outlook for climatic change" (Budyko, 1980).

During the years following this discussion, it was found as often occurs, that both seemingly contrary viewpoints were quite correct. Most researchers now recognize that even the best models of climatic theory cannot be used to predict climatic change without their verification by reliable empirical data on climatic changes. On the other hand, it became clear that climatic changes can be predicted on the basis of generalized empirical data on climatic changes in the past, particularly, on paleoclimatic data.

This possibility also permits detailed verification of climatic theory models and allows us to apply models that agree well with empirical data to forecasting future climate.

In the course of complex studies of climatic change based on theoretical models and empirical data it was possible to solve the problem raised in Chapter 6 of incorporating the climatic effects of altered cloud cover into the prediction of climatic changes. The importance of this problem was mentioned in the study that made the first realistic forecast of mean surface air temperature (Budyko, 1972). In the early 1970s it was known that calculations by the simplest climatic theory models indicated the probable compensation for the influence of cloud cover changes on mean global air temperature due to higher Earth's albedo as cloudiness increases and decreasing outgoing radiation (Budyko, 1971). These first calculations were not very accurate. More reliable data confirming this conclusion were obtained from satellite observations of outgoing radiation (Budyko, 1975; Cess, 1976).

A number of studies used different climatic theory models, including general atmospheric models to solve this problem. They obtained ambiguous results, which can now be attributed to the very high sensitivity of climatic models to comparatively small changes in parametrizing physical processes that determine the development of clouds and their influence on climate.

The most important studies are those that compare calculations by detailed general atmospheric circulation models with empirical data on climatic changes. An example is the aforementioned study by Manabe and Broccoli (1985) which calculates the fields of meteorological elements for the last Ice Age by two versions of climatic models: one considers the cloud cover to be fixed and the other considers it to vary as other meteorological elements change.

The dependence of mean surface air temperature on CO_2 concentration was calculated for each version of the model. Temperature sensitivity of the second version was double that of the first. This result seems to confirm that cloud cover changes have a considerable influence on this dependence, however, this conclusion would be wrong.

The authors of the study found that both versions of their model satisfactorily describe the climate of the Ice Age, and the discrepancy between the results of both calculated distributions of basic meteorological elements were comparatively small. This confirms the rather limited importance of cloud cover variations in the physical mechanism of climatic changes. The conclusion drawn in this study that cloud cover changes have noticeable effect on climatic sensitivity to CO_2 concentration thus seems to be a distinctive feature of this model and not a conclusion about the real laws governing the climatic system confirmed by empirical data.

The most obvious proof of this is the estimates of mean air temperature sensitivity to changing CO_2 concentration obtained from empirical materials, since these estimates found by using different data for various climatic conditions (and, therefore, for various cloud cover distributions) in all cases were essentially the same.

We mention in conclusion that the most recently published authoritative international reports on anthropogenic climatic change recognize that global warming will undoubtedly develop due to the increasing greenhouse effect, give similar estimates of the warming scale, and note the possibility and necessity of predicting basic patterns of climatic change during the warming.

The first such report prepared at the 1981 Soviet-American meeting of experts was published in 1982 (Climatic Effects of Increased Atmospheric Carbon Dioxide..., 1982). This report incorporated maps of expected changes in air temperature and precipitation based on paleoclimatic evidence. Certain information about future climatic conditions is contained in the report of the 1985 Villach meeting (Bolin et al., 1986) and in other reports. There are grounds to believe that the information on future climatic conditions given in this book is more complete than previous published national and international reports.

It should be emphasized that in addition to the viewpoint on anthropogenic warming accepted in modern collective reports, there are some other opinions set forth in a few publications. Their presence can be simply explained: earlier there existed a point of view that no kinds of man's economic activity could exert any noticeable effect on global climate. To reject this concept made certain difficulties for more conservative scientists. As to professional climatologists engaged in this field for many years, not one of them (to their credit, be it said) has doubts now of the hypotheses developed recently concerning anthropogenic climate change. Some American and Soviet specialists have attempted to substantiate the idea that there will be no significant anthropogenic change or that it cannot be estimated yet. Their statements were frequently based on the following.

1. "The accepted estimates of the relationship between mean global temperature and CO_2 concentration exaggerate this dependence several times." This thesis was rejected in a number of publications in other countries, that show that either it follows from explicitly incorrect climatic theory model calculations, where the Earth-atmospheric system energy balance is arbitrarily replaced by the Earth's surface energy balance, or it is substantiated by erroneous

empirical estimates of climatic system sensitivity that ignore the system's thermal inertia during its brief perturbations that cause approximately ten-fold variations in sensitivity compared with its actual value."

The calculations of the mean temperature dependence on CO_2 concentration by climatic theory model with indefinite coefficients also refer to this line of studies. One can obtain any desirable result by choosing indefinite coefficients.

2. "CO_2 concentration cannot increase considerably compared with its pre-industrial level either because of possible unrestricted absorption of CO_2 by marine plankton, or because of its greater expenditure to form carbonate deposits."

Both hypotheses are based on the absolutely arbitrary statements. The first is considered in the Villach conference report where it is noted that it is not confirmed by any empirical data and improbable. The second contradicts detailed studies of carbonate formation as CO_2 concentration rises, that have shown that the time intervals, when considerable changes in carbonate formation could occur, exceed the duration of the present modern climate changes.

3. "It is impossible to reliably predict anthropogenic climatic changes because of large discrepancies in different estimates of a) the rise in fossil fuel consumption; b) carbon cycle components and c) climatic sensitivity to higher CO_2 concentration."

This statement assumes that unsubstantiated published results are equivalent to the conclusions of the most reliable studies. The same is true of any scientific problem that always has some erroneous idea. This does not describe how much the problem has been studied, but the level of its understanding by the authors of this statement.

Foreign reports sometimes discuss similar statements in more detail. Thus, for example, one of the American reports notes that its authors made futile attempts to explain to the authors of the aforementioned concept of anthropogenic warming that their idea was incorrect (Carbon Dioxide..., 1982).

CONCLUSION

Results of investigations. The conclusion drawn in 1972 that considerable global warming due to changing chemical composition of the atmosphere will inevitably occur in the next decades was recently confirmed by the studies of many researchers and in the reports of many national and international scientific organizations. They include, in particular, the report of the Soviet-American meetings of experts held in Leningrad in 1981, 1983 and 1986, as well as the Report of the International Conference on the Assessment of the Role of Carbon Dioxide and of Other Greenhouse Gases in Climate Variations and Associated Impacts (Villach, Austria, 9–15 October 1985).

Since the USSR started studying anthropogenic climatic change before other countries, Soviet scientists now have more detailed information about future climate than that contained in the reports of recent international conferences or surveys on this problem published in other countries.

An important step in studying anthropogenic climatic change made by Soviet scientists was the development of semiempirical and empirical climatic models that allowed them to solve the problems that theoretical climatic models alone could not.

These studies showed that surface air temperature is increasing and is expected to rise in the future as a result of higher concentrations of some atmospheric gas components from industrial emissions, most significantly CO_2 emissions from burning fossil fuel. Changes in atmospheric CO_2 have been recorded since 1958 by fairly precise instruments. It increased from 315 to 343 ppm during the 1958–1984 period. Observations of changes in relative amounts of carbon isotopes ^{12}C, ^{13}C and ^{14}C in the atmosphere, ocean and tree rings, as well as data of monitoring atmospheric CO_2 concentration were used to estimate the CO_2 atmospheric concentration in the mid-nineteenth century at approximately

280 ppm. At the same time, the influx of CO_2 into the atmosphere due to anthropogenic effects on the continental biota was calculated. The biotic emission of CO_2 from 1860 through 1980 did not exceed 90 GTs or no more than 1/2 of the influx into the atmosphere from burning fossil fuel.

The change in future atmospheric CO_2 concentration can be projected on the basis of the predicted rate of annual CO_2 industrial emissions into the atmosphere; possible anthropogenic variations in global biomass (positive or negative) should be considered as an error of this forecast.

If current rates of fossil fuel consumption are maintained to the end of the twentieth century, the atmospheric CO_2 concentration will reach 375–385 ppm by the year 2000. The changing CO_2 concentration in the twenty-first century can be assessed by forecasts of energy development recently made by a number of scientists in different countries.

The calculations of atmospheric CO_2 content in the next century are associated with some uncertainty. The energy development scenarios proposed in different studies do not coincide and there are some discrepancies in the biospheric carbon cycle models because the role of CO_2 biotic sources and sinks in this cycle is not really clear. However, leaving aside the clearly erroneous estimates, it is easy to see that if the forecasts of the expected twenty-first century CO_2 concentrations are based on the most substantiated scenarios of energy development and on independent carbon cycle models, the values of the forecasted concentrations are very similar. In particular, it was found that atmospheric CO_2 concentration compared with pre-industrial level will double in the second half or, possibly, by the middle of the twenty-first century.

Calculations show that if all available fossil fuel resources are used, the CO_2 concentration will increase by a factor of 6 to 10 compared with the pre-industrial level. It is unlikely that fossil fuel resources will be completely exhausted in the future. However, the possibility of a 6 to 10-fold increase in

CO_2 concentration in the more distant future exists, since it is highly probable that new fossil fuel resources will be discovered. This problem has not yet acquired great practical significance, because such an increase in CO_2 concentration cannot occur before the end of the next century.

The modern atmosphere not only contains CO_2, but also a number of gases that, although they comprise a small part of the entire atmospheric volume ("trace gases"), can nevertheless exert a certain influence on climate. Some of them are of anthropogenic origin, and some were present in the earlier atmosphere, although in a very limited amount.

Trace gases include chlorofluorocarbons (CFCs), nitric oxides, methane, and some others. The influence of increasing concentrations of trace gases on climate is similar to that of CO_2, since all these trace gases intensify the greenhouse effect, i.e., lead to climatic warming.

There is no doubt that the quantity of these trace gases in atmospheric air is rising and can cause an acceleration of future climatic change. However, it is harder to forecast concentrations of this trace gas than to forecast concentrations of carbon dioxide.

The forecast of anthropogenic climatic change requires reliable quantitative estimates of climatic sensitivity to altered climate-forming factors.

The ΔT_c-parameter, equal to the change in globally averaged mean surface air temperature with doubled CO_2 concentration, is used as the major characteristic on the Earth's atmospheric global temperature sensitivity to varying CO_2 concentration. The dependence of mean temperature change on CO_2 concentration is assumed to be logarithmic in this case.

Although the ΔT_c-parameter estimates can be obtained by various climatic models (energy-balance, radiative-convective and general atmospheric circulation models), however, because of the more schematic structure of the energy-balance and radiative-convective models, it is sometimes assumed that only the evaluations based on the general atmospheric circulation models are reliable to some extent. Actually, even if the three-dimensional general circulation models could describe fairly completely all the climatic feedbacks known today, they would not guarantee that this description is correct. The most detailed modeling of circulation processes does not realistically incorporate the major feedbacks in the global climatic system. However, the estimates obtained by the best climatic models agree satisfactorily with each other and yield mean ΔT_c equal to about 3°C.

Nevertheless, even consistent theoretical calculations cannot be considered to be quite a reliable solution to the problem in question, since they can contain errors common to all the calculations, because they ignore (or miscalculate) some significant feedbacks for climatic change, and for other reasons. Consequently, a lot of effort has recently been made to develop independent methods of estimating climatic system sensitivity to changing CO_2 concentration. The two most important methods are: one based on generalized observational data on modern natural climatic changes, and the other on evidence on climatic changes in the geological past.

In order to obtain empirical estimates of the ΔT_c-parameter from data on modern mean surface air temperature variations, it was important to find that the major physical mechanism of natural global climatic change is associated with atmospheric transparency fluctuations due to volcanic activity variations. After reliable data were obtained on changing mean annual surface air temperature of the Northern Hemisphere and atmospheric transparency fluctuations over the last 100 years, mean temperature variations due to anthropogenic increase in atmospheric CO_2 concentration were detected and the ΔT_c-parameter was estimated at approximately 3°C.

The atmospheric chemical composition of the geological past was studied in the last decade and showed that the past warm climates and the future warming can mostly be attributed to higher atmospheric CO_2 content. The ΔT_C-estimates based on these data were also close to 3°C.

The best theoretical estimates of the ΔT_c-parameter thus agree with empirical estimates, which shows their validity and the possibility of using them to assess future surface air temperature variations.

Model estimates of climatic sensitivity are associated with invariable conditions, when the climatic system is in equilibrium with external factors. During the expected rapid change in atmospheric gas composition, the actual mean temperature variations will lag behind the equilibrium. We know that the ocean is the major inertial component of the climatic system. By applying models of varying complexity, we can conclude that given the present anthropogenic climatic changes, thermal inertia of the climatic system will diminish the expected mean air temperature variations by about 20–40%.

Individual scientists made the first calculations of future anthropogenic changes in mean air temperature. A joint Soviet-American forecast was published in 1982 and contained the following estimates of mean global temperature variations: 1–2°C by 2000, 2–3°C by 2025, and 3–5°C by 2050.

Subsequent forecasts of the expected mean global temperature increase given in the conclusions of several international meetings agree satisfactorily with the 1982 prediction.

Evaluations of possible regional climatic changes are less certain. Climatic theory models that assess mean temperature sensitivity to the altered climate-forming factors also clarify spatial patterns of possible climatic changes, i.e., the structure of vertical and latitudinal-seasonal distribution of air temperature variations. Less satisfactory results were obtained from modeling the expected precipitation changes.

It should be remembered that paleoclimatic evidence includes data on both surface air temperature and precipitation distributions.

These results made it possible to obtain empirical evaluations of regional changes in some climatic elements at different stages of anthropogenic global warming. These estimates are based on climatic change data during the period of meteorological observations and paleoclimatic reconstructions of the past epochs that were warmer than the present. This approach significantly reduces the uncertainty in the estimate of future climatic conditions.

Until recently the international scientific organizations were rather cautious in their estimates of possible climatic changes. Now the situation has changed significantly. The 1985 Villach (Austria) scientific conference assessed the climatic change effects of higher contents of CO_2 and other greenhouse atmospheric gases. It adopted a statement appealing to all world governments and approved a voluminous report on this problem.

The statement declares that altered atmospheric chemical composition will result in great climatic change in several decades that will be unique in mankind's history. It also mentions the erroneous present practice of making important decisions in economic and social development based on the hypothesis that the past climatic conditions will remain in the future. It points out that this approach cannot be applied to planning irrigation systems for vast territories, power engineering facilities, agricultural production or solving other important problems. It is consequently very urgent and important to obtain estimates of climatic conditions in the near future. Although the conference had only sketchy data on future climate changes, there is no doubt that these changes will have an enormous impact on ecological systems, agriculture, water resources, and sea ice.

The statement focuses on the fact that since the future climatic change is inevitable, the rate and extent of the future warming could depend considerably on energy development projects in different countries, and their governments should incorporate this dependence into these projects.

The statement notes that the governments and the proper agencies should provide more support for research on the forthcoming climatic change. The statement raises the question that in the future a global convention on anthropogenic climatic change should be prepared.

Although we agree with the major conclusions of the Villach conference statement, we regret that it was not adopted by international organizations earlier.

Use of research results. There is no doubt that the problem of anthropogenic climatic change has not yet been fully resolved. The available information about the climatic conditions of the next century is much sketchier than data for the last century obtained at a number of meteorological stations. However, if we recall that only ten years ago almost no one supported the viewpoint that a major warming would inevitably develop in the near future, the progress made in the last ten years is quite considerable.

The most important goal of studying anthropogenic climatic change in the next few years may be a broad synthesis of the results of applying theoretical and empirical estimates of future climatic conditions in different geographical regions. These assessments must include the determination of both changes in mean values of meteorological elements and their variability. The elements of this synthetic approach have already been formed in the most advanced modern studies.

Analysis of the economic consequences of the future climatic change is also an important problem. Current preliminary estimates assume that for the Earth as a whole the developing global warming in the final analysis will have a favorable effect on crop production. However, certain regions will be in a more complicated situation.

If these problems are solved, it may be possible to set up an international scientific center to inform the members of the World Meteorological Organization about expected climate change.

The question of the practical application of estimated future climatic conditions during global warming was discussed in one of the earliest Soviet investigations of this problem published in 1972 which states: "... at current rates of economic development, man's activity could alter global climate in the near future. In 20 or 30 years, these changes can have a pronounced effect on economic activities, in 50 or 80 years they will radically change these conditions in many countries. Taking into account the great practical importance of clarifying future climatic conditions, a detailed study is needed based on the most effective methods of physical climatology.

Although it would be premature to apply the preliminary results presented here to the economic projects, these results can be used to estimate the time during which exact information about future climatic conditions must be obtained. Considering the aforementioned data on possible climatic change rates this time may be no more than ten years."

This was written fifteen years ago (Budyko, 1972), when almost nobody believed that anthropogenic global climatic change was possible. The estimate of the expected climatic change presented in this publication can be compared with the 1985 Villach conference conclusions. The report of this conference notes that higher mean surface air temperature corresponding to a doubled CO_2 concentration will occur in the 2030s. Thus, within 60 years after the publication of the cited work, global climate will change drastically, which conforms to the above prediction.

The available data on future climatic change began to be used to meet the practical demands of our country several years ago, i.e., about ten years after the cited work had appeared.

The third forecast from the above quotation was less accurate. The influence of anthropogenic factors on climate and the national economy has already been detected, i.e., earlier than 20 or 30 years after this work was published. However, it is possible that there are 5 to 15 years remaining before this influence actually becomes easy to notice.

Incorporating the available data, we should assume that there are two aspects to considering the impact of anthropogenic climatic change on economic activity: 1) consideration for the consequences of modern global climatic change, as well as the change that will occur in the near future, approximately by 2000; 2) study of the climatic change effects on man's activities in the more distant future, i.e., in the first half of the next century.

These two aspects of this problem differ both in details and the accuracy of the available information about associated climatic changes and their scale.

We have already noted that the mean global surface air temperature has already increased by approximately 0.5°C as compared to the pre-industrial level.

Although this warming is revealed in detailed analysis of meteorological observations, it is not noticeable for laymen and therefore may seem comparatively insignificant. This would certainly be a gross mistake.

We will merely mention two sectors of the national economy that depend considerably on climate: agriculture and water resources. Both crop yields and river run-off vary as the climate fluctuates slightly. These variations are reflected in the economic indicators that amount to billions of rubles in the USSR.

Changing moisture conditions associated with warming, in particular, increased frequency of droughts in regions with insufficient moisture are of greater practical importance in the next few years than rising surface air temperature. The effect of more frequent droughts on the harvest can be compensated for to a certain extent by the influence of higher CO_2 concentration on crop productivity. This influence is strongest in the years with favorable moisture conditions. As a result of the impact of these two factors, the variability of major crop yield increases from year to year in some regions.

It may be of great economic importance to consider these consequences of global climatic changes.

In passing to anthropogenic climatic changes in the twenty-first century, we note that according to the available estimates they will be very large and can lead to noticeable shifts in the present natural zones, the disappearance of permanent polar sea ice, etc. Although these environmental changes will occur within dozens of years, there are grounds to believe that prediction of these climatic changes is very urgent.

The prediction of climatic changes due to man's activities differs significantly from weather forecasting. Whereas we can limit ourselves to analysis of physical processes in the atmosphere and hydrosphere for the second problem, the first also requires consideration for the temporal changes in the economic indicator.

Consequently, the problem of predicting climatic changes contains two major elements: forecasting the development of some economic aspects (higher fuel consumption that increases the atmospheric CO_2 content, growth in energy production, etc.) and calculation of those climatic changes that will occur as the corresponding types of human activity vary.

This leads to two important features of these predictions. First of all, they will inevitably be conditional. Economic activity is a process that depends on man's impact on climate. Thus, in particular, if this activity results in unfavorable climatic changes, the possibility cannot be excluded that economic activity will be altered before these changes occur. The forecast of possible climatic changes could thus be important in substantiating measures to optimize economic plans.

The second feature of future climatic forecasts is associated with their possible accuracy. Since quantitative prediction of economic development for many years ahead is difficult for a number of reasons, the accuracy of this prediction cannot be high. Therefore, the use of approximate methods to calculate future climate is justified. At the same time, since the estimates of future climatic conditions must be used in long-range economic planning, these assessments obviously have to be more precise. We have already noted that evaluations can be made by applying several independent methods of determining future climatic conditions.

Since even the most reliable data on future climate are very sketchy, we have to develop a technique to use these data most efficiently.

For the next few years information about future climatic conditions and the hydrological cycle should possibly be used as a supplement to the available reference books, that contain generalized data of hydrometeorological observations for the last decades. As research on anthropogenic global climatic change advances this additional information could be improved and worked out in detail.

As already mentioned, we cannot abandon the practical use of the available information on future climatic changes. This would imply that long-range planning would use the hypothesis that modern climatic conditions will be preserved in the future, an idea that is currently not supported by the leading world climatologists. It is easy to understand that if drastic climatic changes are likely, assuming that it is constant can lead to disastrous consequences for development of some economic sectors.

In conclusion we note that for a long time data on past climatic conditions obtained from the meteorological network were viewed as an indefinite forecast of future climatic conditions and were used to solve all the practical climate-dependent problems. Recent studies have shown that this approach to estimating future climate is incorrect. Modern climate already differs noticeably from mean climatic conditions of the past decades, and this difference is rapidly increasing with time. As a result, the problem of determining future climatic conditions becomes the focus of hydrometeorological sciences.

BIBLIOGRAPHY

Adam, R., & West, G. J. 1983. Temperature and precipitation estimates through the last cycle from Clear Lake, California, pollen data. *Science* 219 (4581):168–170.

Agayan, G. M.; Golitsyn, G. S.; & Mokhov, I. I. 1985. Global temperature sensitivity of outgoing thermal radiation. *Bull. Acad. Sci. USSR, Atm. Ocea. Phys.* 21 (6):657–661. (R)

Ajtay, G. L.; Ketner, P.; & Duvigneaud, P. 1979. Terrestrial primary production and phytomass. In *The global carbon cycle.* SCOPE 13. New York: John Wiley and Sons, pp. 129–181.

Alekin, O. A., & Lyakhin, Yu. I. 1984. *Chemistry of the ocean.* Leningrad: Gidrometeoizdat. 343 pp. (R)

Alexandrov, E. L.; Karol, I. L.; Sedunov, Yu. S., et al. 1982. *Atmospheric ozone and global climatic changes.* Leningrad: Gidrometeoizdat. (R)

Alexandrov, E. L.; et al. 1982. *Atmospheric ozone and global climatic changes.* Leningrad: Gidrometeoizdat. 167 pp. (R)

Analysis and interpretation of atmospheric CO_2 data. 1981. WMO/ICSU/UNEP Conference, Bern, 14–18 September. Bern. 327 pp.

Andreae, B. 1980. Brachhaltund in der Weltlandwirtschaft. Extremfruchtfolgen in Marginalzonen des Ackerbaues. *Naturwiss. Rundschau* 34:64–70.

Angell, J. K., & Korshover, J. 1983. Global variation in total ozone and layer-mean ozone: an update through 1981. *J. Climate Appl. Meteorol.* 22 (3):1611–1625.

Aoki, M., & Yabuki, K. 1977. Studies in the carbon dioxide enrichment for plant growth. VII. Change in dry matter production and photosynthetic rate of cucumber during carbon dioxide enrichment. *Agric. Meteorol.* 18.

Apasova, Ye. G., & Gruza, G. V. 1982. *Data on climate structure and variability. Precipitation. The northern hemisphere.* Obninsk: VNIIGMI-MDC. 211 pp. (R)

Ariel, N. Z.; Byutner, E. K.; & Strokina, L. A. 1981. The calculation of the net gas exchange rate through the sea-air interface. *Bull. Acad. Sci. USSR, Atm. Ocea. Phys.* 10:1056–1064. (R)

Ariel, N. Z., & Strokina, L. A. 1982. Dynamic velocity and tangential stress over sea surface. *Meteorol. Hydrol.* 7:59–64. (R)

————. 1985. *Dynamic characteristics of the atmosphere: Sea surface interaction.* Leningrad: Gidrometeoizdat. 112 pp. (R)

————. 1986. The calculation of seasonal changes of gas-exchange rate at different latitudes. *Trans. Main Geophysical Observatory* 54:98–108. (R)

Ariel, N. Z., et al. 1979. The influence of the ocean polluted with oil film on oxygen exchange with the atmosphere. *Meteorol. Hydrol.* 2:57–66. (R)

Armentano, T. V., & Hett (eds.). 1980. *The role of temperate zone forests in the world carbon cycle. Problem definition and research needs.* Washington, D.C.: U.S. Dept. of Energy Publ. 69 pp.

Armentano, T. V., & Ralston, C. W. 1980. The role of temperate zone forests in the global carbon cycle. *Can. J. Forest Res.* 10:53–60.

Arrhenius, S. 1986. On the influence of the carbonic acid in the air upon the temperature of the ground. *Phil. Mag.* 41:237–275.

————. 1908. *Das Werden der Welten.* Leipzig. 108 pp.

Augustsson, T., & Ramanathan, V. 1977. A radiative-convective model study of the CO_2 climate problem. *J. Atm. Sci.* 34 (3):448–451.

Ausubel, J. H. 1983. Can we assess the impacts of climatic changes? *Clim. Change* 5 (1):7–14.

Axelrod, D. I., & Baily, H. P. 1969. Paleotemperature analysis of Tertiary floras. *Palaeogeogr., Palaeoclim., Palaeoecol.* 6.

Babaev, N. S., et al. 1984. Nuclear power, man, and the environment. Moscow: Energoatomizdat. (R)

Bacastow, R. B.; Keeling, C. D.; & Whorf, T. P. 1981. Seasonal amplitude in atmospheric CO_2 concentration at Mauna Loa, Hawaii, 1959–1980, and at Canadian Weather Station P, 1970–1980. In *Analysis and interpretation of atmospheric CO_2 data.* Bern: Proc. Sci. Conf. Bern, 14–18 Sept., pp. 169–176.

Bach, W.; Jung, H.-J.; & Knottenberg, H. 1985. Modeling the influence of carbon dioxide on the global and regional climate. Methodology and results. *Münstersche geogrph. Arbeit* 21. 114 pp.

Balsam, W. H. 1981. Late Quaternary sedimentation in the western North Atlantic: Stratigraphy and paleoceanography. *Palaeogeogr., Palaeoclim., Palaeoecol.* 35 (2–4).

Bandy, O. L.; Casey, R. E.; & Wright, R. C. 1971. Late Neogene planktonic zonation, magnetic reversals and radiometric dates. Antarctic to tropics. *Antarc. Oceanol.* 1:1–26.

Barash, M. S. 1985. Reconstruction of Quaternary oceanic palaeotemperatures from planktonic foraminifera. In *Methods for palaeoclimatic reconstructions.* Moscow: Nauka, pp. 134–142. (R)

Barnett, T. P. 1984. The estimation of "global" sea level change: A problem of uniqueness. *J. Geophys. Res.* 89 (C5):7980–7988.

————. 1986. Long-term changes in precipitation records. In *U. S. Dept. of Energy state of the art report on the detection of climate changes.* Washington, D. C.: U. S. Dept. of Energy.

Barron, E. J., & Washington, W. M. 1984. The role of geographic variables in explaining paleoclimates: Results from Cretaceous climate model sensitivity studies. *J. Geophys. Res.* 89 (C1): 1267–1279.

Bartlein, P. J.; Webb, Th., III; & Fleri, E. 1984. Holocene climatic changes in the northern Midwest: Pollen-derived estimates. *Quatern. Res.* 22 (3):361–374.

Basic hydrological characteristics. 1966–80. Leningrad: Gidrometeoizdat. (R)

Bazilevitch, N. I.; Rodin, L. Ye.; & Rozov, N. N. 1970. Geographical aspects of studying biological productivity. In *Materials of the Vth Congress of the USSR Geograph. Society.* Leningrad: Izd. Geogr. Obschestva SSSR. 27 pp. (R)

Bé, A. W. H. 1977. An ecologic, zoogeographic and taxonomic review of recent planktonic foraminifera. In *Oceanic micropaleontology*, ed. A. T. S. Ramsey, vol. 1. New York: Academic Press.

Berger, A. 1978. Long-term variations of caloric insolation resulting from the Earth's orbital elements. *Quatern. Res.* 9 (2):138–167.

Berger, et al. (eds.). 1984. *Milankovich and climate.* Dordrecht: Reidel.

Berger, W. H. 1982. Increase of carbon dioxide in the atmosphere during deglaciation: The coral reef hypothesis. *Naturwissenschaften* 69.

Barry, J. A., & Raison, J. K. 1981. Responses of macrophytes to temperature. In *Encyclopaedia of plant physiology.* Berlin: Springer-Verlag, vol. 12A.

Bierly, E. W. 1980. Some aspects of economical consequences of atmospheric CO_2 concentration variations. In *Problems of atmospheric carbon dioxide. Proc. Soviet-Amer. Symp.* Leningrad: Gidrometeoizdat, pp. 198–222. (R)

Biske, S. F. 1981. Palaeogene climates in the northeast of Siberia. *Geol. Geophys.* 1:20–26. (R)

Blasing, T. J., & Solomon, A. N. 1983. *Response of North America corn belt to climatic warming.* DOE/NBB-004. Washington, D. C.

Blum, N. S. 1984. Palaeotemperature reconstruction by Pleistocene planktonic foraminifera from different regions of the world ocean. Ph.D. diss. Leningrad. 24 pp. (R)

Bodhaine, B. A. 1983. Aerosol measurements at four background sites. *J. Geophys. Res.* 88 (C15):10753–10768.

Bogorov, V. G. 1974. *Plankton of the world ocean.* Moscow: Nauka. 320 pp. (R)

Bohn, H. L. 1976. Estimate of organic carbon in world soils. *J. Soil Sci.* 40 (3):468–470.

———. 1978. On organic soil carbon and CO_2. *Tellus* 30:472–475.

Bolin, B., 1977. Changes of land biota and their importance for the carbon cycle. *Science* 196:613–615.

———. 1978. Modeling the global carbon cycle. In *Carbon dioxide, climate and society*, ed. J. Williams. New York: Pergamon Press, pp. 41–43.

Bolin, B., & Bishoff, W. 1970. Variation of the carbon dioxide content of the atmosphere in the northern hemisphere. *Tellus* 22:431.

Bolin, B., & Erikson, E. 1959. Changes in the carbon content of the atmosphere and the sea due to fossil fuel combustion. In *The atmosphere and ocean in motion*, ed. B. Bolin. New York: Rockfeller Inst. Press, pp. 130–142.

Bolin, B., et al. (eds.). 1986. *The greenhouse effect, climatic change and ecosystems.* SCOPE 29. New York: John Wiley and Sons. 541 pp.

Bondarenko, I. F., et al. 1982. *Modeling the agroecosystem's productivity.* Leningrad: Gidrometeoizdat. 264 pp. (R)

Borisenkov, E. P., & Altunin, I. V. 1983. Simulation of carbon cycle in the atmosphere-ocean-biosphere system by linear and diffusion models. *Meteorol. Hydrol.* 3:57–63. (R)

Bortkovsky, R. S. 1983. *The atmosphere-ocean heat and moisture exchange during storms.* Leningrad: Gidrometeoizdat. 158 pp. (R)

Borzenkova, I. I. 1980. Moisture fluctuations in the Sahara and adjacent regions during the last 20 Ka. *Bull. Acad. Sci. USSR, Ser. Geogr.* 3:36–46. (R)

———. 1981. On the global temperature trend throughout the Cenozoic. *Meteorol. Hydrol.* 12:25–35. (R)

———. 1987. On land moisture content of the northern hemisphere with global warming (from palaeoclimatic data). *Meteorol. Hydrol.* 10. (R)

Borzenkova, I. I., & Zubakov, V. A. 1984. The Holocene climatic optimum as a global climate model in the early 21st century. *Meteorol. Hydrol.* 8:69–77. (R)

———. 1985. Pliocene climatic optimum as an analogue of the mid-21st century climate. *Trans. State Hydrol. Inst. (GGI)* 339:93–118. (R)

Borzenkova, I. I., et al. 1976. Air temperature change in the northern hemisphere 1881–1975. *Meteorol. Hydrol.* 7:27–35. (R)

Bowen, D. 1981. *Quaternary geology.* Moscow: Mir. 272 pp. (R)

Bowen, H. J. M. 1966. *Trace elements in biochemistry.* Orlando, Fla.: Academic Press. 241 pp.

Bradley, R. S., & Jones, P. D. 1985. Data bases for isolating the effects of the increasing carbon dioxide concentration. In *Detecting the climatic effects of increasing carbon dioxide.* Washington, D.C.: U.S. Dept. of Energy, pp. 29–53.

Brinkman, W. A. R. 1980. Growing season length as an indicator of climatic variations. *Clim. Change* 2 (2):127–138.

Broecker, W. S. 1974. *Chemical oceanography.* Ed. R. S. Deffeyes. Princeton. 212 pp.

———. 1984. Carbon dioxide circulation through ocean and atmosphere. *Nature* 308:602.

Broecker, W. S., & Peng, T.-H. 1982. *Tracers in the sea.* A publication of the Lamont-Doherty Geological Observatory. New York: Columbia Univ. 689 pp.

Brown, S., & Lugo, A. E. 1980. Preliminary estimate of the storage of organic carbon on tropical forest ecosystem. CONF-800350. Washington, D.C.: U.S. Dept. of Energy Publ., pp. 118–139.

———. 1982. The storage and production of organic matter in tropical forests and their role in the global carbon cycle. *Biotropica* 14:161–187.

———. 1984. Biomass of tropical forests: New estimate based on volumes. *Science* 223:1290–1293.

Bryan, K., et al. 1982. Transient climatic response to increasing atmospheric carbon dioxide. *Science* 214:56–58.

Bryson, R. A., & Goodman, B. M. 1980. Volcanic activity and climatic changes. *Science* 207:1041–1044.

Bryson, R. A., & Swain, A. M. 1981. Holocene variations of monsoon rainfall in Rajasthan. *Quatern. Res.* 16:135–145.

Buchardt, B. 1978. Oxygen isotope paleotemperatures from the Tertiary period in the North Sea area. *Nature* 275:121–123.

Budyko, M. I. 1946. Methods for determining natural evaporation. *Meteorol. Hydrol.* 3:3–15. (R)

————. 1950. Climatic factors of external physiographical process. *Trans. Main Geophysical Observatory (MGO)* 19.

————. 1956. *Heat balance of the Earth's surface.* Leningrad: Gidrometeoizdat. 255 pp. (R)

————. 1962. Some ways of influencing the climate. *Meteorol. Hydrol.* 2:3–8. (R)

————. 1964. On the theory of the influence of climatic factors on photosynthesis. *Rep. Acad. Sci. USSR* 158 (2):331–337. (R)

————. 1968. On the origin of ice ages. *Meteorol. Hydrol.* 11:3–12. (R)

————. 1969. *Climatic changes.* Leningrad: Gidrometeoizdat. 35 pp. (R).

————. 1971. *Climate and life.* Leningrad: Gidrometeoizdat. 472 pp. (R) (Eng. translation by D. H. Miller. 1974. New York: Academic Press.)

————. 1972. *Man's impact on climate.* Leningrad: Gidrometeoizdat. 47 pp. (R)

————. 1974. *Climatic change.* Leningrad: Gidrometeoizdat. 280 pp. (R) (Eng. translation: Washington, D.C.: American Geophysical Union.)

————. 1975. The dependence of mean air temperature on solar radiation changes. *Meteorol. Hydrol.* 10:3–10.

————. 1977a. *Global ecology.* Moscow: Mysl. 328 pp. (R) (Eng. translation: Moscow: Progress Publishers.)

————. 1977b. Studies of modern climatic change. *Meteorol. Hydrol.* 11:42–57. (R)

————. 1977c. *Present-day climate change.* Leningrad: Gidrometeoizdat. 47 pp. (R)

————. 1979a. Semi-empirical model of atmospheric thermal conditions and realistic climate. *Meteorol. Hydrol.* 4:5–17 (R)

————. 1979b. *Carbon dioxide problem.* Leningrad: Gidrometeoizdat. 47 pp. (R)

————. 1980. *The Earth's climate past and future.* Leningrad: Gidrometeoizdat. 352 pp. (R) (Eng. translation: 1980. New York: Academic Press.)

————. 1982. *Carbon dioxide problem.* Leningrad: Gidrometeoizdat. 58 pp. (R)

————. 1984. *The evolution of the biosphere.* Leningrad: Gidrometeoizdat. 488 pp. (R) (Eng. translation: 1985. Dordrecht: Reidel.)

Budyko, M. I.; Byutner, E. K.; & Vinnikov, K. Ya. 1986. Detecting anthropogenic climate change. *Meteorol. Hydrol.* 12:5–16. (R)

Budyko, M. I., & Drozdov, O. A. 1976. On the causes of water cycle changes. *Water Res.* 6:35–44. (R)

Budyko, M. I.; & Gates, W. L.; et. al. (eds.). 1984. Anthropogenic climate changes. *Meteorol. Hydrol.* 6:117–122. (R)

Budyko, M. I.; Golitsyn, G. S.; & Izrael, Yu. A. 1986. *Global climatic catastrophes.* Leningrad: Gidrometeoizdat. 156 pp. (R) (Eng. translation: Springer-Verlag.)

Budyko, M. I., & Ronov, A. B. 1979. The evolution of the atmosphere in the Phanerozoic. *Geochimiya* 6:643–653. (R)

Budyko, M. I.; Ronov, A. B.; & Yanshin, A. L. 1985. *History of the Earth's atmosphere.* Leningrad: Gidrometeoizdat. 209 pp. (R) (Eng. translation: Springer-Verlag 1987.)

Budyko, M. I., & Vinnikov, K. Ya. 1973. Present-day changes in climate. *Meteorol. Hydrol.* 9:3–13. (R)

———. 1976. Global warming. *Meteorol. Hydrol.* 7:16–26. (R)

———. 1953. The problem of detecting anthropogenic changes in global climate. *Meteorol. Hydrol.* 9:14–26. (R)

Budyko, M. I.; Vinnikov, K. Ya.; & Yefimova, N. A. 1983. The dependence of air temperature and precipitation on carbon dioxide content of the atmosphere. *Meteorol. Hydrol.* 4:5–13. (R)

Budyko, M. I., & Yefimova, N. A. 1981. Carbon dioxide effects on climate. *Meteorol. Hydrol.* 2:5–17. (R)

———. 1984. Annual variations of meteorological elements as a model of climate changes. *Meteorol. Hydrol.* 1:5–10. (R)

Budyko, M. I., et al. 1978. Expected climate changes. *Bull. Acad. Sci. USSR, Ser. Geogr.* 6:5–10. (R)

Budyko, M. I., et al. 1981. Anthropogenic climatic changes. *Meteorol. Hydrol.* 8:5–14.

Bukshtynov, A. D. 1959. *Forest resources of the USSR and the world.* Moscow. 62 pp. (R)

Bukshtynov, A. D.; Groshev, B. I.; & Krylov, G. V. 1981. *Forests.* Moscow: Mysl. 316 pp. (R)

Burashnikova, G. A.; Muratova, M. V.; Suyetova, I. A. 1982. Climatic model of the USSR territory throughout the Holocene optimum. In *Development of nature in the USSR territory in the Late Pleistocene and Holocene.* Moscow: Nauka. pp. 245–251. (R)

Burgos, J. 1979. Renewable resources and agriculture in Latin America in relation to the stability of climate. In *Proc. World Climate Conf.* Geneva: WMO.

Büringh, P. 1984. Organic carbon soils of the world. In *The role of terrestrial vegetation in the global carbon cycle. Measurement by remote sensing*, ed. G. M. Woodwell. SCOPE 23. New York: John Wiley and Sons, pp. 117–126.

Büringh, P.; Ven Heemst, H.; & Staring, G. 1975. *Computation of the absolute maximum food production of the world.* Wageningen: Agricul. Univ. 55 pp.

Burnett, C. R., & Burnett, E. B. 1984. Observational results on the vertical column abundance of atmospheric hydroxyl: Description of its seasonal behavior, 1977–1982, and of the 1982 El Chichon perturbation. *J. Geophys. Res.* 89 (D6):9603–9611.

Byutner, E. K. 1983. Ocean's response to the change of atmospheric greenhouse effect. *Bull. Acad. Sci. USSR, Atm. Ocea. Phys.* 19 (8):892–895. (R)

———. 1986. *Planetary O_2 and CO_2 gas exchange.* Leningrad: Gidrometeoizdat. 232 pp. (R)

———. 1987. The role of the ocean in anthropogenic climatic changes. In *Theses of papers at the All-Union workshop "Actual problems in oceanology."* Leningrad: Gidrometeoizdat. 132 pp. (R)

Byutner, E. K., & Shabalova, M. V. 1985. Seasonal variations of surface air temperature with atmospheric transparency changes. *Meteorol. Hydrol.* 6:5–12. (R)

Byutner, E. K., & Zakharova, O. K. 1983. Modeling carbon cycle by the method of fractional derivatives. *Meteorol. Hydrol.* 4:14–20. (R)

Byutner, E. K.; Zakharova, O. K.; & Lapenis, A. G. 1986. On the estimation of the preindustrial CO_2 concentration in the Earth's atmosphere. *Bull. Acad. Sci. USSR, Atm. Ocea. Phys.* 22 (8):796–803. (R)

Byutner, E. K., et al. 1981. Anthropogenic changes of atmospheric CO_2 concentration in the nearest 50 years. *Meteorol. Hydrol.* 3:18–31. (R)

Byutner, E. K., et al. 1984. The influence of biomass changes on carbon cycle. *Meteorol. Hydrol.* 3:56–63. (R)

Byutner, E. K., et al. 1985. The analysis of the trend of the amplitudes of atmospheric CO_2 annual variations. *Meteorol. Hydrol.* 2:56–63. (R)

Callendar, G. S. 1938. The artificial production of carbon dioxide and its influence on temperature. *Quart. J. Roy. Met. Soc.* 64 (27):223–240.

The carbon cycle and atmospheric CO_2: Natural variations Archean to present. 1985. *Geophys. Monogr.* 32. 317 pp.

Carbon dioxide and climate: A scientific assessment. 1979. Washington, D.C.: NAS. 22 pp.

Carbon dioxide and climate: A second assessment. 1982. Washington, D.C.: NAS. 72 pp.

Carbon dioxide and climate: Australian research. 1980. Canberra: Australian Academy of Science. 320 pp.

Causes of modern climate change. 1987. *Meteorol. Hydrol.* 1. (R)

Cauwet, G. 1978. Organic chemistry of sea water particulates: Concepts and developments. *Oceanol. Acta* 1:99–105.

Cess, R. D. 1976. Climate change: An appraisal of atmospheric feedback mechanisms employing zonal climatology. *J. Atm. Sci.* 33 (7):1831–1843.

Cess, R. D., & Goldenberg, S. D. 1981. The effect of ocean heat capacity upon global warming due to increasing atmospheric CO_2. *J. Geophys. Res.* 85:498–504.

Cess, R. D., & Hameed, S. 1983. Methane: Greenhouse effect influence on its biospheric sources intensity. *Tellus* 35b (1):1–7.

Cess, R. D., et. al. 1982. Low-latitude cloudiness and climate feedback: Comparative estimates from satellite data. *J. Atm. Sci.* 39 (1):53–59.

Chan, J. H., & Olson, J. S. 1980. Limits on the organic storage of carbon from burning fossil fuels. *J. Environ. Management* 11:147–163.

Chang, C. W. 1975. Carbonic anhydrase and senescence in cotton plants. *Plant Physiol.* 55:515–519.

Changing climate. 1983. Washington, D.C.: NAS. 496 pp.

Chant, V. G. 1981. *Two global scenarios: The evolution of energy use and the economy to 2030.* Laxenburg.

Chen, R. S. 1981. Interdisciplinary research and integration: The case of CO_2 and climate. *Clim. Change* 3 (4): 429–447.

Chepalyga, A. L. 1980. Palaeogeography and palaeoecology of the Black and Caspian seas (Ponto-Caspian) throughout the Plio-Pleistocene. Ph.D. diss. Moscow Inst. Geogr. Acad. Sci. USSR. 48 pp. (R)

Chirkov, Yu. I. 1969. *Agrometeorological conditions and corn productivity.* Leningrad: Gidrometeoizdat. 251 pp. (R)

Chlorofluorocarbons and their effect on stratospheric ozone. 1976. Pollution Paper, no. 5, London.

Chlorofluorocarbons and their effect on stratospheric ozone. Second report. 1979. Pollution Paper, no. 15, London.

Cicerone, R. T.; Walters, S.; & Liu, S. C. 1983. Nonlinear response of stratospheric ozone column to chlorine injections. *J. Geophys. Res.* 88 (66):3647–3661.

Ciesielsky, P. T., & Weaver, T. M. 1974. Early Pliocene temperature changes in the Antarctic seas. *Geol.* 2 (10):511–515.

Clark, W. C., et al. 1982. The carbon dioxide question: Perspectives for 1982. In *Carbon dioxide review.* Oxford and New York: Clarendon Press, pp. 3–54.

CLIMAP Project Members. 1976. The surface of the ice-age Earth. *Science* 191:1131–1138.

———. 1984. The last interglacial ocean. *Quart. Res.* 21 (2):123–224.

Climate change to the year 2000. 1978. Washington, D.C.: Nat. Def. Univ. 109 pp.

Climatic effects of increased atmospheric carbon dioxide. 1982. Proc. Soviet-American meeting on studying climatic effects of increased atmospheric carbon dioxide. Leningrad, June 15–20, 1981. Leningrad: Gidrometeoizdat. 56 pp. (R)

Cooper, C. F. 1982. Food and fiber in a world of increasing carbon dioxide. In *Carbon dioxide review.* New York: Oxford Univ. Press.

Crop yields and climate change: The year 2000. 1978. Selected results from a study conducted by the research directorate of the National Defense University. Washington, D.C.: Fort Lesley J. McNair. 56 pp.

Crop yields and climate change to the year 2000. 1980. Vol. 1. Rept. on the second phase of a climate impact assessment. Washington, D.C.: Fort Lesley J. McNair. 128 pp.

Crutzen, P. J., & Gidel, L. T. 1983. A two-dimensional photochemical model of the atmosphere. 2. The tropospheric budgets of the anthropogenic chlorocarbons, CO, CH_4, CH_3Cl and the effect of various NO_x sources on tropospheric ozone.

Dansgaard, W., et al. 1971. Climatic record revealed by the Camp Century ice core. In *Late Cenozoic glacial ages.* New Haven: Yale Univ. Press, pp. 37–56.

Davis, O. K. 1984. Multiple thermal maximum during the Holocene. *Science* 225 (4662):617–619.

Delcourt, P. A., & Delcourt, H. R. 1983. Late Quaternary vegetational dynamics and community stability reconsidered. *Quantern. Res.* 17 (2):265–271.

Delmas, R.; Ascencio, J.-M.; & Legrand, M. 1980. Polar ice evidence that atmospheric CO_2 20,000 yr B.P. was 50% of present. *Nature* 284:155–157.

Delmas, R., & Boutron, C. 1980. Are the past variations of the stratospheric sulfate burden recorded in central Antarctic snow and ice layers? *J. Geophys. Res.* 85 (C10):5645–5649.

Demchenko, P. F. 1981. Simple statistical model for describing spatial-temporal dependences of mid-latitudinal temperature variations. *Bull. Acad. Sci. USSR, Atm. Ocea. Phys.* 17 (8):805–813. (R)

———. 1982. Estimates of mean hemispheric temperature variance by satellite observations of radiation-balance fluctuations. *Bull. Acad. Sci. USSR, Atm. Ocea. Phys.* 18 (2):138–144. (R)

———. 1984. Analytical model of latitudinal variations of variance and spectra of zonally averaged surface air temperature fluctuations. *Bull. Acad. Sci. USSR, Atm. Ocea. Phys.* 20 (2):144–150. (R)

Demography situation in the countries of the world. Reference book. 1984. Finance and statistics. 446 pp. (R)

Desert dust. 1981. Geol. Soc. Amer. Special Paper, no. 186. 353 pp.

Diester-Haas, L. 1976. Late quaternary climatic variations in Northwest Africa deduced from East Atlantic sediment cores. *Quatern. Res.* 6 (2):299–324.

Dignon, T., & Hameed, S. 1986. A model investigation of the impact of increases in anthropogenic NO_x emissions between 1967 and 1980 on tropospheric ozone. *J. Atm. Chem.*

Dobrodeev, O. P., & Suyetova, I. A. 1976. Living matter of the Earth. In *Problems of general physiography and palaeogeography.* Moscow, pp. 26–58. (R)

Douglas, R. G., & Savin, S. M. 1973. Oxygen and carbon isotope analysis of Tertiary and Cretaceous microfossils from Shatcky Rise and other sites in the North Pacific Ocean. In *Initial reports of the deep sea drilling project*, vol. 32, pp. 509–520.

Downton, W. J. S.; Björkman, O.; & Pike, C. S. 1980. Consequences of increased atmospheric concentration of carbon dioxide for growth and photosynthesis of higher plants. In *Carbon dioxide and climate: Australian research.* Canberra: Australian Academy of Science, pp. 143–151.

Drozdov, O. A. 1966. On changing precipitation in the northern hemisphere with varying temperature of the polar basin. *Trans. Main Geophysical Observatory* 198:3–16. (R)

———. 1981. The formation of land moisture with climatic changes. *Meteorol. Hydrol.* 4:17–28. (R)

———. 1983. On the climate time structure in the Holocene and modern epoch. *Bull. Acad. Sci. USSR, Ser. Geogr.* 4:17–24. (R)

———. 1985. Investigation of the relationship between global temperature and moisture. *Trans. Main Geophysical Observatory* 317:3–22. (R)

Druffel, E. M. 1981. Atmospheric input of CO_2 to the surface ocean. In *Analysis and interpretation of atmospheric CO_2 data.* Bern: WMO/ICSU/UNEP, pp. 201–208.

Duplessy, J. C. 1980. Isotope investigations. In *Climatic Changes.* Leningrad: Gidrometeoizdat, pp. 70–201. (R)

Duplessy, J. C., et al. 1981. Deglacial warming of the northeastern Atlantic Ocean: Correlation with the paleoclimatic evolution of the European continent. *Palaeogeogr., Palaeoclim., Palaeoecol.* 35:121–144.

Dütsch, H. V. 1985. Total ozone trend in the light of ozone soundings: The impact of El Chichon. In *Atmosph. Ozone, Proc. Ozone Sympos. Greece, 1984*, pp. 342–348.

Duvigneaud, P., & Tang, E. 1968. *The biosphere and man's place in it.* Moscow: Progress. 254 pp. (R)

Dvoryashina, E. V.; Dianov-Klokov, V. I.; & Yurganov, L. F. 1984. On variations in carbon monoxide content of the atmosphere, 1970–1982. *Bull. Acad. Sci. USSR, Atm. Ocea. Phys.* 20 (1):40–47. (R)

Dvoryashina, E. V., et al. (eds.). 1985. Carbon dioxide content of the atmosphere in different regions of the world (1970–1984). Preprint. Moscow: Inst. Phys. Atm. Acad. Sci. USSR. 38 pp. (R)

Dymnikov, V. P., et al. 1980. The study of climatic sensitivity to doubling CO_2 by mean-zonal atmospheric general circulation model. In *Mathematical modeling of the dynamics of the atmosphere and ocean*, vol. 2. Novosibirsk, pp. 39–50. (R)

Dzerdzeevsky, B. L. 1975. *Atmospheric general circulation and climate.* Moscow: Nauka. 285 pp. (R)

Edmonds, J., & Reilly, J. M. 1985. Future global energy and CO_2 emissions. In *Atmospheric CO_2 and the global carbon cycle.* Y. Trabalka, Ed. Pages 215–246.

Edmonds, J., & Reilly, S. 1983a. A long-term global energy-economic model of carbon dioxide release from fossil fuel use. *Energy Economics* 5 (2):74–88.

———. 1983b. Global energy and CO_2 to the year 2050. *Energy Journal* 4 (3):21–47.

Edmonds, J., et al. 1984. *An analysis of possible future atmospheric retention of fossil fuel CO_2.* Report No. DEO/OR/21400-1. Washington, D.C.

Edmonds, J., et al. 1987. Special Report DOE/ER-0239. Washington, D.C.

Edvard, E., & David, J. R. 1982. Strategies in fossil fuel technology: Multiple options for unpredicted futures. In *Energy, Resources and Environment.* New York: Pergamon Press.

The effects on the atmosphere of a major nuclear exchange. 1985. Washington, D.C.: Nat. Acad. Sci. Press. 193 pp.

Eicher, U.; Siegenthaler, U.; & Wegmüller, S. 1981. Pollens and oxygen isotope analysis on late and post-glacial sediments of the Tourbiere de Chirens. *Quatern. Res.* 15 (2): 160–170.

Elliot, W. P.; Machta, L.; & Keeling, C. D. 1985. An estimate of the biotic contribution to the atmospheric CO_2 increase based on direct measurements at Mauna Loa Observatory. *J. Geophys. Res.* 90 (D2):3741–3746.

Emiliani, C. 1966. Isotopic paleotemperature. *Science* 154:851–857.

Energy and climate. 1977. Washington, D.C.: NAS. 158 pp.

Enting, I. G.; Pittock, A. B.; & Pearman, G. I. 1984. Comments on "Solar, volcanic and CO_2 forcing of recent climatic changes." *Clim. Change* 6 (4):397–405.

Fabian, P., et al. 1981. Halocarbons in the stratosphere. *Nature* 294 (5843):733–735.

Fairbridge, R. W. 1961. Eustatic changes in sea level. In *Physics and chemistry of the Earth*, vol. 4. New York: Pergamon Press, pp 99–185.

Farman, J. C., et al. 1985. Large losses of total ozone in Antarctic reveal seasonal $C10_x/NO_x$ interaction. *Nature* 315.

Fate of fossil fuel CO_2 in the oceans. 1977. New York: Plenum Press. 780 pp.

Fedorov, E. K. 1979. Climatic changes and the strategy of mankind. *Meteorol. Hydrol.* 7:12–24. (R)

Feister, V., & Warmbt, W. 1985. Long-term surface ozone increase at Arkona (54, 7°N; 134°E). In *Atmosph. Ozone, Proc. Ozone Sympos. Greece, 1984*, pp. 782–787.

Fels, S. B., et al. 1980. Stratospheric sensitivity to perturbations in ozone and carbon dioxide: Radiative and dynamical response. *J. Atm. Sci.* 37:2265–2297.

Feygelson, F. M., & Dmitrieva, L. R. (eds.). 1983. Radiation algorithms in the general atmospheric circulation models. Review information. *Hydrometeorology* 1. Obninsk: VNIIGMI-MDC. 77 pp. (R)

Fisher, B. E. A., & MacQueen, J. F. 1980. The influence of the oceans on the atmospheric burden of carbon dioxide. *Applied Math. Model.* 4:439–448.

Flohn, H. 1981. A climatic feedback mechanism involving oceanic upwelling, atmospheric CO_2 and water vapour. In *Variations in global water budget*, ed. Street-Perrott, Beran & Ratcliffe. Dordrecht: Reidel, pp. 403–414.

————. 1981. *Life on a warmer Earth*. IIASA. 75 pp.

Fogner, H.-H. (ed.). 1983. *IIASA '83 scenario of energy development: Summary*. Laxenburg.

Folland, C. K.; Parker, D. E.; & Kates, F. E. 1984. World marine temperature fluctuations, 1856–1981. *Nature* 670–673.

Food needs of developing countries. Projection of production and consumption to 1990. 1977. Washington, D.C.: NAS. 157 pp.

Frakes, L. A. 1984. The Mesozoic-Cenozoic history of climate change and cause of glaciation. In *27th Int. Geol. Congr. Abstr.* 9 (1):207–208.

Freyer, H. D. 1979a. Atmospheric cycles of trace gases containing carbon. In *The global carbon cycle*, SCOPE 13. New York: John Wiley and Sons, pp. 101–128.

————. 1979b. On the ^{13}C record in tree rings. I. ^{13}C variations in northern hemispheric trees during the last 150 years. *Tellus* 31:124–137.

Freyer, H. D., & Belacy, N. 1983. $^{13}C/^{12}C$ records in northern hemispheric trees during the past 500 years: Anthropogenic impact and climatic superpositions. *J. Geophys. Res.* 88 (C11):6844–6852.

Fukui, H. 1979. Climatic variability and agriculture in tropical moist regions. In *Proc. World Climate Conf.*, Geneva, WMO, February 1979.

Galimov, E. H. 1981. *Nature of biological fractionation of isotopes*. Moscow: Nauka. 245 pp. (R)

Galtsov, A. P. 1961. Meeting on the problem of climate transformation. *Bull. Acad. Sci. USSR, Ser. Geogr.* 5:128–133. (R)

Gammon, R. H.; Cline, J. D.; & Wisegarver, D. J. 1982. Chlorofluoromethanes in the Pacific Ocean. Measured vertical distributions. *J. Geophys. Res.* 87 (C12):9441–9454.

Gammon, R. H., & Steele, L. P. 1984. Global distribution of atmospheric methane determined from the NOAA/GMCC Flask Network. In *Proc. USA/USSR Symp. on Atmosph. Trace Gases Effect Climate*. Vilnius, pp. 49–65.

Garrels, R. M. *Carbon, oxygen and sulphur cycles throughout geological time*. Moscow: Nauka. 48 pp. (R)

Gates, W. C. 1976. Modeling the ice-age climate. *Science* 191: 1138–1144.

Geochronology of the USSR. 1974. Vol. 3. The latest stage. Leningrad: Nedra. 358 pp. (R)

Geophys. monitoring climatic change, 1984–1986. Summ. Rep., no. 12. Boulder, Colo.: NOAA. 184 pp.

Gerasimov, I. P. 1979. Climates of the past geologic epochs. *Meteorol. Hydrol.* 7. (R)

Gerasimov, I. P., & Chichagova, O. A. 1971. Some problems of radiocarbon dating of soil humus. *Potchvovedenie* 10:3–11. (R)

Gibert, G. 1984. La masse forestière congolaise son implantation, ses divers facies. *Bois et forêts des tropiques* 204:3–19.

Gifford, R. M. 1979a. Carbon dioxide and plant growth under water and light stress: Implication for balancing the global carbon budget. *Search* 10:316–318.

———. 1979b. Growth and yield of CO_2-enriched wheat under water-limited conditions. *Aust. J. Plant Physiol.* 6:367–378.

———. 1980. Carbon storage by the biosphere. In *Carbon dioxide and climate: Australian research.* Canberra: Australian Academy of Science, pp. 167–181.

Gilliland, R. L. 1982. Solar, volcanic and CO_2 forcing of recent climatic changes. *Clim. Change* 4 (2):111–131.

Gilliland, R. L., & Schneider, S. H. 1984. Volcanic, CO_2 and solar forcing of northern and southern hemisphere surface air temperatures. *Nature* 310:38–41.

Gilmanov, T. G., & Bazilevitch, N. I. 1983. Quantitative estimate of the sources of humus formation of Russian chernozem. *Bull. Moscow Univ.*, Ser. 17, *Phochvovedenie* 1:9–16. (R)

Girs, A. A. 1974. *Microcirculation method for long-term meteorological forecasts.* Leningrad: Gidrometeoizdat. 488 pp. (R)

Glanz, M. 1979. A political view of CO_2 . *Nature* 280:189.

Global impacts of climate change to the year 2000. 1983. Washington, D.C.: Fort Lesley J. McNair.

Golbert, A. V., et al. 1977. *Palaeoclimates of Siberia during the Cretaceous and Palaeogenic.* Moscow: Nedra. 107 pp. (R)

Goldenberg, J., et al. 1984. Energy for a sustainable world.

Golitsyn, G. S., & Demchenko, P. F. 1980. Statistical properties of a simple energy balance climate model. *Bull. Acad. Sci. USSR, Atm. Ocea. Phys.* 16 (12):1235–1242. (R)

Golitsyn, G. S., & Mokhov, I. I. 1978. Estimates of sensitivity and role of cloudiness in simple climate models. *Bull. Acad. Sci. USSR, Atm. Ocea. Phys.* 14 (5):569–576. (R)

Gonchar-Zaikin, P. P., & Shuravlev, O. S. 1979. A simplified model of the dynamics of humus content of the soil. In *Theoretical foundations and quantitative methods for forecasting crop yields.* Leningrad: Agro-physical Institute, pp. 156–165. (R)

Goodman, H. S. 1980. The $^{13}C/^{12}C$ ratio of atmospheric carbon dioxide at the Australian baseline station, Cape Crim. In *Carbon dioxide and climate: Australian research.* Canberra: Australian Academy of Science, pp. 111–114.

Goody, R. 1958. *Physics of the stratosphere.* Moscow: Mir. 112 pp. (R)

Gornitz, V.; Lebedeff, S.; & Hansen, J. 1982. Global sea level. Trend in the past century. *Science* 215:1611–1614.

Gorshkov, S. P.; Sushckevsky, A. G.; & Shenderuk, G. N. 1980. Organic matter cycle. In *A cycle of a substance in nature and its anthropogenic change.* Moscow, pp. 153–181. (R)

Goudriaan, J., & Ajtay, G. 1979. The possible effects of increased CO_2 on photosynthesis. In *The global carbon cycle.* SCOPE 13. New York: John Wiley and Sons, chap. 8.

Goudriaan, J., & de Ruiter, H. E. 1983. Plant response to CO_2 enrichment, at two levels of nitrogen and phosphorus supply. 1. Dry matter, leaf area and development. *Neth. J. Agric. Sci.* 31:157–169.

Grassl, H., et al. 1984. CO_2 , Kohlenstoff-Kreislauf und Klima. *Naturwissenschaften* 71:129–136.

Grichuk, V. P. 1961. Experience of reconstructing some climate elements in the northern hemisphere throughout the Atlantic period of the Holocene. In *The Holocene.* Moscow: Nauka, pp. 41–57. (R)

————. 1981. The most ancient continental glaciation in Europe: Its indications and stratigraphic position. In *Problems of the Pleistocene palaeogeography of glacial and periglacial regions.* Moscow: Nauka, pp. 7–35. (R)

————. 1985. Reconstruction of scalar climatic indices from floristic materials and the estimation of its accuracy. In *Methods of palaeoclimatic reconstruction.* Moscow: Nauka, pp. 20–28. (R)

Grigoriev, A. A. 1956. *Subarctica.* Moscow: Geographgiz. 233 pp. (R)

Grinchenko, A. M.; Chesnyak, O. A.; & Chesnyak, G. Ya. 1966. The crop-made changes in physico-chemical properties of deep chernozem. *Trans. Kharkov Agric. Inst.* 49. (R)

Grinchenko, A. M.; Mukha, V. D.; & Chesnyak, G. Ya. 1979. Humus transformation with soil cultivation. *Bull. Agric. Sci.* 1:36–40. (R)

Groisman, P. Ya. 1979. Algorithm of estimating linear structural relationship between macroclimatic parameters. *Trans. State Hydrol. Inst.* 257:76–80. (R)

———. 1981. Empirical estimates of the relationship between global warming or cooling and moisture conditions in the USSR territory. *Bull. Acad. Sci. USSR, Ser. Geogr.* 5:86–95. (R)

———. 1983. On the change of some atmospheric circulation characteristics in global warming and cooling. *Meteorol. Hydrol.* 11:26–29. (R)

Grosswald, M. G., & Kotlyakov, V. M. 1978. Forthcoming climatic changes and fate of glaciers. *Bull. Acad. Sci. USSR, Ser. Geogr.* 6:21–32. (R)

Groveman, B. S., & Landsberg, H. E. 1979. *Reconstruction of northern hemisphere temperature: 1579–1880.* Publication N 79-181, Meteorology Program. Univ. of Maryland. 46 pp.

Gruza, G. V., & Rankova, E. Ya. 1980. *Structure and variability of observed climate: Air temperature in the northern hemisphere.* Leningrad: Gidrometeoizdat. 72 pp. (R)

Hafele, W. 1979. A perspective on energy system and carbon dioxide. In *Carbon dioxide, climate and society.* New York: Pergamon Press, vol. 1, pp. 13–34.

Hafele, W. (ed.). 1981. *Energy in a finite world.* Cambridge, Mass.: Ballinger.

Hall, C. S., & Cleveland, C. J. 1981. Petroleum drilling and production in the United States: Yield per effort and net energy analysis. *Science* 211:576.

Halocarbons: Effect on stratospheric ozone. 1976. Panel on atmospheric chemistry. Washington, D.C.

Hamilton, A. 1976. The significance of patterns of distribution shown by forest plants and animals in tropical Africa for the reconstruction of upper Pleistocene palaeoenvironment: A review. *Palaeoecol. of Africa* 9:63–98.

Hammer, C. U. 1984. Traces of Icelandic eruptions in the Greenland ice sheet. *Jökull* 34:51–65.

Hammer, C. U.; Clausen, H. B.; & Dansgaard, W. 1980. Greenland ice sheet evidence of post-glacial volcanism and its climatic impact. *Nature* 288 (5788):230–235.

Hampicke, U. 1979. Net transfer of carbon between the land biota and the atmosphere, induced by man. In *The global carbon cycle.* SCOPE 13. New York: John Wiley and Sons, pp. 219–236.

Hampicke, U., & Bach W. 1980. Die Rolle terrestrischer Öko-Systeme in globalen Kohlenstroffkreislauf. *Münstersche geografische Arbeiten* 6:10–24.

Hansen, J. E., et al. 1979. *Proposal for research in global carbon dioxide sources. Sink budget and climate effects.* New York: Goddard Inst. 60 pp.

Hansen, J. E., et al. 1981. Climate impact of increasing atmospheric carbon dioxide. *Science* 213 (4511):957–966.

Hansen, J., et al. 1983. Efficient three-dimensional global models for climate studies. *Mon. Wea. Rev.* 111:609–662.

Hansen, J. E., et al. 1984. Climate sensitivity: Analysis of feedback mechanisms. In *Climate processes and climate sensitivity. Geophys. Monogr.* 29 (5):130–163.

Harvey, L. D., & Schneider, S. H. 1985. Transient climate response to external forcing on 10^{0}-10^{4} year time scales. *J. Geophys. Res.* 90 (D1): Part 1, pp. 2191–2205; Part II, pp. 2207–2222.

Hasse, L., & Liss, P. S. 1980. Gas exchange across the air-sea interface. *Tellus* 32:470–481.

Headley, M., & Lanley, J.-P. 1983. Ecosystems of a tropical forest: Common features and differences. *Nature and Resources, UNESCO* 19 (1):2–19. (R)

Heath, D. F., & Schlesinger, B. M. 1985. The global response of stratospheric ozone to UV solar flux variations. In *Atmosph. Ozone, Proc. Ozone Sympos. Greece, 1984.*

Held, I. M., & Suarez, M. J. 1974. Simple albedo feedback models of the icecaps. *Tellus* 26 (5):613–629.

Herman, J. H., & McQuillan, G. J. 1985. Atmospheric chlorine and stratospheric ozone nonlinearities and trend detections. *J. Geophys. Res.* 90 (D3):5721–5732.

Heusser, C. J. 1984. Late Glacial-Holocene climate of the Lake district of Chile. *Quatern. Res.* 22 (1):77–90.

Heusser, C. J.; Heusser, L. E.; & Streeter, S. S. 1980. Quaternary temperatures and precipitation for northwest coast of North America. *Nature* 286 (5579):702–704.

Hoffert, M. I., & Flannery, B. P. 1985. Model predictions of the time-dependent response to increasing carbon dioxide. In *Projecting the climatic effects of increasing carbon dioxide.* Washington, D.C.: U.S. Dept. of Energy, pp. 149–190.

Hoffman, J. S.; Keyes, D.; & Titus, J. G. 1983. *Projecting future sea level rise. Methodology, estimates to the year 2100 and research needs.* Washington, D.C.: NAS. 121 pp.

Holloway, J. L., & Manabe, S. 1971. A global general circulation model with hydrology and mountains. *Mon. Wea. Rev.* 99 (5):335–370.

Houghton, R. A., et al. 1985. Net flux of carbon from tropical forests in 1980. *Nature* 316:617–620.

Hrubec, T. C.; Robinson, J. M.; & Donaldson, R. P. 1984. Effect of CO_2 enrichment on soybean leaf and mitochondrial respiration. *Plant Physiol. Suppl.* 75:158.

Huguet, L. 1982. Que penser de la "disparition" des forêts tropicales? *Bois et forêt des tropiques* 195:7–12.

Hunt, B. G., & Wells, N. C. 1979. An assessment of the possible future climatic impact of carbon dioxide increases. *J. Geophys. Res.* 84:787–791.

Imbrie, J.; Van Donk, J.; & Kipp, N. G. 1973. Paleoclimatic investigation of a Late Pleistocene Caribbean deep-sea core: Comparison of isotopic and faunal methods. *Quatern. Res.* 3 (1):10–38.

The impacts of climatic changes on the biosphere: Part 2. Climatic effects. 1975. CIAP Monograph 5. Washington, D.C.: NAS. 547 pp.

Inadvertent climate modification. 1971. Cambridge, Mass., and London: MIT Press. 306 pp.

Isidorov, V. A. 1985. *Organic chemistry of the atmosphere.* Leningrad: Khimia. 265 pp. (R)

Izrael, Yu. A. 1983. Ecological consequences of possible nuclear war. *Meteorol. Hydrol.* 10. (R)

Izrael, Yu. A.; Petrov, V. N.; & Severov, D. A. 1983. On the impact of atmospheric nuclear detonations on ozone content of the stratosphere. *Meteorol. Hydrol.* 9. (R)

Jäger, J., & Kellogg, W. W. 1983. Anomalies in temperature and rainfall during warm Arctic seasons. *Clim. Change* 5:39–60.

Jensen, R. E. 1976. A summary of estimated impacts of climatic change on crop productivity. In *Proc. 4th Conf., CIAP*, February 1975, Springfield, Va., pp. 87–96.

Jensen, W. F. 1978. Limits to growth in world food production. *Science* 201:317–320.

Joint WMO/ICSU/UNEP Meeting of Experts. 1981. *On the assessment of the role of CO_2 on climate variations and their impact.* Villach, Austria, November 1980. Geneva: WMO. 29 pp.

Jones, P. D. 1984. Satellite measurements of atmospheric composition: Three years observations of CH_4 and N_2). *Adv. Space Res.* 4 (4):121–130.

———. 1985. Arctic temperature 1851–1984. *Climate Monitor* 14 (2):43–50.

Jones, P. D., & Kelly, P. M. 1983. The spatial and temporal characteristics of northern hemisphere surface air temperature variations. *J. Climatol.* 3:243–252.

Jones, P. D.; Raper, S. C. B.; & Wigley, T. M. L. 1986. Southern hemisphere surface air temperature variations: 1851–1984. *J. Clim. Appl. Meteorol.* 25 (9):1213–1230.

Jones, P. D.; Wigley, T. M. L.; & Kelly, P. M. 1982. Variations in surface air temperatures: Part 1, Northern hemisphere, 1881–1980. *Mon. Wea. Rev.* 110 (2):59–70.

Jones, P. D.; Wigley, T. M. L.; & Wright, P. B. 1986. Global temperature variations between 1861 and 1984. *Nature* 322 (6078):430–434.

Jones, P. D., et al. 1986. Northern hemisphere surface air temperature variations: 1851–1984. *J. Clim. Appl. Meteorol.* 25 (2):161–179.

Kagan, B. A., & Ryabchenko, V. A. 1981. Nonlinear model of carbon cycle in the ocean. *Rep. Acad. Sci. USSR* 258 (1):212–215. (R)

———. 1982. Numerical experiments on the seasonal evolution of carbon cycle in the ocean. *Bull. Acad. Sci. USSR, Atm. Ocea. Phys.* 18 (4):373–389. (R)

Kagan, B. A.; Ryabchenko, V. A.; & Chalikov, D. V. 1979. Parametrization of the active layer in the model of large-scale interaction between the ocean and atmosphere. *Meteorol. Hydrol.* 12:67–76. (R)

Kagan, R. L. 1979. *Averaging meteorological patterns.* Leningrad: Gidrometeoizdat. 213 pp. (R)

Kagan, R. L., & Lugina, K. M. 1981. On the zonal averaging of meteorological patterns. *Trans. State Hydrol. Inst.* 271:51–61. (R)

Kaplina, T. N., & Lozhkin, A. V. 1982. The history of the plant-cover development in coastal lowlands of Yakutia throughout the Holocene. In *Development of nature in the USSR territory in the late Pleistocene and Holocene.* Moscow: Nauka, pp. 207–220. (R)

Karol, I. L. 1977. The changes of global stratospheric aerosol content and their relation to fluctuations of mean direct solar radiation and temperature at the Earth's surface. *Meteorol. Hydrol.* 3:32–40. (R)

———. 1984. Radiation effects of El Chichon eruption products. *Meteorol. Hydrol.* 3:102–104. (R)

———. 1986. Model estimates of expected changes of mean gas composition and global atmospheric temperature from human activities prior to the year 2000. In *Complex global monitoring of the biosphere state. Trans. III Int. Symp.* Leningrad: Gidrometeoizdat, vol. 1. (R)

Karol, I. L., & Frolkis, V. A. 1984. Energy-balance radiative-convective model of global climate. *Meteorol. Hydrol.* 8:59-67. (R)

Karol, I. L., & Pivovarova, Z. I. 1978. The relationship between stratospheric aerosol concentration and solar radiation fluctuations. *Meteorol. Hydrol.* 9:35–42. (R)

Karol, I. L., & Rozanov, E. V. 1982. Radiative-convective climate models. *Bull. Acad. Sci. USSR, Atm. Ocea. Phys.* 18 (11):1179–1191. (R)

Karol, I. L.; Rozanov, V. V.; & Timofeev, Yu. M. 1983. *Minor gases in the atmosphere.* Leningrad: Gidrometeoizdat. 192 pp. (R)

Karol, I. L., et al. 1986. *Radiative-photochemical models of the atmosphere.* Leningrad: Gidrometeoizdat. 192 pp. (R)

Karpachesky, L. O. 1981. *Forest and forest soils.* Moscow: Forest Industry. 264 pp. (R)

Keeling, C. D. 1973. Industrial production of carbon dioxide from fossil fuels and limestone. *Tellus* 25:174–196.

———. 1979. The Suess effect: ^{13}Carbon-^{14}Carbon interrelations. *Environ. Intern.* 2:229–300.

Keeling, C. D.; Bacastow, R. B.; & Tans, P. P. 1980. Predicted shift in the $^{13}C/^{12}C$ ratio of atmospheric carbon dioxide. *Geophys. Res. Lett.* 7 (7):505–508.

Keeling, C. D.; Mook, W. G.; & Tans, P. P. 1979. Recent trends in the $^{13}C/^{12}C$ ratio of atmospheric carbon dioxide. *Nature* 277:121–123.

Keepin, B. 1984. A technical appraisal of the IIASA Energy Scenarios. *Policy Sciences* 17 (3).

Keepin, B., et al. 1985. *The WMO/ICSU/UNEP International assessment of the impact of an increased atmospheric concentration of carbon dioxide in the environment. Part 2: Emission of CO_2 into the atmosphere.* WMO: Geneva.

Keerberg, O. F., & Viyl, Yu. A. 1982. The system of regulation and energetics of rehabilitative pentozophosphate cycle. In *Photosynthetic physiology.* Moscow: Nauka, pp. 104–118. (R)

Kellogg, W. 1977–78. The impact of human activities on climate. *Bull. WMO* 26 (4):285–299; 27 (1):3–12. (R)

———. 1978. Review of mankind's impact on global climate. In *Multidisciplinary research related to the atmospheric sciences.* Boulder, Colo.: Nat. Cent. for Atm. Res., pp. 64–81.

Kelly, P. M., et al. 1982. Variations in surface air temperature: Part 2, Arctic regions, 1881–1980. *Mon. Wea. Rev.* 110 (2):71–83.

Kelly, P. M., et al. 1985. The extended northern hemisphere surface air temperature record: 1851–1984. In *Extended summaries: Third conference on climate variations and symposium on contemporary climate, 1850–2100.* January 8–11, 1985. Los Angeles: AMS, pp. 35–36.

Kerr, R. A. 1984a. Carbon dioxide and the control of ice ages. *Science* 223:1053–1054.

———. 1984b. Climate since the ice began to melt. *Science* 226:326–327.

Khalil, M. A. K., & Rasmussen, R. A. 1984a. The global increase of carbon monoxide. In *Proc. Spec. Meet. of APCA Conf. on Environ. Impact of Nat. Emissions.* Pittsburgh, Pa. 430 pp.

———. 1984b. Statistical analysis of trace gases in Arctic haze. *Geophys. Res. Lett.* 11 (5):437–440.

———. 1985. Causes of increasing atmospheric methane: Depletion of hydroxyl radicals and the rise of emissions. *Atm. Environ.* 19 (3):397–407.

Khotinsky, N. A. 1977. *The Holocene of northern Eurasia.* Moscow: Nauka. 198 pp. (R)

Khotinsky, N. A., & Savina, S. S. 1985. Palaeoclimatic schemes of the USSR territory in the boreal, atlantic and sub-boreal periods of the Holocene. *Bull. Acad. Sci. USSR, Ser. Geogr.* 4:18–34. (R)

Kimball, B. A. 1983. *Carbon dioxide and agricultural yield: An assemblage and analysis of 770 prior observations.* WCL Report 14, Water Conservation Laboratory, Agricultural Research Service. Phoenix, Ariz.: U.S. Dept. of Agriculture.

Kliege, R. K. 1985. *Global water-exchange alterations.* Moscow: Nauka. 248 pp. (R)

Klimanov, V. A. 1982. Climate of eastern Europe during the Holocene climatic optimum (from palynologic data). In: *The development of nature in the USSR territory in the Late Pleistocene and Holocene.* Moscow: Nauka. pp. 251–258. (R)

Klimanov, V. A., & Nikolskaya, M. V. 1983. The analysis of subrecent spore-pollen spectra and some climatic indices of the Holocene for the north of Siberia. In: *Palaeogeographical analysis and stratigraphy of the Anthropogene for the Far East.* Vladivostok, pp. 27–49. (R)

Kobak, K. I. 1964. Some problems of CO_2 supply of forest biogeocenoses. In *Problems of ecology and physiology of forest plants,* vol. 2. Leningrad, pp. 61–98. (R)

Kobak, K. I., & Kondrasheva, N. Yu. 1985. Anthropogenic impacts on forest ecosystems and the role of these impacts in the global carbon cycle. *Bot. J.* 70 (3):305–313. (R)

———. 1986. The distribution of organic carbon in soils over the globe. *Trans. State Hydrol. Inst.* 320:61–76. (R)

Kobak, K. I.; Yatsenko-Khmelevsky, A. A.; & Kondrasheva, N. Yu. 1980. Carbon dioxide balance in high- and low-productive plant associations. In *Problems of atmospheric CO_2*. Leningrad: Gidrometeoizdat, pp. 252–264. (R)

Kobak, K. I., et al. 1985. Present-day estimates of the role of forests in carbon cycle. *Trans. State Hydrol. Inst.* 339:3–37. (R)

Komhyr, W. D., et al. 1983. *Global atmospheric CO_2 distribution and variations from 1968–1982.* NOAA/GMCC Flask Samples Data. Summary of U.S. NOAA Monitoring Program. 36 pp.

Kondratjev, K. Ya. 1980. *Radiation factors of current climatic changes.* Leningrad: Gidrometeoizdat. 279 pp. (R)

Kondratjev, K. Ya., & Moskalenko, N. I. 1984. *Greenhouse effect of the atmosphere and climate. Results of science and technology.* Moscow: VINITI. 262 pp. (R)

Kondratjev, K. Ya.; Moskalenko, N. I.; & Pozdnyakov, D. V. 1982. *Atmospheric aerosols.* Leningrad: Gidrometeoizdat. 226 pp. (R)

Kononova, M. M. 1976. Humus formation in the soil and its decomposition. *Achievements in Microbiology* 11:134–151. (R)

———. 1984. Organic matter and soil fertility. *Pochvovedenie* 8:6–20. (R)

Kononova, M. M., & Alexandrova, I. V. 1974. Humus formation as a link of carbon cycle in the soil. In *Trans. Int. Congress of Pedologists.* Moscow: Nauka, vol. 2, pp. 81–90. (R)

Konstantinov, A. R.; Zoidze, E. K.; & Smirnova, S. I. 1981. *Soil-climatic resources and crop location.* Leningrad: Gidrometeoizdat. 278 pp. (R)

Köppen, W. 1873. Über mehrjährige Perioden der Witterung. *Zeitschrift der Osterreichischen Gesellschaft für Meteorologie* 8 (16): 241–248; (17):257–267.

Kotlyakov, V. M., & Gordienko, F. G. 1982. *Isotope and geochemical glaciology.* Leningrad: Gidrometeoizdat. 288 pp. (R)

Koval, L. A.; Menzhulin, G. V.; & Savvateev, S. P. 1983. On the principles of building parameterized models of crop productivity. *Trans. State Hydrol. Inst.* 280:119–129. (R)

Koval, L. A.; Nikolaev, M. V.; & Savvateev, S. P. 1985. On the problem of the comparative estimation of climatic wheat productivity in the USSR, western Europe and North America. *Trans. State Hydrol. Inst.* 339:82–92. (R)

Koval, L. A., & Savvateev, S. P. 1982. The use of productivity models for estimating the influence of climate changes on crop productivity. In *Problems of land hydrology. Papers at the conference of young scientists and specialists.* Leningrad: Gidrometeoizdat, pp. 219–223. (R)

Kovda, V. A. 1973. *Fundamentals of soil theory.* Vol. 1. Moscow: Nauka. 447 pp. (R)

————. 1977. *Aridization of land and drought-fighting.* Moscow: Nauka. 272 pp. (R)

Kovyneva, N. P. 1982. Statistical investigation of modern changes in surface air temperature patterns and atmospheric pressure. In *Problems of land hydrology.* Leningrad: Gidrometeoizdat, pp. 200–210. (R)

————. 1984. Modern changes in surface air temperature and atmospheric precipitation patterns. *Bull. Acad. Sci. USSR, Ser. Geogr.* 6:29–39. (R)

Kozuto, N. A.; Lugina, K. M.; & Severina, N. G. 1981. Spatial variability of mean annual values of air temperature. *Trans. State Hydrol. Inst.* 271:62–71. (R)

Krishtofovitch, A. N. 1957. *Palaeobotany.* Moscow: Gostoptekchizdat. 646 pp. (R)

Kulik, M. S. 1971. *Methodology handbook on long-range agrometeorological forecasts of mean regional productivity of winter cereals in the non-chernozem zone.* Leningrad: Gidrometeoizdat. 24 pp. (R)

Kutzbach, J. E. 1983. Monsoon rains of the late Pleistocene and early Holocene: Patterns, intensity and possible causes of changes. In *Variations in global water budget.* Dordrecht: Reidel, pp. 371–389.

Kutzbach, J. E., & Otto-Bliesner, B. S. 1982. The sensitivity of the African-Asian monsoonal climate to orbital parameter changes for 9000 years B.P. in a low-resolution general circulation model. *J. Atm. Sci.* 39:1177–1188.

Kutzbach, J. E., & Street-Perrott, F. A. 1985. Milankovitch forcing of fluctuations in the level of tropical lakes from 18 to 0 Kr B.P. *Nature* 317 (6033):130–134.

Kuzakova, L. I. 1975. *Talks about nature, society and man.* Moscow: Znanie. 190 pp. (R)

Kuzmin, I. I.; Romanov, S. V.; & Chernoplekov, A. N. 1984. On a quantitative approach to the safety assessment. Preprint of I. V. Kurchatov. Inst. of Atomic Energy, no. 4011. (R).

Kuzmin, I. I., & Stolyarevsky, A. Ya. 1984. The forecast of the global energy balance. *Energy: Economics, Technology, Ecology* 11:30–42. (R)

Lacis, A., et al. 1981. Greenhouse effect of trace gases, 1970–1980. *Geophys. Res. Lett.* 8 (10):1035–1038.

Lagan, J. A. 1985. Tropospheric ozone: Seasonal behavior trends and anthropogenic influence. *J. Geophys. Res.* 90 (D6): 10463–10482.

Lamb, H. H. 1970. Volcanic dust in the atmosphere, with a chronology and assessment of its meteorological significance. *Phil. Trans. Roy. Soc., London, A* 266 (1178):425-533.

————. 1974. *The current trend of world climate: A report on the early 1970s and a perspective.* Norwich: Univ. of East Anglia, Climatic Research Unit, School of Environmental Sciences. 28 pp.

Landsberg, H. E.; Groveman, B. S.; & Hakharinen, I. M. 1978. A simple method for approximating the annual temperature of the northern hemisphere. *Geophys. Res. Lett.* 5 (6):505–506.

Lanly, J. P. 1982. Extrait de la forestry planning news letter publieé par le département des forêts de la FAO. Projet FAO/PNUE d'évolution des ressources forestières tropicales. *Bois et forêts des tropiques* 195:22–31.

Lanly, J. R. (ed.). 1982. *Les ressources forestières tropicales. Organization des nations unies pour l'alimentation et l'agriculture.* Rome. 113 pp.

Laysk, A. Kh. 1977. *Kinetics of photosynthesis and photorespiration of C_3-plants.* Moscow: Nauka. 196 pp. (R)

————. 1981. Investigation of kinetics of photosynthesis and photorespiration of leaves. In *Results of investigations of photosynthesis and productivity in Estonian SSR.* Tallinn-Tartu, pp. 32–51. (R)

————. 1982. Correspondence of photosynthetic system to environmental conditions. In *Photosynthetic physiology.* Moscow: Nauka. pp. 221–234. (R)

Legasov, V. A., & Kuzmin, I. I. 1981. Energetics problems. *Priroda* 2:8–23. (R)

Legasov, V. A.; Kuzmin, I. I.; & Chernoplekov, A. N. 1984. The impact of energetics on climate. *Bull. Acad. Sci. USSR, Atm. Ocea. Phys.* 20 (20):1089–1106. (R)

Lerman, J. C.; Mook, W. C.; & Vogel, J. C. 1970. ^{14}C in tree rings from different localities. *Nobel Symp.* 12:275–282.

Levin, F. I. 1983. *Problems of cultivation, degradation and rise of fertility of arable soils.* Moscow. 93 pp. (R)

Levitus, S. 1982. *Climatological atlas of the world ocean.* NOAA Prof. Pap. 13. Rockville. 173 pp.

Lian, M. S., & Cess, R. D. 1977. Energy balance climate models: A reappraisal of ice-albedo feedback. *J. Atm. Sci.* 34:1058–1062.

Liberman, A. A.; Muratova, M. V.; & Suytova, I. A. 1985. Application of linear interpolation for building palaeoclimatic models. In *Palaeoclimatic reconstruction methods.* Moscow: Nauka, pp. 48–53. (R)

Lieth, H.; Seeliger, I.; & Zimmermeyer, G. 1980. Die CO_2-Frage aus geoökologische and energiewirtschaftlicher Sieht. In *Proc. 11th World Energy Conf.*, vol. 3, Münich.

Liou, K. N. 1984. *Fundamentals of radiation processes in the atmosphere.* Leningrad: Gidrometeoizdat. 376 pp. (R)

Liss, P. S. 1973. Processes of gas exchange across an air-water interface. *Deep Sea Res.* 20:221–238.

Logan, J. A. 1985. Tropospheric ozone: Seasonal behavior trends and anthropogenic influence. *J. Geophys. Res.* 90 (D6): 10463–10482.

Loginov, V. F.; Pivovarova, Z. I.; & Kravchuk, E. G. 1983. Variability of direct solar radiation and temperature in the northern hemisphere due to volcanic eruptions. *Bull. All-Union Geogr. Soc.* 5:401–411. (R)

Lokshina, I. Yu. 1986. Plant-cover productivity in the Pliocene. *Bull. Acad. Sci. USSR, Ser. Geogr.* 3. (R)

Long-term series of mean regional combined meteorological parameters for the main agricultural zone of the USSR. 1985. Leningrad. 324 pp. (R)

Lorius, C., et al. 1985. A 150,000-year climate record from Antarctic ice. *Nature* 316 (6029):991–996.

Loucks, P. L. 1980. Recent results from studies of carbon cycling in the biosphere. Washington, D.C.: U.S. Dept. of Energy Publ., pp. 3–42.

Lough, J. M., et al. 1983. Climate and climate impact scenarios for Europe in a warmer world. *J. Clim. Appl. Meteorol.* 22 (10):1673–1684.

Lovins, A. B., et al. 1981. *Least cost energy: Solving the CO_2 problems.* Andover: Brick House.

Lowe, J. J.; Gray, J. M.; & Robinson (eds.). 1980. *Studies in the Late-glacial of Northwest Europe.* Oxford: Pergamon Press. 205 pp.

Lugina, K. M., & Speranskaya, N. A. 1984. Variability of mean annual surface air temperature in high latitudes of the northern hemisphere. *Trans. State Hydrol. Inst.* 295:87–97. (R)

Lugo, A. E.; Brown, S.; & Hall, Ch. 1980. The role of tropical forests in the carbon balance of the world. In *Carbon dioxide effects research and assessment program. Proc. of carbon dioxide and climate research program conference,* ed. L. Schmitt. Washington, D.C.: pp. 261–273.

Lysak, G. N. 1980. Agricultural ecology and soil erosion. In *Ecology and agriculture.* Moscow: Nauka, pp. 106–113. (R)

Lysgaard, L. 1950. On the present climatic variation. In *Centenary Proc. Roy. Met. Soc.,* pp. 206–211.

MacCracken, M. C.; Cess, R.; & Rotter, G. L. 1986. The climatic effects of Arctic aerosols: Illustration of climate feedback mechanisms with one- and two-dimensional climate models. *J. Geophys. Res.* 91.

MacCracken, M. C., & Luther, F. M. (eds.). 1985. *Projecting the climatic effects of increasing carbon dioxide.* DOE/ER-0237. Livermore, 381 pp.

McDonald, A. 1981. *Energy in a finite world.* IIASA, A-2361. Laxenburg.

MacDowell, E. D. H. 1973. Growth of Harquis wheat II. Carbon dioxide dependence. *Can. J. Bot.* 50:883–889.

McKenna, M. C. 1980. Eocene paleoaltitude, climate and mammals of Elsmere Islands. *Palaeogeogr., Palaeoclim., Palaeo- ecol.* 30 (3–4):349–362.

McQuigg, J. 1979. Climatic variability and agriculture in the temperate regions. In *Proc. World Climate Conf.,* Geneva, WMO, February 1979. Geneva: WMO.

Madden, R. A., & Ramanathan, V. 1980. Detecting climate change due to increasing carbon dioxide. *Science* 209 (4458):763–768.

Mahlman, T. D.; Levy, H.; & Moxim, W. T. 1980. Three-dimensional tracer structure and behavior as simulated in two ozone precursor experiments. *J. Atm. Sci.* 37 (3).

Makukhin, A. N. 1985. Ecological strategy. *Energy: Economics, Technology, Ecology* 7:2–9. (R)

Mamedov, Z. D. 1982. Pluvials and arids in the Late Pleistocene and Holocene history of USSR deserts and adjacent countries. In *Development of nature in the USSR territory in the Late Pleistocene and Holocene.* Moscow: Nauka, pp. 94–99. (R)

Manabe, S. 1970. The dependence of atmospheric temperature on the concentration of carbon dioxide. In *Global effects of environment pollution.* Dordrecht: Reidel.

Manabe, S., & Broccoli, A. J. 1985. A comparison of climate model sensitivity with data from the last glacial maximum. *J. Atm. Sci.* 42 (23):2643–2651.

Manabe, S., & Bryan, K. 1985. CO_2-induced change in a coupled ocean-atmosphere model and its paleoclimatic implications. *J. Geophys. Res.* 90 (C11):11689–11707.

Manabe, S., & Stouffer, R. J. 1980. Sensitivity of a global climate model to an increase of CO_2 concentration in the atmosphere. *J. Geophys. Res.* 85 (C10):5529–5553.

Manabe, S., & Wetherald, R. T. 1967. Thermal equilibrium of the atmosphere with a given distribution of relative humidity. *J. Atm. Sci.* 24 (3):241–259.

———. 1975. The effects of doubling the CO_2 concentration on the climate of a general circulation model. *J. Atm. Sci.* 32 (1):3–15.

———. 1980. The distribution of climate change resulting from increase in CO_2 content of the atmosphere. *J. Atm. Sci.* 37:99–118.

Manabe, S., et al. 1981. Summer dryness due to an increase of atmospheric CO_2 concentration. *Clim. Change* 3 (4):347–386.

Managing climatic resources and risks. 1981. Washington, D.C: Nat. Acad. Press. 51 pp.

Mankin, W. G., & Coffey, M. T. 1984. Increased stratospheric hydrogen chloride in the El Chichon cloud. *Science* 266:170–172.

Mann, D. 1982. Research on political institutions and their response to the problem of increasing CO_2 in the atmosphere. In *Social science research and climatic change: An interdisciplinary appraisal*, ed. R. S. Chen, E. M. Boulding & S. H. Schneider. Boston: Reidel.

Marchuk, G. I. 1979. Modeling climatic changes and the problem of long-range weather forecasting. *Meteorol. Hydrol.* 7. (R)

Markov, K. K. 1941. On the multiplicity of glaciations. *Bull. Acad. Sci. USSR, Ser. Geogr., Geophys.* 2:203–207. (R)

Marland, G., & Rotty, R. M. 1984. Carbon dioxide emissions from fossil fuels: A procedure for estimation and results for 1950–1982. *Tellus* 368:232–261.

Mattei, F. 1979. Climatic variability and agriculture in the semiarid tropics. In *Proc. World Climatic Conf.*, Geneva, WMO, February 1979. Geneva: WMO.

Mauney, J. R.; Fry, K. E.; & Guinn, G. 1978. Relationship of photosynthetic rate to growth and fruiting of cotton, soybean, sorghum, and sunflower. *Crop Sci.* 18:259–263.

Meier, M. F. 1984. Contribution of small glaciers to global sea level. *Science* 226 (4681):1418–1421.

Melentyev, L. A. 1984. Long-range energetics program in the USSR. *Energy: Economics, Technology, Ecology* 4:2–12. (R)

Meleshko, V. P. 1980. The calculation of global distribution of three-layer large-scale cloudiness. *Meteorol. Hydrol.* 9:12–33. (R)

Menzhulin, G. V. 1976. The influence of climate changes on crop productivity. *Trans. Main Geophysical Observatory* 365:41–48. (R)

———. 1984. The influence of current climate changes and CO_2 concentration on agricultural plant productivity. *Meteorol. Hydrol.* 4:95–101. (R)

Menzhulin, G. F.; Nikolaev, M. N.; & Savvateev, S. P. 1983. The estimates of economic and weather components of cereal-yield variability. *Trans. State Hydrol. Inst.* 280:111–119. (R)

Menzhulin, G. V., & Savvateev, S. P. 1980. The influence of modern climate changes on crop productivity. In *Problems of atmospheric carbon dioxide. Proc. Soviet-Amer. Symp.* Leningrad: Gidrometeoizdat, pp. 186–197. (R)

———. 1981. Modern climate changes and crop productivity. *Trans. State Hydrol. Inst.* 271:90–103. (R)

Mercer, J. H. 1978. West Antarctic ice sheet and CO_2 greenhouse effect: A threat of disaster. *Nature* 271:321–325.

Methods for paleoclimatic reconstructions. 1985. A. Velichko and Yl. Gurtovaya, Eds. Moscow: Nauka. 287 pp. (R)

Meyer-Abich, K. M. 1980. Socioeconomic impacts of CO_2-induced climatic changes and the comparative changes of alternative political … .

Miles, M. K., & Gildersleeves, P. B. 1977. A statistical study of the likely causative factors in the climatic fluctuation of the last 100 years. *Met. Meg.* 106:314–322.

Miller, J. R., et al. 1983. Annual oceanic heat transports computed from an atmospheric model. *Dynam. Atm. Oceans* 7 (1):95–109.

Miller, P. C. (ed.). 1980. *Carbon balance in northern ecosystems and the potential effect of carbon-dioxide-induced climatic change.* Washington, D.C.: U.S. Dept. of Energy Publ. 109 pp.

Mintz, Y. 1984. The sensitivity of numerically simulated climates to land-surface boundary conditions. In *The global climate*, ed. J. T. Houghton. Cambridge: Cambridge Univ. Press, 79–105.

Mitchell, J. 1975. A reassessment of atmospheric pollution as a cause of long-term changes of global temperature. In *The changing global environment*, ed. S. F. Singer. Dordrecht and Boston: Reidel, pp. 149–173.

Mitchell, J., & Lupton, G. 1984. A 4 × CO_2 integration with prescribed changes in sea surface temperatures. *Progress in Biometeorol.* 3:353–374.

Moiseytchik, V. A. 1975. *Agrometeorological conditions and winter crops under winter conditions.* Leningrad: Gidrometeoizdat. 295 pp. (R)

Mokhov, I. I. 1981. On the impact of CO_2 on thermal conditions of the Earth's climatic system. *Meteorol. Hydrol.* 4:24–34. (R)

————. 1982. On the cloudiness-temperature relationship with spatial averaging. *Meteorol. Hydrol.* 10:334–45. (R)

————. 1983. Vertical temperature gradient in the troposphere and its correlation with surface air temperature by empirical data. *Bull. Acad. Sci. USSR, Atm. Ocea. Phys.* 19 (9):913–919. (R)

————. 1984. Antiscreening cloudiness effect of outgoing heat radiation. *Bull. Acad. Sci. USSR, Atm. Ocea. Phys.* 20 (3):244–254. (R)

Mokhov, I. I., & Petukhov, V. K. 1978. Parameterization of outgoing long-wave radiation for climatic models. Preprint. Moscow: Inst. Phys. Atm. Acad. Sci. USSR. 234 pp. (R)

Molina, L. T., et al. 1985. *J. Phys. Chem.* 89.

Monin, A. S., & Obukhov, A. M. 1954. Basic features of turbulent mixing in the surface air layer. *Trans. Geophys. Inst. Acad. Sci. USSR* 24 (151):163–187. (R)

Monsi, M., & Saeki, T. 1953. Über den Lichtfaktor in den Pflanzengesellschaften und wiseine Bedeutung für die Stoffproduction. *Jap. J. Bot.* 14 (1):22–52.

Monthly climatic data for the world. 1960–80. Ashville.

Montreal Protocol on Substances That Deplete the Ozone Layer. 1987. UNEP.

Mook, W. G.; Bommerson, J. C.; & Stoverman, W. H. 1974. Carbon isotope fractionation between dissolved bicarbonate and gaseous carbon dioxide. *Earth Planet. Sci. Lett.* 22 (2):169–176.

Mook, W. G.; Keeling, C. D.; & Herron, A. 1981. Seasonal and secular variations in the abundance and $^{13}C/^{12}C$ ratio of atmospheric CO_2. In *Analysis and interpretation of atmospheric CO_2 data.* Bern: WMO et al., pp. 195–198.

Moore, B., et al. 1980. *A simple model for analysis of the role of terrestrial ecosystems in the global carbon budget.* Durham: Univ. New Hampshire. 21 pp.

Moore, B., et al. 1981. A simple model for analysis of the role of terrestrial ecosystems in the global carbon budget. Chichester: John Wiley and Sons, pp. 365–385.

Mörner, N.-A. 1980. A 10,000-year temperature record from Gotland Pleistocene-Holocene boundary events in Sweden. *Boreas* 9 (4):283–287.

Müller, D. 1960. Kreislauf des Kohlenstoffes. *Handbuch der Pflanzenphysiol.*, Springer-Verlag, 12 (2):934–938.

Munk, W. H. 1966. Abyssal recipes. *Deep Sea Res.* 13:707–721.

Murzaeva, V. E., et al. 1984. Pluvial conditions of the Late Pleistocene and Holocene in the arid zone of Asia and Africa. *Bull. Acad. Sci. USSR; Ser. Geogr.* 4:15–25. (R)

Namias, J. 1980. Some concomitant regional anomalies associated with hemispherically averaged temperature variations. *J. Geophys. Res.* 85 (C3):1585–1590.

The natural matter cycle and its change due to the influence of human activities. 1980. Moscow. 272 pp. (R)

Neales, T. F., & Nickolls, A. O. 1978. Growth responses of young wheat plants to a range of ambient CO_2 levels. *Aust. J. Plant Physiol.* 5:45–49.

Neftel, A., et al. 1982. Ice core sample measurements give atmospheric CO_2 content during the past 40,000 yr. *Nature* 295:220–223.

Neftel, A., et al. 1985. The increase of atmospheric CO_2 in the last two centuries. Evidence from polar ice cores. *Nature* 315:43–45.

Newman, J. E. 1980. Climate change impacts on the growing season of the North American corn belt. *Biometeorology* 7 (2):128–142.

———. 1982. Impacts of a rising atmospheric carbon dioxide level on agricultural growing seasons and crop water use efficiencies. In *Environmental and Social Consequences of a Possible CO_2 -Induced Climate Change*, vol. 11, part 8. DOE/EV/10019-8. Washington, D.C.: U.S. Dept. of Energy.

Nikolaev, M. V. 1985a. On economic and weather climatic constituents of crop production in the USA. *Trans State Hydrol. Inst.* 339:48–60. (R)

———. 1985b. On estimating crop productivity variation using economic indices of agricultural production. In *Problems of land hydrology. Papers at the conference of young scientists and specialists.* Leningrad: Gidrometeoizdat, pp. 208–213. (R)

Nikolaev, M. V.; Menzulin, G. V.; & Savvateev, S. P. 1985. Some features of crop productivity variations in the USSR and USA. *Trans. State Hydrol. Inst.* 339:61–81. (R)

Nikolaev, M. V., & Savvateev, S. P. 1982. On economic and climatic variations of crop productivity. In *Problems of land hydrology. Papers at the conference of young scientists and specialists.* Leningrad: Gidrometeoizdat, pp. 224–227. (R)

Nikolskaya, M. V. 1982. Palaeobotanic and palaeoclimatic reconstructions of the Holocene for Taimir. In *The anthropogene in Taimir.* Moscow: Nauka, pp. 148–157. (R)

Nordhause, W. D., & Yoe, G. 1983. Future paths of energy and carbon dioxide emissions. In *Changing climate.* Washington, D.C.: NAS.

North, G. R. 1975. Theory of energy balance climate models. *J. Atm. Sci.* 32 (1):3–15.

North, G. R.; Calahan, R. F.; & Coakley, J. A. 1981. Energy-balance climate models. *Rev. Geophys. Space Phys.* 19:91–122.

North, G. R.; Mengel, J. G; & Short, D. A. 1983. Simple energy balance model resolving the seasons and the continents: Application to the astronomical theory of the ice ages. *J. Geophys. Res.* 88 (C10):6576–6586.

Nozaki, Y., et al. 1978. A 200 year record of carbon-13 and carbon-14 variations in a Bermuda coral. *Geophys. Res. Lett.* 5:825–828.

Nuclear power, the environment and man. 1982. Vienna: International Atomic Energy Agency.

Nyers, N. 1980. The conversion of tropical forests. *Environment* 22 (6):24–29.

Oberländer, H., & Roth, K. 1968. Transformation of ^{14}C-labelled plant material in soils under field conditions. In *Isotopes and radiation in soil organic matter studies.* Vienna: International Atomic Energy Agency, p. 241.

Oeschger, H.; Siegenthaler, U.; & Heimann, U. 1980. The carbon cycle and its perturbation by man. In *Interaction of energy and climate.* Münster, pp. 107–128.

Oeschger, H., et al. 1975. A box diffusion model to study the carbon dioxide exchange in nature. *Tellus* 27:168–192.

Olson, J. S. 1963. Energy storage and the balance of producers and decomposers in ecological systems. *Ecology* 44 (2):322–331.

————. 1978. *Changes in the global carbon cycle and the biosphere.* Oak Ridge Nat. Lab. 169 pp.

———. 1982. Earth's vegetation and atmospheric carbon dioxide. In *Carbon dioxide review 1982*, ed. M. W. Clark. Oxford and New York: Clarendon Press, pp. 388–398.

Omer, St. L., & Horvath, S. M. 1983. Elevated carbon dioxide concentration and whole plant senescence. *Ecology* 64:1311–1314.

Oram, P. A. 1985. Sensitivity of agricultural production to climatic change. *Clim. Change* 7 (1):129–152.

Outline plan and basis for the world climate programme, 1980–1983. 1979. Rept. no. 540. Geneva: WMO. 64 pp.

Palaeogeography of Europe during the last 100 thousand years. Atlas monograph. 1982. Moscow: Nauka. 155 pp.

Paleoclimates in the Late Glacial and Holocene. 1989. N. Khotinskyi, Ed. Moscow: Nauka. (R)

Parry, M. L. 1985. Estimating the sensitivity of natural ecosystems and agriculture to climatic change. *Clim. Change* 7 (1):1–4.

Parry, M. L., & Carter, T. R. 1985. The effect of climatic variations on agricultural risk. *Clim. Change* 7 (1):95–110.

Parry, M. L., & Carter, T. L. (eds.). 1984. *Assessing the impact of climatic change in cold regions.* Summer Rept. Luxenburg, Austria: IIASA. 40 pp.

Pasov, V. M. 1973a. Crop yield variability in different climatic zones of the USSR. *Meteorol. Hydrol.* 7:82–86. (R)

———. 1973b. Climatic variability of winter wheat yields. *Meteorol. Hydrol.* 2:94–103. (R)

Pavlova, V. N., & Sirotenko, O. P. 1985. On the use of dynamic models for estimating the influence of possible climate changes and fluctuations on crop productivity. *Trans. All-Union Inst. Agr. Machine Build.* 10:81–90. (R)

Pearman, G. I. 1980a. *Atmospheric CO_2 concentration measurements. A review of methodologies, existing programmes and available data.* WMO Rep., no. 3. 27 pp.

———. 1980b. Preliminary studies with a new global carbon cycle model. In *Carbon dioxide and climate: Australian Research.* Canberra: Australian Academy of Science, pp. 79–90.

Ped', D. A. 1975. On drought and excessive moisture content indices. *Trans. Hydrometeorol. Center* 156:19–39. (R)

Peng, T.-H. 1985. Atmospheric CO_2-variations based on the tree-ring ^{13}C record. In *The carbon cycle and atmospheric CO_2: Natural variations, Archean to present. Geophys. Monogr.* 32:123–131.

Peng, T.-H., et al. 1983. A deconvolution of the tree ring based on $\delta^{13}C$ record. *J. Geophys. Res.* 88 (C6):3609–3620.

Petit-Maire, N., & Riser, J. (eds.). 1982. *Sahara ou Sahel? Quaternaire recent du Bassin de Taoudenni (Mali).* 473 pp.

Petukhov, V. K., & Manuilova, N. I. 1984. Estimating some climate-forming factors in a simple thermodynamic climate model. *Meteorol. Hydrol.* 10:31–37. (R)

Physical basis of climate and climate modeling. 1975. Rep. of the Int. Study Conf., no. 16. Stockholm: JARP.

Pielke, R.; Styles, T.; & Biondini, R. 1979. Changes in the growing season. *Weatherwise* 10:207–210.

Pittock, A. B., & Salinger, J. M. 1982. Towards regional scenarios for a CO_2-warmed Earth. *Clim. Change* 4:23–40.

Pivovarova, Z. I. 1977. *Radiation characteristics of climate in the USSR.* Leningrad: Gidrometeoizdat. 335 pp. (R)

Popov, N. I.; Fedorov, K. N.; & Orlov, V. M. 1979. *Sea water.* Moscow: Nauka. 327 pp. (R)

Post, W. M., et al. 1982. Soil carbon pools and world life zones. *Nature* 298 (5870):156–158.

Potential climatic impacts of increasing atmospheric CO_2 with emphasis on water availability and hydrology in the United States. 1984. New York: NASA. 96 pp.

Prinn, R. G., et al. 1983. The atmospheric lifetime experiment. I–IV. *J. Geophys. Res.* 88 (C13):8353–8519.

Problems of atmospheric CO_2. 1980. Leningrad: Gidrometeoizdat. 310 pp. (R)

Pulwarty, R. S., & Cohen, S. J. 1984. Possible effects of CO_2-induced climate change on the world food system: A review. *Climatol. Bull.* 18 (2):33–48.

Punning, Ya. M. K., & Raukas, A. V. 1985. *Palaeogeography of the Late Quaternary. Results of science and technology.* Vol. 2. Moscow: VINITI. (R)

Ramanathan, V. 1976. Radiative transfer within the Earth's troposphere and stratosphere: A simplified radiative-convective model. *J. Atm. Sci.* 33 (7):1330–1346.

———. 1981. The role of ocean-atmosphere interactions in the CO_2 climate problem. *J. Atm. Sci.* 38 (8):918–930.

Ramanathan, V., & Coakley, J. A. 1978. Climate modeling through radiative-convective models. *Rev. Geophys. Space Phys.* 16 (3):465–489.

Ramanathan, V.; Lian, M. S.; & Cess, R. D. 1979. Increased atmospheric CO_2: Zonal and seasonal estimates of the effect. *J. Geophys. Res.* 84:4949–4958.

Ramanathan, V., et al. 1979. Greenhouse effect due to climatic chlorofluorocarbons: Climatic implications. *Science* 190:50.

Ramanathan, V., et al. 1985. Trace gas trends and their potential role in climate change. *J. Geophys. Res.* 90 (D3):5547–5566.

Ramanathan, V., et al. 1986. Climate-chemical interactions and effects of changing atmospheric trace gases. *Rev. Geophys.* 24.

Ramirez, J. M.; Sakamoto, C. M.; & Jensen, R. E. 1975. Wheat. In *Impacts of climatic change on the biosphere.* CIAP Monogr. 5, p. 2. U.S. Dept. of Transportation. Washington, D.C.

Raper, S. C. B., et al. 1984. Variations in surface air temperatures. Part 3. The Antarctic, 1957–1982. *Mon. Wea. Rev.* 112 (7):1241–1353.

Raschke, K. 1975. Stomatal action. *Ann. Rev. Plant Physiol.* 26:309–340.

Rasmussen, R. A.; & Khalil, M. A. K. 1981. Atmospheric methane (CH_4): Trends and seasonal cycles. *J. Geophys. Res.* 86 (C10):9826–9832.

Rauner, Yu. L. 1981. *Climate and crop productivity.* Moscow: Nauka. 163 pp. (R)

Rauner, Yu. L., & Lazovskaya, L. A. 1978. Variations of productivity of wheat and corn in the grain-growing zone of North America. *Bull. Acad. Sci. USSR, Ser. Geogr.* 1:90–101. (R)

Reister, D. B. 1984. *An assessment of the contribution of gas to the global emissions of carbon dioxide.* Final Rep. GRI-84/003. Chicago, Ill.: Gas Res. Inst.

Report of the 5th session of Coordination Committee on Ozone Layer. 1981. Copenhagen: UNEP. (R)

Report of the meeting of experts. 1982. NCAR-Boulder, Sept. 1982. WMO Global Ozone Res. Monitor, Project Rep. no. 14. Geneva: WMO. 95 pp.

Report of the study conference on sensitivity of ecosystems and society to climate change. 1984. Geneva: WMO.

Revelle, R. R. 1983. Probable future changes in sea level resulting from increased atmospheric carbon dioxide. In *Changing climate report of the carbon dioxide assessment committee.* Washington, D.C.: NAS, pp. 433–448.

Revelle, R., & Munk, W. 1977. The carbon dioxide cycle and the biosphere. In *Energy and climate.* Washington, D.C.: NAS, pp. 140–158.

Revelle, R., & Suess, H. E. 1957. Carbon dioxide exchange between atmosphere and ocean and the question of an increase of atmospheric CO_2 during the past decades. *Tellus* 9:18.

Ritchie, J. C.; Cwynuar, L. C.; & Spear, R. M. 1983. Evidence from northwest Canada for the early Holocene Milankovitch thermal maximum. *Nature* 305 (5930):126–128.

Robock, A. 1983. Ice and snow feedbacks and the latitudinal and seasonal distribution of climate sensitivity. *J. Atm. Sci.* 40:986–997.

Roche, L. 1984. Une autre opinion sur la "disparition" des forêts tropicales: Extrait de "Botanistes, zoologistes et défense des écosystèmes des forêts tropicales humides." *Bois et forêts des tropiques* 204:3–19.

Rodin, L. E., & Basilevitch, N. I. 1965. *Organic matter's dynamics and biological cycle of ash elements and nitrogen in basic vegetation types of the world.* Moscow and Leningrad: Nauka. 254 pp. (R)

Romankevitch, E. A. 1977. *Geochemistry of organic matter in the ocean.* Moscow: Nauka. 256 pp. (R)

Ronov, A. B. 1976. Volcanism, carbon accumulation, life. *Geochimiya* 8:1252–1277. (R)

Rose, R. J.; Miller, M. M.; & Agnew, G. 1983. *Global energy futures and CO_2-induced climate change.* Cambridge, Mass.: MIT Energy Lab.

Rosenberg, N. J. 1981. The increasing CO_2 concentration in the atmosphere and its implications for agricultural productivity. 1. Effects on photosynthesis, transpiration and water use efficiency. *Clim. Change* 2:387–409.

———. 1982. The increasing CO_2 concentration in the atmosphere and its implications for agricultural productivity. II. Effects through CO_2-induced climate change. *Clim. Change* 4:239–254.

Ross, Yu. K. 1964. On the mathematical theory of plant-cover photosynthesis. *Rep. Acad. Sci. USSR* 157 (5):1239–1242. (R)

Rossignol-Strick, M. 1985. Mediterranean Quaternary sapropels, an immediate response of the African monsoon to variation of insolation. *Palaeogeogr., Palaeoclim., Palaeoecol.* 49 (314):237–265.

Rowntree, P., & Walker, J. 1978. The effects of doubling the CO_2 concentration on the radiative-convective equilibrium. In *Carbon dioxide, climate and society,* ed. J. Williams. New York: Pergamon Press, pp. 181–192.

Rozov, N. N., & Stroganova, M. N. 1979. *Soil cover of the world.* Moscow. 287 pp. (R)

Rubinshtein, E. S., & Polozova, L. G. 1966. *Modern climatic change.* Leningrad: Gidrometeoizdat. 268 pp. (R)

Ruddiman, W. F., & Duplessy, J. C. 1985. Conference on the last deglaciation: Timing and mechanism. *Quatern. Res.* 23 (1):1–17.

Ryabchikov, A. M. 1968. Hydrothermal conditions and productivity of phytomass in main landscape zones. *Vestnik Moscov. Univers., Geograph.* 5: 41-48.

Safarova, S. A.; Nikolaev, V. N.; & Blum, N. S. 1984. Palaeogeography of the Pacific basin 18 Ka B.P. *Oceanology* 24 (4):643–648. (R)

Sagan, C.; Toon, O. B.; & Pollack, J. B. 1979. Human impact on climate of global significance since the domestication of fire. *Science* 204 (4425):1363–1368.

Sakamoto, C., et al. 1980. Climate and global grain yield variability. *Clim. Change* 2 (4):349–361.

Santer, B. 1985. The use of general circulation models in climate impact analysis: A preliminary study of the impacts of a CO_2 -induced climatic change on West European agriculture. *Clim. Change* 7 (1):71–94.

Savin, S. M. 1982. Stable isotopes in climatic reconstructions. In *Climate in Earth history*. Washington, D.C.: NAS, pp. 164–171.

Savin, S. M.; Douglas, R. G.; & Stehli, F. G. 1975. Tertiary marine paleotemperatures. *Geol. Soc. Bull.* 86:1499.

Savina, S. S., & Khotinsky, M. A. 1982. Zonal method of palaeoclimatic reconstructions for the Holocene. In *The development of nature in the USSR territory in the Late Pleistocene and Holocene*. Moscow: Nauka, pp. 231–244. (R)

Savvateev, S. P. 1985. On the problem of computing changes of climatic crop productivity by parameterized models. In *Problems of land hydrology. Papers at the conference of young scientists and specialists*. Leningrad: Gidrometeoizdat, pp. 213–216. (R)

Schlesinger, M. E. 1982. *CO_2-induced climatic warming*. Rept. no. 36. Clim. Res. Inst. 25 pp.

———. 1983. A review of climate models and their simulation of CO_2-induced warming. *Int. J. Environ. Studies* 20 (2):103–114.

———. 1984. Climate model simulations of CO_2-induced climatic change. *Adv. Geophys.* 26:141–235.

———. 1985. *The role of the ocean in CO_2 -induced climate change*. Rept. no. 60. Clim. Res. Inst., Oregon State Univ. 39 pp.

Schlesinger, M. E., et al. 1985. *The role of the ocean in CO_2-induced climate change: Preliminary results from OSU coupled atmosphere-ocean general circulation model.* Rept. no. 60. Clim. Res. Inst., Oregon State Univ.

Schmitt, L. E. (ed.). 1980. *Proceedings of the carbon dioxide and climate research program conference.* Washington, D.C. 287 pp.

Schneider, S. N. 1977. Climate change and the world predicament: A case study for interdisciplinary research. *Clim. Change* 1 (1):21–43.

Schneider, S. W., & Dickinson, R. E. 1974. Climate modeling. *Rev. Geophys. Space Phys.* 12 (3):447–493.

Schnell, R. C. 1984. Arctic haze and the Arctic gas and aerosol sampling program (AGASP). *Geophys. Res. Lett.* 11 (5):361–364.

Schönwiese, C. D. 1984. Northern hemisphere temperature statistics and forcing. Part B: 1579–1980 A.D. *Arch. Met. Geophys. Biol.*, Ser. B35:155–178.

Schwarz, M., & Gale, J. 1984. Growth response to salinity at high levels of carbon dioxide. *J. Exp. Bot.* 35:193–196.

Scientific plan for world climate research programme. 1984. WCRP Publ. Ser. no. 2. Geneva: WMO/ICSU.

Seidel, S., & Keyes, D. 1983. *Can we delay a greenhouse warming?* Washington, D.C.: Environmental Protection Agency.

Seiler, W. 1984. Contribution of biological processes to the global budget of the CH_4 in the atmosphere. In *Current perspectives in microbiology and ecology.* Am. Soc. Microbiol. Washington, D.C.: NAS, pp. 468–477.

Seiler, W., et al. 1984. The seasonality of CO abundance in the southern hemisphere. *Tellus* 36B (4):219–231.

Sellers, W. D. 1969. A global climate model based on the energy balance of the Earth-atmosphere system. *J. Appl. Meteorol.* 8:392–400.

Shackleton, N. J., & Kennett, J. P. 1975. Paleotemperature history of the Cenozoic and initiation of Antarctic glaciation: Oxygen and carbon isotope analyses in DSDP sites 277, 279, 281. In *Initial reports of the deep-sea drilling project*, vol. 29, pp. 743–755.

Shackleton, N. J., & Pisias, N. S. 1985. Atmospheric carbon dioxide orbital forcing and climate. In *The carbon cycle and atmospheric CO_2: Natural variations, Archean to present*, ed. E. T. Sungquist & W. S. Broecker. *Geophys. Monograph* 32:303–317.

Shaw, C. E. 1985. On the climatic relevance of Arctic haze: Static energy balance considerations. *Tellus* 37B (1):50–52.

Shoeberl, M. R., & Kruger, A. J. 1986. Overview of the Antarctic ozone depletion issue. *Geophys. Res. Lett.* 13 (12):1191.

Shotton, F. M. (ed.). 1978. *British Quaternary studies.* Oxford.

Shukla, J., & Mintz, Y. 1982. Influence of land-surface evapotranspiration on the Earth's climate. *Science* 215:1498–1501.

Siegenthaler, U. 1981. Uptake of excess CO_2 calculated by an outcrop diffusion model of the ocean. In *Analysis and interpretation of atmospheric CO_2 data.* Bern: WMO/ICSU/ UNEP, pp. 169–176.

———. 1984. Nineteenth-century measurements of atmospheric CO_2: A comment. *Clim. Change* 6:409–411.

Siegenthaler, U.; Heimann, M.; & Oeschger, H. 1978. Model responses of the atmospheric CO_2 level and $^{13}C/^{12}C$ ratio to biogenic CO_2 input. In *Carbon dioxide, climate and society.* Oxford: Pergamon Press, pp. 79–88.

Siegenthaler, U., & Oeschger, H. 1978. Predicting future atmospheric carbon dioxide levels. *Science* 199 (4327):388–395.

Siegenthaler, U., & Wenk, T. H. 1984. Rapid atmospheric CO_2 variations and ocean circulation. *Nature* 308:624–626.

Sinitsyn, V. M. 1965. *Ancient climates of Eurasia. Part 1.* Leningrad: Leningrad Univ. Publ. 287 pp. (R)

———. 1966. *Ancient climates of Eurasia. Part 2.* Leningrad: Leningrad Univ. Publ. 166 pp. (R)

———. 1967. *Introduction to palaeoclimatology.* Leningrad: Nedra. 232 pp. (R)

———. 1970. *Ancient climates of Eurasia. Part 3.* Leningrad: Leningrad Univ. Publ. 133 pp. (R)

———. 1976. *Climate of laterite and boxite.* Leningrad: Nedra. 151 pp. (R)

———. 1980. *Environment and climate in the USSR territory in the Early and Middle Cenozoic.* Leningrad: Leningrad Univ. Publ. (R)

Sionit, N.; Hellmers, H.; & Strain, B. R. 1982. Interaction of atmospheric CO_2 enrichment and irradiance on plant growth. *Agron. J.* 74:721–725.

Sionit, N.; Mortenson, D. A.; Strain, B. R.; & Hellmers, H. 1981. Growth responses of wheat to CO_2 enrichment at different levels of mineral nutrition. *Agron. J.* 1024–1027.

Sionit, N.; Strain, B. R.; & Backford, H. A. 1981. Environmental controls on the growth and yield of okra. I. Effects of temperature and of CO_2 enrichment at cool temperature. *Crop Sci.* 21:885–888.

Sirotenko, O. D. 1981. *Mathematical modeling of water-thermal conditions and productivity of agroecosystems.* Leningrad: Gidrometeoizdat. 167 pp. (R)

Sirotenko, O. D.; Abashina, E. V.; & Pavlova, V. N. 1984. Estimation of the impact of possible fluctuations and changes of climate on agricultural productivity. *Bull. Acad. Sci. USSR, Atm. Ocea. Phys.* 20 (11):1104–1110. (R)

Sirotenko, O. D., & Boiko, A. P. 1980. The use of simulation model of the soil-plant-atmosphere system for estimating the influence of CO_2 concentration on agrocenoses productivity. In *Problems of atmospheric carbon dioxide. Proc. Soviet-Amer. Symp.* Leningrad: Gidrometeoizdat, pp. 243–251. (R)

Skopintsev, B. A. 1971. Modern achievements in studying organic matter of oceanic waters. *Oceanology* 11 (6):939–956. (R)

————. 1977. Oxygen consumption in deep waters. *Ambio. Special Report* 5:103–105. (R)

Smith, G. I., & Street-Perrott, F. A. 1983. Pluvial lakes of the western United States. In *Late Quaternary environments of the U.S.*, ed. H. E. Wright, Jr. Vol. 1. *The Late Pleistocene.* Minneapolis: Univ. of Minnesota Press, pp. 190–212.

Solar cycle may cause ozone hole. 1986. *EOS Trans. Amer. Geophys. Union* 67.

Sommer, A. 1976. Attempt at an assessment of the world's tropical moist forests. *Unasylva* 28:5–25.

Sorkina, A. I. 1972. *Long-term fluctuations of mean monthly values of the intensity and mean monthly geographical locations of atmospheric action centers in the northern hemisphere.* Moscow, Obninsk: USSR Hydrometeorological Center. VNIIGMN-MDC. 35 pp. (R)

Soviet-American workshop on nature-climatic changes in the Pleistocene and Holocene. 1977. *Meteorol. Hydrol.* 5:121–123. (R)

Spelman, M. J., & Manabe, S. 1984. Influence of oceanic heat transport upon the sensitivity of a model climate. *J. Geophys. Res.* 89 (C1):571–586.

Stauffer, B., et al. 1985. Increase of atmospheric methane recorded in Antarctic ice core. *Science* 229 (4720):1386–1388.

Stepanov, V. N. 1974. *World ocean.* Moscow: Nauka. 255 pp. (R)

Strat. ozone. 1984. *Causes and effects of stratospheric ozone reduction: Update 1983.* Washington, D.C.: Nat. Acad. Press.

Stratospheric Ozone. 1988. UK Stratospheric Ozone Review Group. London: HUSO. 71 pp.

Street, F. A., & Grove, A. T. 1979. Global maps of lake-level fluctuations 30,000 yr B.P. *Quatern. Res.* 12:83–118.

Street-Perrott, F., & Roberts, N. 1981. Fluctuations in closed-basin lakes as an indicator of past atmospheric circulation patterns. In *Variations in the global water budget*, pp. 331–341.

Strokina, L. A. 1963. Heat balance of the surface of the ocean. *Meteorol. Hydrol.* 1:25–34. (R)

————. 1967. Determination of changing heat content of the ocean. *Trans. Main Geophysical Observatory* 209.

————. 1982. Mean latitudinal values of air and water temperature for the world ocean. *Meteorol. Hydrol.* 4:50–55. (R)

————. 1986. Heat balance and evaporation from the world ocean's surface. In *Theses of papers of the 5th All-Union Hydrological Congress.* Leningrad: Gidrometeoizdat, pp. 68–69. (R)

Stuiver, M. 1978. Atmospheric carbon dioxide and carbon reservoir changes. *Science* 199 (4328):253–258.

Stuiver, M., & Quay, P. D. 1980. Changes in atmospheric carbon-14 attributed to a variable Sun. *Science* 207 (4426):11–19.

————. 1981. Atmospheric ^{14}C changes resulting from fossil fuel CO_2 release and cosmic ray flux variability. *Earth and Planet. Sci. Lett.* 53:349–362.

Styrikovitch, M. A.; Sinyak, Yu. B.; & Chernavsky, S. Ya. 1981. Longterm prospects of the development of world energetics. *Achievements and prospects*, vol. 14, *Energetics, Fuel*, no. 3. (R)

Suess, H. E. 1955. Radiocarbon concentration in modern wood. *Science* 122:415–417.

Sundquist, E. T. 1985. Geological perspectives on carbon dioxide and carbon cycle. In *The carbon cycle and atmospheric CO_2: Natural variations, Archean to present. Geophys. Monogr.* 32, ed. W. S. Broecker, pp. 5–59.

Swaminathan, M. 1979. Global aspects of food production. *Proc. World Climate Conf.* Geneva: WMO.

————. 1984. Climate and agriculture. In *Climate and development*, ed. A. K. Biswas. Natural Resources and the Environment Series, vol. 13. Dublin: Tycooly International, chap. 3, pp. 65–95.

Synoptic bulletin. The northern hemisphere. 1961–83. Moscow, Obninsk: USSR Hydrometeorological Center. VNIIGMI-MDC. (R)

Syuetova, I. A. 1973. Quantitative distribution of continental and ocean's biomass. *Bull. State Moscow Univ., Ser. Geogr.* 6:20–46. (R)

Taira, K. 1979. Holocene migrations of warm-water front and sea-level fluctuations in the North-Western Pacific. *Palaeogeogr., Palaeoclim., Palaeoecol.* 28 (3–4):197–204.

Talrose, V. L., & Larin, I. K. 1982. Atmospheric ionization and ozone. In *Complex global monitoring of natural environmental pollution. Trans. II Int. Symp.* Leningrad: Gidrometeoizdat. (R)

Tenyakov, V. A., & Yasamanov, N. A. 1981. Phanerozoic formation of boxites: evolution of some atmospheric parameters. *Rep. Acad. Sci. USSR* 188 (2):342–344. (R)

Terjung, W. H., et al. 1984. Climatic change and water requirements for grain corn in the North American Great Plains. *Clim. Change* 6 (2):193–220.

Thomas, R. H., et al. 1979. Effect of climatic warming on the West Antarctic ice sheet. *Nature* 277:355–358.

Thompson, I. 1976. Living with climatic change: Phase II. *Symp. Rep.* Mitre Corp. 14 pp.

Thompson, L. M. 1975. Weather variability, climatic change and grain production. *Science* 188:535–541.

Trabalka, J. R. (ed.). 1985. *Atmospheric carbon dioxide and the global carbon cycle.* U.S. Dept. of Energy, Oak Ridge Nat. Lab. 315 pp.

Turchinovitch, I. E. 1983a. Distribution of radioactive carbon ^{14}C of anthropogenic origin between the atmosphere and ocean after 1963. *Meteorol. Hydrol.* 3:65–70. (R)

———. 1983b. The Suess effect. *Meteorol. Hydrol.* 1:99–102. (R)

Turchinovitch, I. E., & Vager, B. G. 1985. Model of distribution of ^{12}C and ^{13}C carbon isotopes between the atmosphere and ocean. *Meteorol. Hydrol.* 8:60–68. (R)

Turco, R. P., et al. 1983. Nuclear winter: Global consequences of multiple nuclear explosions. *Science* 222:1283–1292.

Uchijima, Z. 1976. Long-term change and variability of air temperature above 10°C in relation to crop production. In *Climatic change and food production.* Int. Symp. on Recent Climatic Change and Food Production. October 1976, Taukuba-Tokyo, pp. 217–229.

Ulanova, E. S. 1975. *Agrometeorological conditions and winter wheat productivity.* Leningrad: Gidrometeoizdat. 302 pp. (R)

UNEP Expert Group Meeting on Climate Impact Studies. 1980. *Rept. Meeting of Experts.* February 1980, Nairobi. 27 pp.

UNESCO. 1978. *Tropical forest ecosystems: A state-of-knowledge report.* Natural Resources Research 14. Paris: UNESCO/UNEP/FAO.

Vager, B. G., & Turchinovitch, I. E. 1987. The estimation of biogenic CO_2 emission into the atmosphere 1860–1981. *Meteorol. Hydrol.* 9:17–28. (R)

Van Loon, H., & Williams, J. 1974. The connection between trends of mean temperature and circulation at the surface: Part I, Winter. Part II, Summer. Part III, Spring and Autumn. *Mon. Wea. Rev.* 104 (4):365–380; (8):1003–1011; (12):1592–1596.

Varushchenko, A. N.; Varushchenko, S. I.; & Kliege, R. K. 1980. The change of the Caspian Sea level in the Late Pleistocene-Holocene. In *Fluctuations of moisture conditions in the Aral-Caspian region during the Holocene.* Moscow: Nauka, pp. 79–90. (R)

Varushchenko, S. I. 1984. Does central Asia dry up? *Bull. Moscow Univ. Ser. 5, Geography,* 1:51–58. (R)

Velichko, A. A. 1973. *Natural process in the Pleistocene.* Moscow: Nauka. 254 pp. (R)

———. 1985. Empirical palaeoclimatology (principles and extent of accuracy). In *Methods of palaeoclimatic reconstructions.* Moscow: Nauka, pp. 7–20.

Velichko, A. A. (ed.). 1984. *Late Quaternary environments of the Soviet Union.* Minneapolis: Univ. of Minnesota Press. 327 pp.

———. 1987. The structure of paleoclimatic Mezo-Cenozoic thermal variations based on data from eastern Europe. In *Earth's climates in the geologic past.* Moscow: Nauka, pp. 5–41. (R)

———. 1987. Relationship of climatic changes in high and low latitudes of the Earth during the late Pleistocene and Holocene. In *Paleogeography and loess.* Budapest: Akadémiae Kiadó.

Velichko, A. A., et al. 1982. Palaeoclimatic reconstructions for the Miculino interglacial optimum in the European territory. *Bull. Acad. Sci. USSR, Ser. Geogr.* 1:15. (R)

Velichko, A. A., et al. 1983. Palaeoclimate of the USSR territory during the last (Miculino) Interglacial. *Bull. Acad. Sci. USSR, Ser. Geogr.* 6:30–45. (R)

Velichko, A. A., et al. 1984. Climate of the northern hemisphere during the last Miculino interglacial. *Bull. Acad. Sci. USSR, Ser. Geogr.* 1:5–18. (R)

Vinnikov, K. Ya. 1985. *Modern changes of global climate.* Obninsk. 52 pp. (Review information. *Ser. Meteor.* VNIIGMI-MDC, vol. 8.) (R)

———. 1986. *Climate sensitivity. Empirical investigations of modern climate changes.* Leningrad: Gidrometeoizdat. 224 pp. (R)

Vinnikov, K. Ya., & Groisman, P. Ya. 1979. Empirical model of modern climatic changes. *Meteorol. Hydrol.* 3:25–36. (R)

————. 1981. Empirical analysis of CO_2-induced changes in mean annual surface air temperature in the northern hemisphere. *Meteorol. Hydrol.* 11:30–43. (R)

————. 1982. Empirical investigation of climate sensitivity. *Bull. Acad. Sci. USSR, Atm. Ocea. Phys.* 18 (11):1159–1169. (R)

Vinnikov, K. Ya; Groisman, P. Ya; and Lugina, K. M. 1989. Empirical data on modern climate changes (temperature and precipitation). *Journal of Climate* (in press).

Vinnikov, K. Ya., & Kovyneva, N. P. 1983. On the distribution of climate changes with global warming. *Meteorol. Hydrol.* 5:10–19. (R)

Vinnikov, K. Ya., & Lugina, K. M. 1982. Some problems of monitoring global thermal regime of the northern hemisphere. *Meteorol. Hydrol.* 11:5–14. (R)

Vinnikov, K. Ya., et al. 1980. Modern climate changes in the northern hemisphere. *Meteorol. Hydrol.* 6:5–17. (R)

Vinnikov, K. Ya., et al. 1987. Mean annual air temperature change in the northern hemisphere during the period 1841 to 1985. *Meteorol. Hydrol.* 1:45–55. (R)

Vitels, L. A., & Drozdov, O. A. 1975. Manifestation of secular variations of precipitation in the USA through the recurrence of large precipitation deficit and comparison with analogous conditions in arid regions of the USSR. *Trans. Main Geophysical Observatory* 354:76–80. (R)

Voitkevitch, G. V., et al. 1977. *Brief reference book on geochemistry.* Moscow: Nedra. 106 pp. (R)

Volcanoes, stratospheric aerosol and climate of the Earth. 1986. Leningrad: Hydrometeoizdat. (R)

Volkova, V. S. 1977. *Stratigraphy and the history of vegetation development in western Siberia during the Late Cenozoic.* Moscow: Nauka. 236 pp. (R)

Voskresensky, A. I., & Marshunova, M. S. 1982. The dynamics of modern climate of polar regions. *Bull. Acad. Sci. USSR, Atm. Ocea. Phys.* 18 (12): 1269–1277. (R)

Wagener, K. 1978. Total anthropogenic CO_2 production during the period 1800–1935 from carbon-13 measurement in tree rings. *Radiat. Environ. Biophys.* 15:101–111.

Waggoner, P. E. 1983. Agriculture and a climate changed by more carbon dioxide. In *Changing climate report of the carbon dioxide.* Assessment Committee. Washington, D.C.: Nat. Acad. Press.

Wang, W. C., et al. 1976. Greenhouse effects due to man-made perturbations of trace gases. *Science* 194:685.

Warrick, R. A., & Riebsame, W. E. 1981. Societal response to CO_2 -induced climate change: Opportunities for research. *Clim. Change* 3 (4):387–428.

Washington, W. M., & Meehl, G. A. 1983. General circulation model experiments on the climatic effects due to a doubling and quadrupling of carbon dioxide concentration. *J. Geophys. Res.* 88:6600–6610.

————. 1984. Seasonal cycle experiment on the climate sensitivity due to a doubling of CO_2 with an atmospheric general circulation model coupled to a simple mixed layer ocean model. *J. Geophys. Res.* 89:9475–9503.

Watts, Ju. A. 1982. The carbon dioxide question. In *Carbon dioxide*, ed. W. S. Clark. New York: Clarendon Press, pp. 4431–4441.

Watts, W. A. 1983. Vegetational history of the eastern United States 25,000 to 10,000 years ago. In *Late Quaternary environment of the United States*, ed. H. E. Wright, Jr. Vol. 1. *The Late Pleistocene*. Minneapolis: Univ. of Minnesota Press, pp. 290–310.

WCP-46: Report of the JSC study conference of land surface processes in atmospheric general circulation models. Greenbelt, Md., January 1981. Geneva: WMO.

WCP-76: Report of the meeting of experts on the design of a pilot atmospheric hydrological experiment of the WGO. Geneva, 28 November–2 December 1983. Geneva: WMO.

Webb, Th., III; Street, F. A.; & Howe, S. 1980. *Precipitation and lake-level changes in the West and Midwest over the past 10,000 to 24,000 years. A final report.* Lawrence Livermore Laboratory, Univ. of Calif.

Weber, K.-H., & Flohn, H. 1984. Oceanic upwelling and air-sea-exchange of carbon dioxide and water vapor as a key for large-scale climate change? *Bonner Meteorol. Abhandlungen* 31:74–107.

WEC. 1983. *Energy 2000–2020: World prospects and regional stresses.* Ed. J. R. Frish. World Energy Conference Conservation Commission and Oil Substitution. World Outlook to 2020. London: Oxford Univ. Press.

Weiss, R. F. 1981. The temporal and spatial distribution of tropospheric nitrous oxide. *J. Geophys. Res.* 86 (C8):7185–7195.

Wetherald, R. T., & Manabe S. 1975. The effects of changing the solar constant on the climate of a general circulation model. *J. Atm. Sci.* 32:1485–1510.

————. 1980. Cloud cover and climate sensitivity. *J. Atm. Sci.* 37 (7):1485–1510.

————. 1981. Influence of seasonal variation upon the sensitivity of a model climate. *J. Geophys. Res.* 86 (C2):1194–1204.

Whittaker, R. H. 1975. *Communities and ecosystems.* New York: Macmillan. 286 pp.

Whittaker, R. H., & Likens, G. E. 1973. Carbon and the biota. In *Carbon and the biosphere*, ed. G. M. Woodwell & E. V. Pecan. Springfield: U.S. Atomic Energy Commission.

————. 1975. The biosphere and man. In *Primary productivity of the biosphere*, ed. H. Lieth & R. H. Whittaker. New York: Springer-Verlag.

Wigley, T. M. L. 1984. Carbon dioxide, trace gases and global warming. *Climate Monitor* 13 (5):133–148.

Wigley, T. M. L.; Jones, P. D.; & Kelly, P. M. 1980. Scenario for a warm high-CO_2 world. *Nature* 283:17–21.

Wigley, T. M. L., & Schlesinger, M. E. 1985. Analytical solution for the effect of increasing CO_2 on global mean temperature. *Nature* 315:649–652.

Williams, J. 1980. Anomalies in temperature and rainfall during warm Arctic seasons as a guide to the formulation of climatic scenarios. *Clim. Change* 2:249–266.

Williams, J. W. (ed.). 1979. *Carbon dioxide, climate and society.* Pergamon Press, vol. 1.

Williams, M. A. J., & Faure, H. (eds.). 1980. *The Sahara and the Nile.* Rotterdam: A. A. Balkema. 607 pp.

Wit, C. T. de. 1965. Photosynthesis of leaf canopies. *Agric. Res. Rep.* no. 663. Wageningen, pp. 1–57.

Wit, C. T. de; Broiwer, R.; & Penning de Vries, F. W. T. 1971. A dynamic model of plant and crop growth. In *Potential crop production: A case study*, ed. P. P. Wareing, Jr. London: Cooper, pp. 117–142.

Wittwer, S. H. 1980. Environmental and societal consequences of a possible CO_2-induced climate change on agriculture. In *Annual Meeting of the American Association for the Advancement of Science: The Effect of Increasing Atmospheric Carbon Dioxide on Carbon Storage in Various Ecosystem Types. Rongemont*: N.S., March 1980.

Woillard, C. M., & Mook, W. G. 1982. Carbon-14 dates of Grande Pile: Correlation of land and sea chronologies. *Science* 215:159–161.

Wolfe, J. A. 1980. Tertiary climates and floristic relationships at high latitudes in the northern hemisphere. *Palaeogeogr., Palaeoclim., Palaeoecol.* 30 (3–4):313–325.

Wong, S. C. 1980a. Elevated atmospheric partial pressure of CO_2 and plant growth. I. Interactions of nitrogen nutrition and photosynthetic capacity in C_3 and C_4 plants. *Oecologia* 44:68–74.

————. 1980b. Effects of elevated partial pressure of CO_2 on rate of CO_2 assimilation and water-use efficiency in plants. In *Carbon dioxide and climate: Australian research.* Canberra: Australian Academy of Science, pp. 159–166.

Woodwell, G. M. 1978. The carbon dioxide question. *Sci. Amer.* 238:34–43.

————. 1983. Biotic effect on the concentration of atmospheric carbon dioxide: A review and projection. In *Changing climate.* Washington, D.C.: Nat. Acad. Press, pp. 216–241.

Wood, B., & Dubeaut, R. 1975. *The Earth is only one.* Moscow: Nauka. 217 pp. (R)

World and regional demography forecasts. 1974. UN document for the World Demography Conference. E/CONF, 60/CBP/ 15. 36 pp. (R)

The world food situation and prospects to 1985. 1975. Foreign Agric. Econ. Rept., no. 98. Washington, D.C. 90 pp.

World water balance and water resources of the Earth. 1974. Leningrad: Gidrometeoizdat. 640 pp. (R)

World weather records. 1881–1920, 1921–30, 1931–40, 1941–50, 1951–60. Washington, D.C.

Wright, H. E. (ed.). 1983. *Late-Quaternary environment of the United States.* Vol. 1. *The Late Pleistocene.* Minneapolis: Univ. of Minnesota Press.

Wuebbles, D. J.; Luther, F. M.; & Penner, J. E. 1983. Effect of coupled anthropogenic perturbations on stratospheric ozone. *J. Geophys. Res.* 88 (C2):1444–1456.

Wuebbles, D. J.; McCracken, M. C.; & Luther, F. M. 1984. *A proposed reference set of scenarios for radiatively active atmospheric constituents.* Lawrence Livermore Nat. Lab. 51 pp.

Yaglom, A. M. 1977. Comments on wind and temperature flux-profile relationships. *Boundary Layer Meteorol.* 11 (1): 89–102.

Yasamanov, N. A. 1978. *Topography climatic conditions of the Jurassic, Cretaceous and Palaeogene of the south of the USSR.* Moscow: Nedra. 223 pp. (R)

———. 1985. *Ancient climates of the Earth.* Leningrad: Gidrometeoizdat. 292 pp. (R)

Yeh, T.-C., et al. 1983. A model study of the short-term climatic and hydrologic effects of sudden snow-cover removal. *Mon. Wea. Rev.* 111:1013–1024.

Yung, J. L., et al. 1976. Greenhouse effect due to atmospheric nitrous oxide. *Geophys. Res. Lett.* 3:619.

Zakharov, V. F., & Strokina, L. A. 1978. Present-day changes in ice cover of the Arctic Ocean. *Meteorol. Hydrol.* 7:35–43. (R)

Zakharova, O. K., & Byutner, E. K. 1985. The influence of physical and hydrochemical factors on the absorption of anthropogenic CO_2 by the ocean. *Bull. Acad. Sci. USSR, Atm. Ocea. Phys.* 21 (5):739–747. (R)

Zavelsky, F. S. 1975. Radiocarbon dating and theoretical models of carbon cycle in soils. *Bull. Acad. Sci. USSR, Ser. Geogr.* 1:27–34. (R)

Zinke, P. Z., et al. 1984. *Worldwide organic soil carbon and nitrogen data.* Rep. ORN4/TM-8857. Oak Ridge, Oak Ridge Nat. Lab. 141 pp.

Zubakov, V. A., & Borzenkova, I. I. 1983. *Palaeoclimates in the Late Cenozoic.* Leningrad: Gidrometeoizdat. 214 pp. (R)

Zubenok, L. I. 1976. *Evaporation under natural conditions.* Leningrad: Gidrometeoizdat. 264 pp. (R)

ABOUT THE EDITORS

M. I. Budyko and Yu. A. Izrael are colleagues at the State Hydrological Institute of the Soviet Union, where Budyko serves as director of the Climate Change and Atmospheric Water Cycle Research Department. Budyko is the chief Soviet scientist involved in the joint American-Soviet study of the greenhouse effect, and his research has earned him the International Meteorological Organization prize, the Lenin Prize, the Gold Medal of the Soviet Geographical Society, and the Vinogradov Prize of the Soviet Academy of Science. His books available in English include *The Evolution of the Biosphere* (1986), *History of the Earth's Atmosphere* (1987), and *Global Climatic Catastrophes* (1988).